# Simulation with Entropy Thermodynamics

# Simulation with Entropy Thermodynamics

Editor

**Christophe Goupil**

MDPI • Basel • Beijing • Wuhan • Barcelona • Belgrade • Manchester • Tokyo • Cluj • Tianjin

*Editor*
Christophe Goupil
University of Paris
France

*Editorial Office*
MDPI
St. Alban-Anlage 66
4052 Basel, Switzerland

This is a reprint of articles from the Special Issue published online in the open access journal *Entropy* (ISSN 1099-4300) (available at: https://www.mdpi.com/journal/entropy/special_issues/Simulation_Entropy_Thermodynamics).

For citation purposes, cite each article independently as indicated on the article page online and as indicated below:

LastName, A.A.; LastName, B.B.; LastName, C.C. Article Title. *Journal Name* **Year**, *Volume Number*, Page Range.

**ISBN 978-3-0365-0114-7 (Hbk)**
**ISBN 978-3-0365-0115-4 (PDF)**

© 2021 by the authors. Articles in this book are Open Access and distributed under the Creative Commons Attribution (CC BY) license, which allows users to download, copy and build upon published articles, as long as the author and publisher are properly credited, which ensures maximum dissemination and a wider impact of our publications.

The book as a whole is distributed by MDPI under the terms and conditions of the Creative Commons license CC BY-NC-ND.

# Contents

**About the Editor** . . . . . . . . . . . . . . . . . . . . . . . . . . . . . . . . . . . . . . . . . . . . . . . . . . . . . . . . . . . . . . . vii

**Preface to "Simulation with Entropy Thermodynamics"** . . . . . . . . . . . . . . . . . . . . . . . . . ix

**Ioulia Chikina, Valeri Shikine and Andrey Varlamove**
The Ohm Law as an Alternative for the Entropy Origin Nonlinearities in Conductivity of Dilute Colloidal Polyelectrolytes
Reprinted from: *Entropy* **2020**, *22*, 225, doi:10.3390/e22020225 . . . . . . . . . . . . . . . . . . . . . . . . 1

**Pablo Eduardo Ruiz-Ortega, Miguel Angel Olivares-Robles and Olao Yair Enciso-Montes de Oca**
Segmented Thermoelectric Generator under Variable Pulsed Heat Input Power
Reprinted from: *Entropy* , *21*, 929, doi:10.3390/e21100929 . . . . . . . . . . . . . . . . . . . . . . . . . . . 11

**Armin Feldhoff**
Power Conversion and Its Efficiency in Thermoelectric Materials
Reprinted from: *Entropy* **2020**, *22*, 803, doi:10.3390/e22080803 . . . . . . . . . . . . . . . . . . . . . . . . 27

**Prasanna Ponnusamy, Johannes de Boor and Eckhard Müller**
Discrepancy between Constant Properties Model and Temperature-Dependent Material Properties for Performance Estimation of Thermoelectric Generators
Reprinted from: *Entropy* **2020**, *22*, 1128, doi:10.3390/e22101128 . . . . . . . . . . . . . . . . . . . . . . . 67

**Mario Wolf, Alexey Rybakov, Richard Hinterding and Armin Feldhoff**
Geometry Optimization of Thermoelectric Modules: Deviation of Optimum Power Output and Conversion Efficiency
Reprinted from: *Entropy* **2020**, *22*, 1233, doi:10.3390/e22111233 . . . . . . . . . . . . . . . . . . . . . . . 85

**Akihiro Nishiyama, Shigenori Tanaka and Jack A. Tuszynski**
Non-Equilibrium Quantum Brain Dynamics: Super-Radiance and Equilibration in 2 + 1 Dimensions
Reprinted from: *Entropy* **2019**, *21*, 1066, doi:10.3390/e21111066 . . . . . . . . . . . . . . . . . . . . . . 109

**Ilia A. Luchnikov, Alexander Ryzhov, Pieter-Jan Stas, Sergey N. Filippov and Henni Ouerdane**
Variational Autoencoder Reconstruction of Complex Many-Body Physics
Reprinted from: *Entropy* **2019**, *21*, 1091, doi:10.3390/e21111091 . . . . . . . . . . . . . . . . . . . . . . 137

**Gianpiero Colonna and Annarita Laricchiuta**
Thermodynamic and Transport Properties of Equilibrium Debye Plasmas
Reprinted from: *Entropy* , *22*, 237, doi:10.3390/e22020237 . . . . . . . . . . . . . . . . . . . . . . . . . . 159

**Nicolás Pérez, Constantin Wolf, Alexander Kunzmann, Jens Freudenberger, Maria Krautz, Bruno Weise, Kornelius Nielsch and Gabi Schierning**
Entropy of Conduction Electrons from Transport Experiments
Reprinted from: *Entropy* **2020**, *22*, 244, doi:10.3390/e22020244 . . . . . . . . . . . . . . . . . . . . . . . 169

**Mateusz Korpyś, Anna Gancarczyk, Marzena Iwaniszyn, Katarzyna Sindera, Przemysław J. Jodłowski and Andrzej Kołodziej**
Analysis of Entropy Production in Structured Chemical Reactors: Optimization for Catalytic Combustion of Air Pollutants
Reprinted from: *Entropy* **2020**, *22*, 1017, doi:10.3390/e22091017 . . . . . . . . . . . . . . . . . . . . . . 177

**Christophe Goupil and Eric Herbert**
Adapted or Adaptable: How to Manage Entropy Production?
Reprinted from: *Entropy* **2020**, 22, 29, doi:10.3390/e22010029 . . . . . . . . . . . . . . . . . . . . **191**

# About the Editor

**Christophe Goupil** is a Professor at the University of Paris. Specialized in condensed matter physics, he successively contributed to the study of the pinning and the dynamics of vortices in superconductors, then directed his work towards the processes of energy conversion, in particular by the thermoelectric way. From 2010, he contributed to the foundation of the "Laboratoire Interdisciplinaire des Energies de Demain", UMR 8236, of which he is the Deputy Director. As the head of the DyCo team, he develops an out-of-equilibrium thermodynamic approach of coupled transport processes. The conversion of energy and matter are central to his work and are now applied in different domains of physics, including thermoelectricity, muscle physiology, and sport performance.

# Preface to "Simulation with Entropy Thermodynamics"

Entropy is the strange state function which extends the description of energy by adding to it the unique property in physics that is quality. Indeed, in addition to being quantified by the first law of thermodynamics, energy is qualified. Beyond this rather formal description, it is indeed on the emergence of a new tool that we should focus our attention. If, thanks to a variational principle, analytical mechanics triumphs in predicting the evolution of the systems it is able to describe, it nevertheless leaves aside whole sections of the description of processes, where the dispersion of energy and/or matter, in degrees of freedom made accessible, apparently makes it impossible to apply an extremal principle. It is here, with subtlety and elegance, that the entropic approach can be inserted. Whether it concerns the dimensioning of thermoelectric systems, the study of the stability of thermodynamic parameters, the out-of-equilibrium behavior of plasmas, or the transport of charge carriers in different states of matter, the entropic approach often proves to be the only closed relationship that allows us to understand the fate of a system in contact with a reservoir of matter and a reservoir of energy. The world of engineering has a lot of experience and pragmatism on these subjects, only a part of which has today found a fundamental analytical formulation. Far from claiming to answer the open-ended problems of maximizing or minimizing the production of entropy in a general framework, this book wants to offer the reader, through very different systems, a look at the different uses that can be made of entropy, as a measurement of the dispersion of energy and matter. This book brings together the variety of uses of entropy by its authors in their respective works. We would like to thank them all warmly for their work.

Christophe Goupil
*Editor*

Article

# The Ohm Law as an Alternative for the Entropy Origin Nonlinearities in Conductivity of Dilute Colloidal Polyelectrolytes

Ioulia Chikina [1], Valeri Shikin [2] and Andrey Varlamov [3,*]

[1] LIONS, NIMBE, CEA, CNRS, Université Paris-Saclay, CEA Saclay, 91191 Gif-sur-Yvette, France
[2] ISSP, RAS, Chernogolovka, 142432 Moscow, Russia
[3] CNR-SPIN, c/o DICII-Universitá di Roma Tor Vergata, Via del Politecnico, 1, 00133 Roma, Italy
[*] Correspondence: andrey.varlamov@spin.cnr.it

Received: 10 January 2020; Accepted: 13 February 2020; Published: 17 February 2020

**Abstract:** We discuss the peculiarities of the Ohm law in dilute polyelectrolytes containing a relatively low concentration $n_\odot$ of multiply charged colloidal particles. It is demonstrated that in these conditions, the effective conductivity of polyelectrolyte is the linear function of $n_\odot$. This happens due to the change of the electric field in the polyelectrolyte under the effect of colloidal particle polarization. The proposed theory explains the recent experimental findings and presents the alternative to mean spherical approximation which predicts the nonlinear I–V characteristics of dilute colloidal polyelectrolytes due to entropy changes.

**Keywords:** polyelectrolytes; Ohm law; colloids

## 1. Introduction

Polyelectrolytes are polymers whose repeating units contain a group of electrolytes. These groups dissociate in aqueous solutions, making the polymers charged. Polyelectrolyte properties resemble those of both electrolytes and polymers, and, like salts, their solutions are electrically conductive. The incorporation of the nano- and micro-meter-sized charged colloidal particles can dramatically change the electrical and heat transport properties of such systems. For instance, the authors of Ref. [1] study the electrical transport in charged colloidal suspensions of iron oxide nanoparticles (maghemite) dispersed in an aqueous medium, while in Ref. [2], the thermal and electrical transport is investigated in ionically stabilized magnetic nanoparticles dispersed in aqueous potassium ferro/ferricyanide electrolytes. Both groups report the unusual effect of multiply charged colloidal particles on conductivity of the dilute polyelectrolytes. It turns out that the latter grows linearly with an increase of colloidal particle concentration.

This finding seems to be non-trivial from the point of view of the percolation theory (see, for example, [3]). Indeed, in accordance with the latter, the conductivity of a mixture between dielectric (in our case water molecules) and conducting (colloidal particle with counter-ions coat) components remains minute until the fraction of the conducting phase approaches the percolation threshold, and only in the vicinity of the latter, the conductivity growths smoothly have a value of dielectric component that is similar to to that of a metallic one.

Before discussing this contradiction, let us make an excursus into the physics of semiconductors. In the theory of semiconductors [3], the regions of weak and strong doping (i.e., introduction of charged impurities or structural defects with the purpose of changing the electrical properties of a semiconductor) are distinguished. In the low doping regime, the impurity concentration $n_\odot$ is so small that the distances between them significantly exceed the Debye length $\lambda_0$ and the bare radius of the colloidal particle $R_0$, i.e.

$$n_\odot (\lambda_0 + R_0)^3 \ll 1, \quad R_0 \leq \lambda_0, \tag{1}$$

and the intrinsic charge carriers of semiconductor completely screen the electric fields produced by the charged impurities (see Figure 1). In the strong doping regime, when criterion (1) is violated, the fields produced by the dopants are screened only partially and their interaction becomes significant.

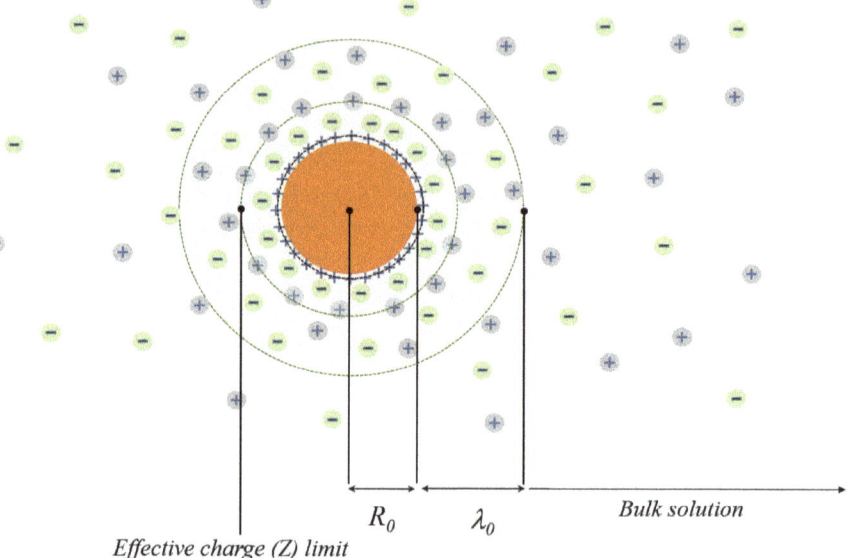

**Figure 1.** The schematic presentation of the multiply charged colloidal particle surrounded by the cloud of counter-ions.

Returning to the case of the dilute colloidal polyelectrolyte, one can map its properties to the ones of the weak doped semiconductor and identify $n_\odot$ with the concentration of the colloidal particles, while $\lambda_0$ should be related to their characteristic size. The latter is determined by the known concentration $n_0$ of the counterions of the electrolyte hosting charged colloidal particles.

The criterion (1) is in a reasonable agreement to the common concepts of the physics of dilute polyelectrolytes developed in the 40s of the last century by Derjaguin, Landau, Verwey, and Overbeek [4,5] and known as DLVO formalism. Namely, if the colloidal particles are neutral, they are not stationary in dilute solution and coagulating due to van der Waals forces acts between them. In order to prevent such coagulation processes, one should immerse individual colloidal particles in the electrolyte specific for each sort of them. The latter are called stabilizing electrolytes.

Being immersed (or synthesized within) in an electrolyte solution, the nanoparticles acquire surface ions (e.g., hydroxyl groups, citrate, etc. [6–8]) resulting in a very large structural charge $eZ$ ($|Z| \gg 10$). Its sign can be both positive or negative, depending on the surface group type. The latter, in return, attracts counterions from the surrounding solvent creating an electrostatic shielding coat of the size $\lambda_0$ with an effective charge $-eZ$. In these conditions, nano-particles approaching between them to the distances $r \leq \lambda_0$ begin to repel each other without flocculation [4,5,9]. The region of an essential interaction between them in terms of the criterion (1) corresponds to the condition

$$n_\odot^c (\lambda_0 + R_0)^3 \sim 1. \tag{2}$$

In Ref. [1,2], the massive multiply charged colloidal particles are surrounded by the clouds of counter-ions screening their positive charge. Such formations, according to Ref. [3], should not affect the conductivity of the dilute polyelectrolyte until the shells of the neighbor charged complexes do not overlap among themselves (see Equation (2)). The results of both Ref. [1] and [2] demonstrate the opposite: the conductivity of dilute colloidal polyelectrolyte grows linearly with increase of concentration already in the range $n_\odot \ll n_\odot^c$, where there is not yet place for percolation effects.

This contradiction can be removed by noticing that the presence of the multiply charged colloidal particles has an effect not only on the value of conductivity of a solution but also on the local value of the electric field:

$$j(n_\odot) = \sigma(n_\odot)E(n_\odot). \qquad (3)$$

It is important to note that the factors in Equation (3) are affected by the presence of the multiply charged colloidal particles in different ways. While the conductivity of the electrolyte at low concentrations of multiply charged colloidal particles ($n_\odot \leq n_\odot^c$) remains almost unchanged, their effect on the local electric field in this range of concentrations is essential. This happens due to polarization of the colloidal particles by an external electric field which, in accordance with the Le Chatelier's principle, results in the decrease of the effective value of the field. Consequently, the growth of conductivity [1,2] as a function of concentration $n_\odot$ is observed in experiments. When the concentration of multiply charged colloidal particles reaches the percolation threshold ($n_\odot = n_\odot^c$), the role of the factors in Equation (3) is reversed. Here, the subsystem of colloidal particles forms clusters and cannot be considered more as the gas of polarized highly conducting particles. Yet, in this range of concentrations, the new channel of percolation charge transfer is opened and the total conductivity of the electrolyte growth further increase by $n_\odot$.

The state-of-the-art in transport phenomena in polyelectrolytes was recently reviewed in Ref. [1]. Focusing mainly on the results of the microscopic approach (so called mean spherical approximation theory (MSA)) [10–12], the authors discuss mobility, diffusion coefficient, and the effective charge space distribution of the colloidal particles as the function of their concentration. Yet, in Ref. [1], there is not any information concerning the effect of clusters of polarization on the charge transfer process in such complex systems. This aspect of the problem is the subject of our work.

## 2. Effective Electric Field in Bulk of Colloidal Polyelectrolyte

The colloidal polyelectrolyte is a weakly conducting liquid with the small but finite fraction of relatively highly (due to $Z \gg 1$) conducting inclusions, i.e., colloidal particles. The collective polarization of these inclusions occurs when the external electric field $E_0$ is applied. This phenomenon is analogous to polarization of neutral atoms in gas. The only difference is that the neutral atoms reside in vacuum, while the charged conducting clusters of colloidal polyelectrolyte are immersed in a less, but still conducting, medium. Hence, our goal is to account for this peculiarity and find the effective field which governs the charge transport in such a complex system.

### 2.1. Electric Field in Absence of Current

The space distribution of the effective electric field of the colloidal particle is determined by the Poisson equation (see [3,9])

$$\Delta \varphi = \frac{4\pi}{\epsilon}\rho(r), \quad \rho(r) = |e|[n_+(r) - n_-(r)], \qquad (4)$$

where $\epsilon$ is the dielectric permittivity of stabilizing electrolyte.

The concentrations of the screening counterions $n_\pm(r)$ is determined self-consistently via the value of local electrostatic potential

$$n_\pm(r) = n_0 \exp[e_\pm \varphi(r)/T], \qquad (5)$$

$n_0 = n_0^+ = n_0^-$ is the counterions bare concentration, occuring due to the complete dissociation of the electrolyte which stabilizes the gas of colloidal particles.

In assumption $e\varphi(r) < T$ the Poisson equation can be linearized and takes form

$$\Delta\varphi = \varphi/\lambda_0^2, \quad \lambda_0^{-2} = \frac{8\pi e^2}{\epsilon T} n_0. \tag{6}$$

This equation should be solved accounting for the boundary conditions

$$r\varphi(r)|_{r \to R_0} \to Z|e|, \quad \varphi(r)|_{r \to \infty} \to 0, \tag{7}$$

what results in the standard screened Coulomb potential:

$$\varphi(r) = Ze \frac{\exp(-\frac{r}{\lambda_0})}{r}. \tag{8}$$

The values $Z, R_0$ and $n_0$ of the electrolyte, which stabilizes the colloidal solution can be determined by independent experiments (for example, by measurements of the electrophoretic forces, osmotic pressure, etc. [1]).

One should remember that even strongly diluted polyelectrolytes can undergo the transition to the state of a Wigner crystal in the case of strongly charged colloidal particles ($Z \gg 1$). For description of this, observed experimentally [13–15], phenomenon the authors of [16] assumed that the interaction between two colloidal particles has the same form of Yukawa potential (8), yet with the renormalized effective charge $Z^* \ll Z$, explicitly depending on the colloidal particles density $n_\odot$. The value of $Z^*$ is determined in the Wigner-Seitz model from the new boundary condition

$$\frac{\partial \varphi}{\partial r}\Big|_{r \to n_\odot^{-1/3}} = 0$$

replacing that ones, valid for the isolated charged particle in the screening media (see Equation (7)). For some range of the colloidal particles densities $n_\odot$ the conditions $Z \gg 1$ and $Z^* \ll 1$ can be satisfied simultaneously. The former characterizes the properties of the multiply charged colloidal particles, while the latter is determined by the strength of their interaction and $n_\odot$. In the range of densities $n_\odot$ satisfying Equation (1), the effect of the effective charge $Z^*$ on the Ohmic transport is negligible.

## 2.2. Electric Field in Presence of Current

When a stationary current flows through the polyelectrolyte, an internal electric field $\vec{E}$ appears in it. In the approximation of a very diluted solution, one can start considerations from the effect of presence of the isolated colloidal particle on a flowing current. Namely, one should find the perturbation of the internal electric field which would provide the homogeneity of the transport current far from the colloidal particle. A corresponding problem recalls that one of classic hydrodynamics: calculus of the associated mass of the particle moving in the ideal liquid [17].

We choose the center of spherical coordinates coinciding with the colloidal particle and direct the $z$-axis along the electric field $\vec{E}_0$. We assume that the conductivity of the electrolyte in the absence of colloidal particles is $\sigma_0$. The highly charged colloidal particle we will model as the conducting solid sphere of the radius $R \simeq (R_0 + \lambda_0)$ (see Figure 1) with conductivity $\sigma_\odot > \sigma_0$. Analysis of the charge transport in multi-phase systems (see [18]) is based on the requirements

$$div\vec{j} = 0, \quad \vec{j} = \sigma\vec{E}. \tag{9}$$

When the medium conductivity is invariable in space the constancy of the current, this automatically means the homogeneity of the electric field. The situation changes when the system is inhomogeneous and $\sigma \neq const$. The continuity Equation (9) in this case should be solved with the boundary conditions accounting for the current flow through the boundaries between domains of diverse conductivity. According to Ref. [18,19], the tangential components of electric field intensity at the boundary must be continuous, while the normal ones provide the continuity of the charge transfer. Applying these rules to our simple model of the highly charged colloidal particle in the less conductive medium, one can write

$$j_n^0 = J_n^\odot, \quad \text{or} \quad \sigma_0 E_0 = \sigma_\odot E_\odot. \tag{10}$$

Solution of the system of Equations (9) and (10) for the electrostatic potential in the vicinity of the colloidal particle ($r \geq R$) acquires the form:

$$\varphi(r,\theta) = -E_0 r \cos\theta + \left(\frac{\gamma-1}{\gamma+2}\right) E_0 \frac{R^3}{r^2} \cos\theta, \tag{11}$$

with $\gamma = \sigma_\odot/\sigma_0$. In the limit $\gamma \to 1$ the electric field remains unperturbed, $\vec{E} = -\nabla\varphi \to \vec{E}_0$. In the opposite case, $\gamma \gg 1$, the dipole perturbation takes the form corresponding to the case of metallic inclusion of the radius $R$ in the weakly conducting environment (Ref. [18]):

$$\varphi(r,\theta) = -E_0 r \cos\theta \left(1 - \frac{R^3}{r^3}\right). \tag{12}$$

One can see that in accordance with the intuitive expectations, the presence of an isolated colloidal particle in an electrolyte leads to the appearance of the local perturbation of the electric field of the dipole type $\nabla\varphi \propto r^{-3}$ with the value of the dipole moment of one colloidal particle

$$p_\odot = \left(\frac{\gamma-1}{\gamma+2}\right) R^3 E_0. \tag{13}$$

Returning to the initial problem of the rarefied gas of colloidal particles of concentration $n_\odot$ in the electrolyte media, one can introduce the effective dielectric permittivity $\epsilon_\odot$. It can be related to the dipole moment (13) by means of the Clausius–Mossotti relation (see Ref. [18]) and in terms of the material parameters of the problem which is read as:

$$\epsilon_\odot = 1 + 4\pi \left(\frac{\gamma-1}{\gamma+2}\right) R^3 n_\odot. \tag{14}$$

One can try to make the model of colloidal particles more realistic assuming that the latter has the structure of a thick-walled sphere; a "nut" with the conducting shell and the insulating core of the bare radius $R_0$. This intricacy leads to the change in the expression for the corresponding dipole momentum: instead of Equation (13) it takes the form (see Ref. [18])

$$\tilde{p}_\odot = \frac{(2\gamma+1)(\gamma-1)}{(2\gamma+1)(\gamma+2) - 2(\gamma-1)^2 R_0^3/R^3} \left(R^3 - R_0^3\right) E_0. \tag{15}$$

This formula contains two geometrical parameters: $R$ and $R_0$. The latter should be determined from some independent measurements. The difference $R - R_0$ can be identified with the Debye length $\lambda_0$ or to consider it as the fitting parameter.

## 3. Ohmic Transport in a Weak Colloidal Polyelectrolyte

Equation (14) demonstrates that growth of the nano-particle concentration $n_\odot$ leads to increase of the dielectric constant $\epsilon_\odot$, which, in its turn, results in the decrease of the effective electric field in an electrolyte. The latter, in conditions of the fixed transport current, is perceived as the growth of conductivity with an increase of the colloidal particles concentration:

$$\sigma(n_\odot) = j\epsilon_\odot/E_0 = \sigma_0 \left[1 + 4\pi n_\odot \frac{p_\odot(E_0)}{E_0}\right]. \tag{16}$$

## 3.1. Approximation of the Conducting Spheres

Substituting the dipole moment taken in the approximation of Equation (13) in Equation (16) one finds

$$\frac{\Delta\sigma_{CP}(n_\odot)}{\sigma_0} = \frac{\sigma(n_\odot) - \sigma_0}{\sigma_0} = 4\pi n_\odot \left(\frac{\gamma - 1}{\gamma + 2}\right) R^3, \qquad (17)$$

where $\Delta\sigma_{CP}$ is the excess conductivity due to the presence of colloidal particles. The left-hand-side of this equation can be extracted from the data presented in Figure 2. Indeed, in the interval of the nanoparticles concentrations $0 \leq \phi \leq 0.6\%$ the behavior of conductivity $\sigma(n_\odot)$ is almost linear and $\sigma(n_\odot)/\sigma_0 - 1 = 0.7$. In turn, the concentration $\varphi = 0.6\%$ corresponds to $n_\odot^{(1)} = 5.45 \times 10^{15}$ cm$^{-3}$.

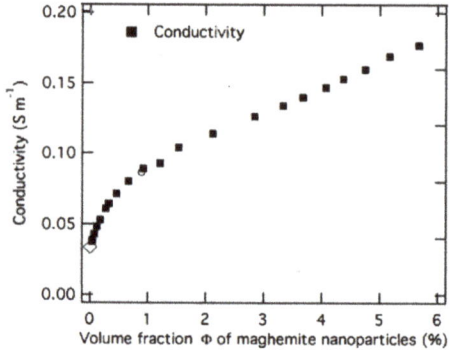

**Figure 2.** Experimental values of electrical conductivity of water based polyelectrolyte solution as a function of colloidal concentrations. Measurements were performed in pH = 3.1 solutions containing maghemite nanoparticles with an average diameter of 12 nm. More detailed information on the colloidal solution preparation methods and the nature of other ions is found in Ref. [1,2].

For further estimations, it will be crucial that Equation (17) is sensitive to the value of $\gamma$ only when it is not very large. When $\gamma \gg 1$ (we will justify this limit below) the combination $(\gamma - 1)/(\gamma + 2) \to 1$ and it ceases to influence the evaluations based on Equation (17). This allows us to find this limit

$$R_{exp}^{(1)} = 2.17 \times 10^{-6} cm,$$
$$n_\odot^{(1)} \left[R_{exp}^{(1)}\right]^3 = 0.055 \ll 1. \qquad (18)$$

One can see that these values, together with the nanoparticle concentration $n_\odot^{(1)}$, confirm the validity of the assumed above approximation (1). The plausible reasons for the discovered considerable difference between $R_{exp}^{(1)}$ and the value of bare radius $R_0^{(1)} = 6 \times 10^{-7}$ cm given in Ref. [1] will be discussed below.

The above found conductivity correction $\Delta\sigma_{CP}(n_\odot) \propto n_\odot R^3$ (see Equation (17)) caused by presence of nanoparticles in electrolyte can be confidently distinguished from the standard Onsager–Debye conductivity ($\sigma_{OD}$) of the diluted 1:1 electrolyte [20–22]. Indeed, first of all, the concentration dependencies of these conductivities are different: $\Delta\sigma_{CP}(n_\odot) \propto n_\odot$ while $\sigma_{OD}(n_\odot) \propto \sqrt{n_\odot}$.

Let us focus on the unusual dependence of the excess conductivity (17) of the nanoparticle size: $\Delta\sigma_{CP}$ growths with increase of $R$. Usually, this dependence is supposed to be opposite (the larger radius of the sphere in Stokes viscous law, the lower its mobility, and hence, the conductivity).

One can analyze the available experimental data on the conductivity of the stabilized diluted colloidal solution [1,2] in the conditions described by Equation (2). In accordance with Equation (17), the excess conductivities for different sizes of nanoparticles in assumption of the same concentration should scale as $[R_0^{(1)}/R_0^{(2)}]^3$. Taking the value $R_0^{(1)} = 6\ nm$ from [1] and $R_0^{(2)} = 3.8\ nm$ from [2] one finds that the ratio

$$\frac{\Delta\sigma_{CP}^{(1)}}{\sigma_0^{(1)}} \Big/ \frac{\Delta\sigma_{CP}^{(2)}}{\sigma_0^{(2)}} = \left(\frac{6}{3.8}\right)^3 \approx 4 \qquad (19)$$

Experimental data for this value give even more striking difference:

$$\frac{\Delta\sigma_{CP}^{(1)}}{\sigma_0^{(1)}} \Big/ \frac{\Delta\sigma_{CP}^{(2)}}{\sigma_0^{(2)}} = \frac{0.7}{0.06} \approx 11.7. \qquad (20)$$

## 3.2. Approximation of the Conducting Thick-Walled Spheres

Here, it is necessary to note that the value $R_{exp}^{(1)}$ obtained in the simple approximation of Equations (13) and (16) and the measured in Ref. [1] bare radius of the colloidal particle $R_0$ form a relatively small numerical parameter, $[R_0/R_{exp}^{(1)}]^3 \simeq 0.02$. It makes sense to improve the experimental data proceeding replacing the value $p_\odot$ in Equation (16) by the two parametric expressions (15). Tending $\gamma \to \infty$ in it one finds

$$\frac{\sigma(n_\odot) - \sigma_0}{\sigma_0} = 4\pi n_\odot [R_{exp}^{(1)}]^3 \left[1 - \frac{3}{\gamma} - \frac{9}{2\gamma}\frac{R_0^3}{\left([R_{exp}^{(1)}]^3 - R_0^3\right)}\right] \qquad (21)$$

From this expression, it is clear that the approximation (17) is valid when $\gamma \gg 1$.

The parameter $\gamma$ requires special discussion. In the DLVO colloidal model, it is assumed that some bare core exists which is able to cause the van der Waals forces between colloidal particles in dilute, non-stabilizing solutions. The conducting properties of this core is not so essential. For example, one can suppose this bare core of the radius $R_0$ to be a semiconductor possessing its intrinsic charge carriers which are confined in its volume. If the solvent possesses the stabilizing properties its own mobile charge carriers, counterions have the same properties as the intrinsic charge carriers of the bare core. The requirement of electrochemical potential constancy leads to the charge exchange between the bare core and the solvent. Such exchange results in the formation of the Debye shell (see Equations (4)–(8)), where the concentration of counterions considerably exceeds that in the solvent bulk. We assumed above that the value of corresponding conductivity $\sigma(n_\odot)$ considerably exceeds $\sigma_0$ of the electrolyte conductivity in absence of the nanoparticles. This assumption ($\gamma \gg 1$) breaks when the average value of electrochemical potential in the Debye shell $e\phi_\odot$ exceeds the temperature. The authors of Ref. [23] state that in these conditions the Debye shell of the DLVO colloid can crystallize due to Coulomb forces and the latter becomes an insulator with $\sigma(n_\odot) \leq \sigma_0$.

## 4. Conclusions

The main result of this work consists of the proposition of an alternative scenario explaining the linear growth of the polyelectrolyte conductivity versus the concentration of colloidal particles observed in Ref. [1,2] in the conditions of the validity of Equation (1). It drastically differs from the existing ideas of the transport in electrolytes resulting in the empirical Kohlrausch's law (see [22,24])

$$\Delta\sigma \sim \sqrt{n_\odot}. \qquad (22)$$

The speculations justifying Equation (22) were firstly proposed in early papers such as Ref. [20,21] and the recent efforts to improve this mechanism were undertaken in Ref. [25].

The fact of the observation of the Ohmic transport in strong electrolytes (Ref. [1,2]) denies the applicability of Kohlrausch's law in the interval of a very low concentration of the colloidal particles. Conversely, the mechanism proposed above, based on the analogy to the percolation mechanism of conductivity occurring in doped semiconductors, allows to get an excellent agreement in the observed linear dependence. Moreover, it also provides very reasonable values of the microscopic parameters of the problem.

One can believe that the validity of Kohlrausch's law is restored in the domain of higher concentrations and the crossover point between the two regimes (16) and (22) is determined by the condition (2), as is shown in Figure 2. One can find the pro-arguments for this statement in the experimental curve shown in Figure 2 of Ref. [1], where the regimes are changed in the vicinity of the concentration $n_\odot^{(1)} = 5.45 \times 10^{15}\ cm^{-3}$.

The question that arises is why such linear growth below the percolation threshold was never reported in measurements performed on semiconductors. The answer probably consists of the overwhelming supremacy of the colloidal particle dipole momentum in comparison to that of the dopant in semiconductors.

It would be interesting to compare the values of effective charge $Z$ extracted from the experiments on conductivity of [2] and the review article [1]. Unfortunately, this is not easy to do because of the analysis of the data for different $Z$ results in very different values of $R_0$. It is why one cannot judge the influence of the effective charge $Z$ on the bare radius of the colloidal particle $R_0$.

The relative insensibility of the polyelectrolyte conductivity on the value of parameter $Z$ is not extended on the Seebeck coefficient. The measurements of [2] demonstrate the existence in its kinetics of the two different phases; the initial and steady ones. The authors dealt with two types of colloids; one is almost electroneutral ($Z \geq 1$) and the other is supposed to have $Z \gg 1$.

**Author Contributions:** I.C.: Methodology, Formal analysis, Writing—review and editing, Visualization, Funding acquisition. V.S.: Methodology, Formal analysis, Writing—review and editing, Visualization. A.V.: Methodology, Formal analysis, Writing—review and editing, Visualization, Funding acquisition. All authors have read and agreed to the published version of the manuscript.

**Funding:** This research was supported by the European Union's Horizon 2020 Research and Innovation Programme: under grant agreement No.731976 (MAGENTA).

**Acknowledgments:** The authors acknowledge multiple and useful discussions with Sawako Nakamae.

**Conflicts of Interest:** The authors declare no conflict of interest.

## References

1. Lucas, I.T.; Durand-Vidal, S.; Bernard, O.; Dahirel, V.; Dubois, E.; Dufrêche, J.F.; Gourdin-Bertin, S.; Jardat, M.; Meriguet, G.; Roger, G. Influence of the volume fraction on the electrokinetic properties of maghemite nanoparticles in suspension. *Mol. Phys.: Int. J. Interface Chem. Phys.* **2014**, *112*, 1463–1471. [CrossRef]
2. Salez, T.J.; Huang, B.; Rietjens, M.; Bonetti, M.; Wiertel-Gasquet, C.; Roger, M.; Filomeno, C.L.; Dubois, E.; Perzynski, R.; Nakamae, S. Can charged colloidal particles increase the thermoelectric energy conversion efficiency? *Phys. Chem. Chem. Phys.* **2017**, *19*, 9409–9416; doi:10.1039/C7CP01023K. [CrossRef] [PubMed]
3. Shklovski, B.I.; Efros, A.L. *Electronic Properties of Doped Semiconductors*, 1st ed.; Springer-Verlag: Berlin, Germany, 1984; pp. 94–107.
4. Derjaguin, B.V.; Landau, L.D. Theory of the stability of strongly charged lyophobic sols and of the adhesion of strongly charged particles in solutions of electrolytes. *Acta Phys. Chem. URSS* **1941**, *14*, 633–662. [CrossRef]
5. Verwey, E.; Overbeek, J. *Theory of the Stability of Lyophobic Colloids*, 1948 ed.; Elsevier: Amsterdam, The Netherlands, 1948; pp. 131–136.
6. Riedl, J.C.; Akhavan Kazemi, M.A.; Cousin, F.; Dubois, E.; Fantini, S.; Lois, S.; Perzynski, R.; Peyre, V. Colloidal dispersions of oxide nanoparticles in ionic liquids : Elucidating the key parameters. *Nanoscale Adv.* **2020**. [CrossRef]
7. Bacri, J.C.; Perzynski, R.; Salin, D.; Cabuil, V.; Massart, R. Ionic ferrofluids: A crossing of chemistry and physics. *J. Magn. Magn. Mater.* **1990**, *85*, 27–32. [CrossRef]
8. Dubois, E.; Cabuil, V.; Boué, F.; Perzynski, R. Structural analogy between aqueous and oily magnetic fluids. *J. Chem. Phys.* **1999**, *111*, 7147–7160. [CrossRef]
9. Landau, L.D.; Lifshitz, E.M. Statistical Physics. In *Course of Theoretical Physics*, 3rd ed.; Elsevier: Amsterdam, The Netherlands, 2011; Volume 5, pp. 276–278.

10. Rotenberg, B.; Dufrêche, J.F.; Turq, P. Frequency-dependent dielectric permittivity of salt-free charged lamellar systems. *J. Chem. Phys. B* **2005**, *123*, 154902–154903. [CrossRef] [PubMed]
11. Durand-Vidal, S.; Jardat, M.; Dahirel, V.; Bernard, O.; Perrigaud, K.; Turq, P. Determining the radius and the apparent charge of a micelle from electrical conductivity measurements by using a transport theory: Explicit equations for practical use. *J. Chem. Phys. B* **2006**, *110*, 15542–15547. [CrossRef] [PubMed]
12. Jardat, M.; Dahirel, V.; Durand-Vidal, S; Lucas, I.; Bernard, O.; Turq, P. Effective charges of micellar species obtained from Brownian dynamics simulations and from an analytical transport theory. *Mol. Phys.* **2004**, *104*, 3667–3674. [CrossRef]
13. Heltner, P.; Papir, Y.; Krieger, I. Diffraction of light by nonaqueous ordered suspensions. *J. Phys. Chem.* **1971**, *75*, 1881–1886. [CrossRef]
14. Kose, A.; Ozake, T.; Takano, K.; Kobayschi, Y.; Hachisu, S. Direct observation of ordered latex suspension by metallurgical microscope. *J. Colloid Interface Sci.* **1973**, *44*, 330–338. [CrossRef]
15. Williams, R.; Crandall, R. The structure of crystallized suspensions of polystyrene spheres. *Phys. Lett. A* **1974**, *48*, 225–226. [CrossRef]
16. Alexander, S.; Chaikin, P.; Grant, P.; Morales, G.; Pincus, P. Charge renormalization, osmotic pressure, and bulk modulus of colloidal crystals: Theory. *J. Chem. Phys.* **1984**, *80*, 5776–5781. [CrossRef]
17. Landau, L.D.; Lifshitz, E.M. *Course of Theoretical Physics: Vol. 6, Fluid Mechanics*, 2nd ed.; Elsevier: Amsterdam, The Netherlands, 2013; p. 46.
18. Jackson, J.D. *Classical Electrodynamics*, 3rd ed.; John Wiley & Sons, Inc.: New York, NY, USA, 1999; p. 114.
19. Dykhne, A.M. Conductivity of a Two-Dimensional Two-Phase System. *Sov. JETP* **1971**, *32*, 63–65.
20. Onsager, L. On the theory of electrolytes. *Physica Z* **1927**, *28*, 277–298.
21. Debye, P.; Huckel, E. The theory of the electrolyte II-The border law for electrical conductivity. *Physica Z* **1923**, *24*, 305–325.
22. Lifshitz, E.M.; Pitaevskii, L.P. *Physical Kinetics: Course of Theoretical Physics—Volume 10*, 1st ed.; Butterworth-Heinenann Ltd.: London, UK, 2002; p. 125.
23. Grosberg, A.; Nguyen, T.; Shklovskii, B. The physics of charge inversion in chemical and biological systems. *Rev. Mod. Phys.* **2002**, *74*, 329–345. [CrossRef]
24. Robinson, R.; Stokes, R. *Electrolyte Solutions*, 1959 ed.; Butterworths Scientific Publications: London, UK, 1959; p. 119.
25. Lizana, L.; Grossberg, A. Exact expressions for the mobility and electrophoretic mobility of a weakly charged sphere in a simple electrolyte. *Europhys. Lett.* **2013**, *104*, 68004–68009. [CrossRef]

© 2020 by the authors. Licensee MDPI, Basel, Switzerland. This article is an open access article distributed under the terms and conditions of the Creative Commons Attribution (CC BY) license (http://creativecommons.org/licenses/by/4.0/).

Article

# Segmented Thermoelectric Generator under Variable Pulsed Heat Input Power

**Pablo Eduardo Ruiz-Ortega [1,†], Miguel Angel Olivares-Robles [2,\*,†] and Olao Yair Enciso-Montes de Oca [2,†]**

1. Instituto Politecnico Nacional, Depto. Ingenieria Bioquimica, ENCB, Ciudad de Mexico 07738, Mexico; eduardo29491@gmail.com
2. Instituto Politecnico Nacional, SEPI ESIME Culhuacan, Ciudad de Mexico, Coyoacan 04430, Mexico; olaoyairenciso1991@gmail.com
\* Correspondence: olivares@ipn.mx; Tel.: +52-555-729-6000 (ext. 73262); Fax: +52-555-656-2058
† These authors contributed equally to this work.

Received: 16 August 2019; Accepted: 22 September 2019; Published: 24 September 2019

**Abstract:** In this paper, we consider the transient state behavior of a segmented thermoelectric generator (STEG) exposed to a variable heat input power on the hot side while the transfer of heat on the cold side is by natural convection. Numerical analysis is used to calculate the power generation of the system. A one-dimensional STEG model, which includes Joule heating, the Peltier effect with constant properties of materials, is considered and governing equations are solved using the finite differences method. The transient analysis of this model is typical for energy harvesting applications. A novel design methodology, formulated on the ratio of the figure of merit of the thermoelectric materials, is developed including segmentation on the legs of the thermoelectric generator, which does not consider previous studies. In our approach, the figure of merit is an advantageous parameter to analyze its impact on thermal and electrical efficiency. The transient state of the thermoelectric generator is analyzed, considering two and three heat input sources. We obtain the temperature profiles, voltage generation, and efficiency of the STEG under pulsed heat input power. The results showed that the temperature drop along the semiconductor elements was more considerable when three pulses were applied, and when the thermal conductivity in the first segment was higher than that of the second segment. Furthermore, we show that the generated voltage and the maximum efficiency in the system occur when the value of the figure of merit in the first segment, which is in contact with the temperature source, is lower than the figure of merit for the second thermoelectric segment of the leg. The model investigated in this paper offers an essential guide on the thermal and electrical performance behavior of the system under transient conditions, which are present in many variable thermal phenomena such as solar radiation and the normalized driving cycles of an automotive thermoelectric generator.

**Keywords:** segmented thermoelectric generator; pulsed heat; transient

---

## 1. Introduction

Power generation based on thermoelectric effects, which use new thermoelectric materials technology, is of interest for researchers because it converts thermal energy into electricity and is utilized as a new way to harvest clean energy. To satisfy world energy demands, it is vital to investigate new areas in energy conversion technologies for power generation. Recent research into the improvement of efficiency and reduced costs of thermoelectric materials make essential contributions towards harvesting clean energy [1,2]. Therefore, advances in the development of new applications to take advantage of renewable energy sources have increased in recent years and are being carried out as a new solution to reduce the use of fossil fuels [3]. Thermoelectric energy harvesting employs thermoelectric generators. A thermoelectric generator (TEG) is composed of several pairs of p-type and n-type semiconductor elements, known as legs, connected electrically in a series and thermally in parallel.

Thermoelectric generators (TEGs) generate electricity directly from thermal energy in a closed-circuit when there exists a temperature difference between the hot side and the cold side of the device, which has no moving parts. The governing thermoelectric effects in the energy conversion are the Seebeck, Peltier, Thomson, and Joule effects [4–6]. TEGs provides several advantages against other power generating systems such as gas-free emissions, endless shelf life, no noise, a simple structure, maintenance-free operation, no pollution [7–9], and have been used as a harvester of waste heat from power plants. Therefore, most of the new research focuses on improving the efficiency of TEG to reach high energy conversion [10]. Some groups of researchers have made an effort to investigate new methods for energy harvesting with several numerical models to optimize the performances of TEGs have been proposed [11,12]. Thereby, the development of promising industry and daily life applications of TEGs is expected. In the improvement of TEGs energy conversion efficiency, it is necessary to use new techniques that can be focused through device design or developing new semiconductor materials according to its thermoelectric figure of merit [13]. Geometry optimization for the legs of a thermoelectric generator has been previously analyzed by Ma et al. [14], who propose a generator with minimized thickness for maximum power output. Meng [15] showed the effects of thermocouple physical size on the performance of a thermoelectric heat pump driven by a TEG. The segmentation of the semiconductor elements improves the performance of TEGs as demonstrated by the results shown in previous research [16,17]. A segmented thermoelectric generator (STEG) contains legs composed of two thermoelectric materials and is used to take advantage of the working conditions at different temperature gradients. The efficiency of a STEG increases if the two joined thermoelectric materials are compatible, which may lead to higher efficiency. Ming et al. [18] conducted a thermal analysis, using a three-dimensional finite element model on a segmented thermoelectric generator and results indicated that maximum efficiency increased 11.2% when the load resistance value was very nearly to the internal resistance value. Shu et al. [19] propose a thermoelectric generator for engine waste heat recovery, using a three-dimensional numerical model and results showed that the output power was higher than that of a non-segmented TEG by 13.4%.

It is well known that the study of TEGs operating with pulsed heat input is more challenging in a numerical simulation than thermoelectric coolers (TECs) because both the temperature gradient along the semiconductors and the electric currents vary through time. The dynamic behavior of thermoelectric devices has been studied and their importance has been reported in works such as Paul E. Gray's book [20]. The application of pulsed heat input is a novel method to enhance the maximum performance of thermoelectric devices and the understanding of its transient behavior is essential for optimizing energy harvesting from waste heat. Crane [21] studied the differences between TEG behavior at steady-state and transient models in a MATLAB/Simulink environment where the devices and systems modeled were optimized according to an advanced multiparameter optimization technique. Mahmoudinezhad et al. [22] studied the performance of a STEG with self-adhesive graphite sheet attached to the hot surface, under variable solar radiation at high operation temperatures using a numerical simulation by the finite volume method. Results showed that the graphite absorber had an effect on the power generation by the enhancement of absorbed radiation. Samson et al. [23] investigated a segmented asymmetrical thermoelectric generator (SASTEG) under transient conditions and the results showed that by using an asymmetrical leg the thermal stress reduced by 39.21% compared to the symmetrical leg geometry.

In transport vehicles, several studies using TEGs have been performed for waste heat recovery. For example, in driving the exhaust temperature and gas flow rate vary depending on the engine operating conditions. High temperatures are achieved in the exhaust line recovery and a portion of this wasted energy can be converted into electricity, which is advantageous. Transient tests have demonstrated that the overall power generation of a TEG can be improved by controlling the hot-side temperature [24]. In addition to this, the placement of TEGs taking into account the interaction with the internal combustion engine has also been investigated [25]. This kind of research can be carried out by using a numerical simulation. Research has also been done on the variation of solar radiation, power generation, and efficiency of a TEG under the transient condition where it has been shown the impact of the thermal contact resistance on the temperature profile and system efficiency [26]. It is very important to study the transient behavior of a TEG when the heat source is variable and even more so when using segmented materials, which is known to increase the efficiency of devices. Thermoelectric devices and thermal collectors have been studied for energy harvesting, steady-state systems, and different types of practical applications [27,28].

The previously presented literature shows the optimization of thermoelectric generators using segmented legs. Measurements of figure of merit has been investigated using different techniques and by also considering different leg materials (inhomogeneous or of an irregular shape), as well as configurations of TEGs, observing an increase in the figure of merit due to the bulk thermoelectric effect [29–31]. Previous studies do not take into account the relationship of the figure of merit between two different materials using variable pulsed heat input to optimize TEG performance. This paper focuses on the application of the segmented thermoelectric materials of a TEG when exposed to a variable pulsed heat input for different energy harvesting applications.

## 2. One-Dimensional Model of a Segmented Thermoelectric Generator (STEG)

In this work, numerical analysis based on the finite differences method is developed to study a STEG's transient characteristics for example, spatial temperature profiles, voltage output, and efficiency. The proposed model for the segmented legs of the STEG is shown in Figure 1, where $A$ is the cross-sectional area, $L_1$ and $L_2$ are the lengths of the first and second segment, respectively, $L = L_1 + L_2$ is the total length, $T_c$ and $T_h$ are the temperatures at the cold and heat ends of the thermoelectric element, and $T_m$ is the temperature at the junction of the two segments. Numerical solutions for the spatial temperature profiles, power output, and efficiency are carried out considering variable heat input pulses when the STEG is exposed to (a) two heat input sources, and (b) three heat input sources.

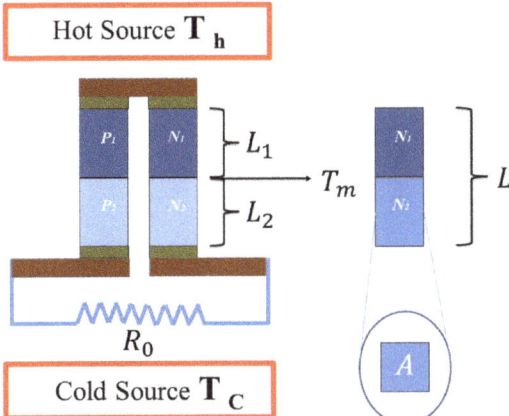

**Figure 1.** 1D model of a segmented thermoelectric generator (STEG) under variable heat input power. $L_1$ and $L_2$ are the length for the first and second segment of the leg, respectively. $P_1$, $N_1$ and $P_2$ $N_2$ are the p-type and n-type elements for the first and second segment, respectively, and $R_0$ is the load resistance.

## 3. Numerical Model

The thermoelectric phenomena are described using (a) the energy balance equation,

$$DC_p \frac{\partial T}{\partial t} + \nabla \cdot q'' = Q' \tag{1}$$

where $C_p$ is the specific heat capacity, $D$ is the density, $t$ is the time, and $T$ is the temperature. $Q'$ is the Joule heating energy,

$$Q' = JV \tag{2}$$

$q''$ is the input heat flux,

$$q'' = -\kappa \nabla T + \Pi J \tag{3}$$

$\Pi$ is the Peltier coefficient, $V$ is the difference in electric potential, and $J$ is the electric current flux. (b) is the current density continuity,

$$\frac{\partial \rho_c}{\partial t} = \nabla \cdot J \tag{4}$$

and $\rho_c$ is the charge density.

It is well known that,

$$\Pi = \alpha T \tag{5}$$

and

$$J = -\sigma \nabla E - \sigma \alpha \nabla T \tag{6}$$

where $\sigma$ is the electrical conductivity, $\kappa$ is the thermal conductivity, $E$ is the electric field, and $\alpha$ is the Seebeck coefficient.

The energy balance to the thermoelement (n-type and p-type) using the 1D unsteady-state heat transfer modeling is given by:

$$\frac{\partial^2 T}{\partial x^2} + \frac{\rho_{1,2} I^2}{\kappa_{1,2} A^2} = \frac{D_{1,2} C_{p1,p2}}{\kappa_{1,2}} \frac{\partial T}{\partial t} \tag{7}$$

where $\rho$, $I$, and $x$ are the electrical resistivity, the electric current flowing in a closed-circuit, and spatial coordinate, respectively. The subscripts determine the type of material, where 1 is for the $CoSb_3$ and 2 is for the $Bi_2Te_3$.

Equation (8) represents the energy balance for $CoSb_3$ and $Bi_2Te_3$ at the interface. Equation (9) represents the energy balance for the cold side of the thermoelement exposed to natural convection and is expressed as follows:

$$\frac{\Delta x A D_1 C_{p1} + \Delta x A D_2 C_{p2}}{2} \frac{\partial T}{\partial t} = -\alpha_1 TI + \kappa_1 A \frac{\partial T}{\partial x} + \alpha_2 TI - \kappa_2 A \frac{\partial T}{\partial x} \tag{8}$$

$$\frac{\Delta x A D_2 C_{p2}}{2} \frac{\partial T}{\partial t} = -\alpha_2 TI + \kappa_2 A \frac{\partial T}{\partial x} + hA(T - T_a) \tag{9}$$

where $h = 2.5 \frac{W}{mK}$ is the natural convection coefficient and $T_a$ is the ambient temperature [21].

### 3.1. Initial Conditions

The initial conditions on the lower and upper surface of the thermoelement are as follows:

$$T_h(t, L) = T_h(t, 0) = 388 \text{ K} \qquad T_c(t, L) = T_c(0, L) = 298 \text{ K} \tag{10}$$

### 3.2. Transient State Equations Solution by the Finite Differences Method

The finite difference method has been used to solve the differential equation with the transient term (7) since this method is stable and more accurate than other methods. Further details about this numerical technique to solve the finite difference equations are in: Numerical Methods for engineers, Steven C. Chapra (2007). All different temperature profiles for the transient state are obtained by applying the finite differences method to the energy balance equations. The thermoelement have been divided into $N$ number of equal parts with equivalent length ($\Delta x$) and node numbers from $i = 0$ to $i = N+1$, where $N = 110$.

Solving Equations (8)–(10), we obtain,

$$T_{i-1}^{n+1}\left(\frac{\kappa_{1,2}\Delta t}{C_{p1,p2}D_{1,2}\Delta x^2}\right) - T_i^{n+1}\left(\frac{2\kappa_{1,2}\Delta t}{C_{p1,p2}D_{1,2}\Delta x^2} + 1\right) + T_{i+1}^{n+1}\left(\frac{\kappa_{1,2}\Delta t}{C_{p1,p2}D_{1,2}\Delta x^2}\right) + T_i^n + \frac{\rho_{1,2}I^2}{A} = 0 \quad (11)$$

Equation (11) describes the temperature distribution for $CoSb_3$ and $Bi_2Te_3$.

$$T_{i-1}^{n+1}\left(\frac{\kappa_1 A\Delta t}{\psi \Delta x}\right) - T_i^{n+1}\left(\frac{\alpha_2 I \Delta t}{\psi} + \frac{\kappa_2 A \Delta t}{\psi \Delta x} + \frac{\kappa_1 A \Delta t}{\psi \Delta x} - \frac{\alpha_1 I \Delta t}{\psi} - 1\right) - T_{i+1}^{n+1}\left(\frac{\kappa_2 A \Delta t}{\psi \Delta x}\right) + T_i^n = 0 \quad (12)$$

Equation (12) describes the temperature $CoSb_3$ and $Bi_2Te_3$ at interface.

$$T_{i-1}^{n+1}\left(\frac{2\kappa_2 A\Delta t}{C_{p2}D_2\Delta x^2}\right) - T_i^{n+1}\left(\frac{2\kappa_2\Delta t}{C_{p2}D_2\Delta x^2} + \frac{2h\Delta t}{D_2 C_{p2}\Delta x} - \frac{2\alpha_2 I \Delta t}{D_2 A C_{p2}\Delta x} - 1\right) + T_i^n - \frac{2hT_a\Delta t}{D_2 C_{p2}\Delta x} = 0 \quad (13)$$

where $\psi$ is given as:

$$\psi = \frac{\Delta x A D_1 C_{p1} + \Delta x A D_2 C_{p2}}{2} \quad (14)$$

Equation (13) describes the temperature of the surface exposed to the natural convection of the $Bi_2Te_3$ material.

### 3.3. Electrical Performance Equations

In closed-circuit mode, the generated output voltage is given by:

$$V_{cc} = V_{oc} - IR \quad (15)$$

where $V_{oc}$ is the open-circuit voltage and can be determined with the following equation:

$$V_{oc} = T_0 \alpha_E - T_L \alpha_E \quad (16)$$

The effective coefficient of Seebeck is calculated for a segmented semiconductor material [32] and is defined in Equation (17) as follows:

$$\alpha_E = \alpha_1 \frac{T_0 - T_{\frac{N}{2}+1}}{T_h - T_c} + \alpha_2 \frac{T_{\frac{N}{2}+1} - T_{N+1}}{T_h - T_c} \quad (17)$$

where $T_0$, $T_{\frac{N}{2}+1}$ and $T_{N+1}$ are the hot side constant temperature, interface temperature through time, and cold side temperature through time, respectively.

The electrical resistance, $R$, of the thermoelement is given by:

$$R = \frac{\rho_1 L}{A} + \frac{\rho_2 L}{A} \quad (18)$$

where $L$ is the length of the thermoelement of the STEG. Considering a resistive load $R_0 = 1.5\,\Omega$ [33], the electric current can be determined with the following equation:

$$I = \frac{V_{oc}}{R_0 + R} \quad (19)$$

In a thermoelectric power generation device, efficiency is given by,

$$\eta = \frac{W}{Q_h} = \frac{I(\alpha_E \Delta T - IR)}{K_{th}\Delta T + \alpha_E I T_h - \frac{1}{2}I^2 R} \quad (20)$$

where $W$ is the power delivered by the TEG system, $Q_h$ is the heat flow from the heat source to the sink, and $K_{th} = \kappa A/L$ is the thermal conductance.

### 3.4. The Figure of Merit

To achieve a high value of the figure of merit, the dimensionless parameter that determines the performance of a thermoelectric material requires high electrical conductivity, low thermal conductivity, and a high Seebeck coefficient.

The figure of merit is defined as follows:

$$Z_{1,2} = \frac{\alpha_{1,2}^2}{\rho_{1,2}\kappa_{1,2}} \quad (21)$$

Equation (21) is used to know the merit figure of material 1 and 2.

$$Z_r = \frac{Z_1}{Z_2} \quad (22)$$

where $Z_r$ is the ratio of material 1's figure of merit to material 2's figure of merit.

## 4. Material Properties and Geometry Description

In this paper, we assume that the thermoelectric materials' properties are independent of temperature and the constant parameters and dimensions of the semiconductor elements are given in Table 1. It has been proven in other works that by using constant material properties for calculations, thermal and electrical characterization can be matched with experimental data [33]. It has been proven that the Thomson effect does not affect the temperature profile in TEGs, but the output voltage is impacted by the Thomson effect and even more so by the load resistance. Here we focus on optimization according to the ratio of figure of merit $Z_r$ since the value of the figure of merit changes, in small ranges, due to the variation of the average temperature. Therefore, the temperature gradient along the semiconductor's elements is the most dominant factor and hence thermal and electrical characterization can be achieved [34]. Optimization that takes into account the Thomson effect is presented in this study. A novel design methodology using a computational tool that focuses on examining the influence of segmentation is proposed. Depending on the particular application for energy harvesting, TEG performance varies mainly due to $\alpha$, $\kappa$, $\tau$, and $\rho$ which in this paper are considered in $Z_r$ and $Z_{r,\tau}$, without the Thomson effect and with the Thomson effect, respectively. Therefore, and according to this last statement, only a resistive load $R_0$ is considered to show the thermal and electrical response of a STEG considering different values of the figure of merit. Thomson coefficient values are assumed as constant and Seebeck is related to the Thomson effect as follows:

$$\alpha_{1,2} = (\alpha_{p(1,2)} - \alpha_{n(1,2)}) + (\tau_{p(1,2)} - \tau_{n(1,2)})\ln\left(\frac{T_{avg}}{T_{ref}}\right) \quad (23)$$

where $T_{ref}$ is the room temperature (298 K) and $T_{avg}$ is the average temperature when the heat fluxes are applied, i.e., 468 K.

Table 1. Constant properties of the thermoelectric materials [22,33].

| Property | Material 1 $CoSb_3$ | Material 2 $Bi_2Te_3$ | Unit |
|---|---|---|---|
| $\alpha$ | $459 \times 10^{-6}$ | $512 \times 10^{-6}$ | V K$^{-1}$ |
| $\kappa$ | 3.22 | 3.518 | W m$^{-1}$ K$^{-1}$ |
| $\rho$ | $1.01 \times 10^{-5}$ | $4.378 \times 10^{-5}$ | $\Omega$m |
| $D$ | 7582 | 8160 | Kg m$^3$ |
| $T_a$ | 298 | 298 | K |
| $A$ | $2.25 \times 10^{-6}$ | $2.25 \times 10^{-6}$ | m$^2$ |
| $L$ | $2.2 \times 10^{-3}$ | $2.2 \times 10^{-3}$ | m |
| $C_p$ | 238.7 | 155 | J kg$^{-1}$ K$^{-1}$ |
| $\tau$ | $157 \times 10^{-6}$ | $22.394 \times 10^{-6}$ | V K$^{-1}$ |

## 5. Heat Input Power Effect on Performance

A thermoelectric generator consists of copper conductors that connect the p-type and n-type semiconductor materials and facilitates electrical conductivity within the TEG. In this work, these copper conductors as well as heat losses due to radiation and transverse convection are not taken into account, but an external load resistor, and thus voltage output of the STEG, can be measured. For calculations, all materials are assumed homogeneous, and property materials such as thermal conductivities, electrical resistivity, the Thomson coefficient, and materials' specific heat capacities are assumed constant and do not change throughout time or temperature. We considered a length of the semiconductor element in the STEG of $L_{1,2} = 1.1 \times 10^{-3}$ m for the first and second segment. Variable heating conditions are applied to the hot side of the TEG while the cold side changes through time. We calculated the spatial temperature profiles along the STEG elements, the cold side temperature, and output voltage during pulse heat operation under the following conditions:

(a) The system begins at room temperature at 298 K;
(b) The heat pulses are input to the system, alternating the temperature sources $T_1 = 388$ K for some time of $t = 0$–5 s and then;
(c) A source of $T_2 = 500$ K for a period of time of $t' = 5$–10 s, as shown in Figure 2.

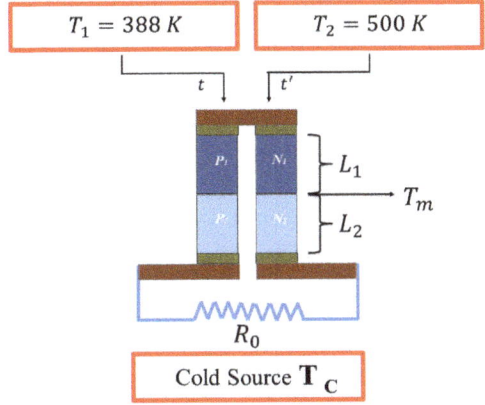

Figure 2. Two temperature sources are applied on the thermoelements at time $t$ and $t'$.

This process is repeated up to a total time of 35 s. For the second case (b), firstly a temperature source of $T_1 = 388$ K is applied, then a temperature source of $T_2 = 468$ K, and finally a temperature source of $T_3 = 548$ K, with each source applied in a period of time of $t = 0$–5 s, $t = 5$–10 s, and $t = 10$–15 s, respectively, as shown in Figure 3. This process is repeated up to a total time of 50 s. In both cases, $CoSb_3$ is placed in the first segment and $Bi_2Te_3$ in the second segment. The load resistance is kept constant throughout this study. Rectangular heat input flux function is used to model the pulse in all cases. In our simulation model, the relevant constant parameters used in the numerical simulation are listed in Table 1.

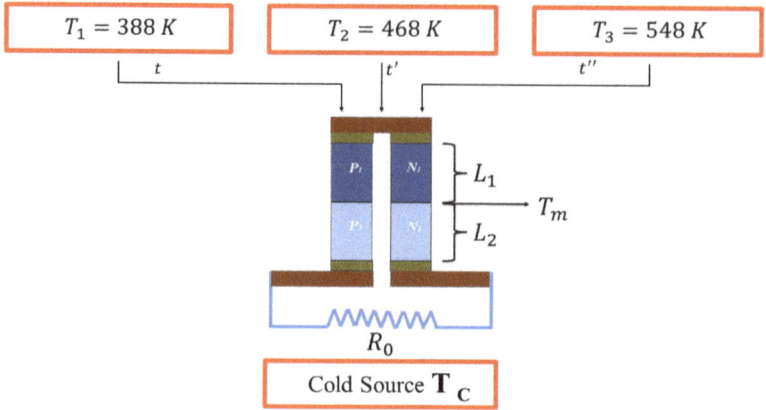

Figure 3. Three temperature sources are applied on the thermoelements at time $t$, $t'$, and $t''$.

## 5.1. Two Heat Input Pulses: Thermal Behavior

In particular, the output voltage and the temperature profiles across a STEG under periodic pulse heating were compared for different $Z_r$ and $Z_{r,\tau}$ values, for both case (a) and (b). In all temperature and voltage profiles, solid lines do not take into account the Thomson effect and dashed lines are for results taking into account the Thomson effect. According to Equations (11)–(13), Figure 4 shows the spatial temperature profile evaluated when the heat pulse of 500 K was applied in a period of time of $t = 5$ s, along the STEG elements until the system reached a steady-state distribution for case (a) as previously described. Here the temperature profile behavior was similar in all heat pulses for all cases, and so there was no need to include all the figures. For this result a temperature difference of $\Delta T = 112$ K when the heat input pulse of 500 K was applied is used, and the temperature distribution was evaluated until the temperature in the segment reached a steady-state distribution.

It can be seen that the temperature increases on the hot side, i.e., in the first segment, at the moment when the heat pulse was applied which caused an increase in the temperatures in a specific length and here no change in the interface was observed for the first instants of time. After 0.25 s there was a change at the interface because the value of the thermal conductivity of the $CoSb_3$ was higher than the material $Bi_2Te_3$, $\kappa_1 = 3.09$ W/mK and $\kappa_2 = 6.523$ W/mK respectively, and therefore heat was preferentially conducted toward the cross-sectional area easily in the first segment. Notice that the heat absorbed in the first segment was due to $CoSb_3$ material, and the heat released in the second segment was due to $Bi_2Te_3$.

In this paper, the thermal characterization was made considering open-circuit conditions. The application of the two transient rectangular pulsed heat power is shown in Figure 5 which shows the transient behavior of the cold side temperature, $T_c$, for different $Z_r$ and $Z_{r,\tau}$ values shown in the figure. As shown from these results, the temperature profiles were not affected by the Thomson effect as is well known from the literature and only output voltage results change considering the Thomson effect, as demonstrated in the next section. To show the periodic behavior of the system when the pulses were repeatedly applied, the total time in all simulation was set to 35 s. Simulations were performed by varying the thermal conductivity of the $Bi_2Te_3$ material to obtain the different values of $Z_r$ and this range spanned the cases when $\kappa_1 < \kappa_2$ to $\kappa_1 > \kappa_2$. It can be seen from these results that the cold side temperature $T_c$ values were very similar for all $Z_r$ values where the thermal conductivity, electrical resistivity, and Seebeck coefficient of the $CoSb_3$ material remained as constant values, i.e., $Z_1 = cte$, and only material properties of $Bi_2Te_3$ were changed in order to vary $Z_2$ values.

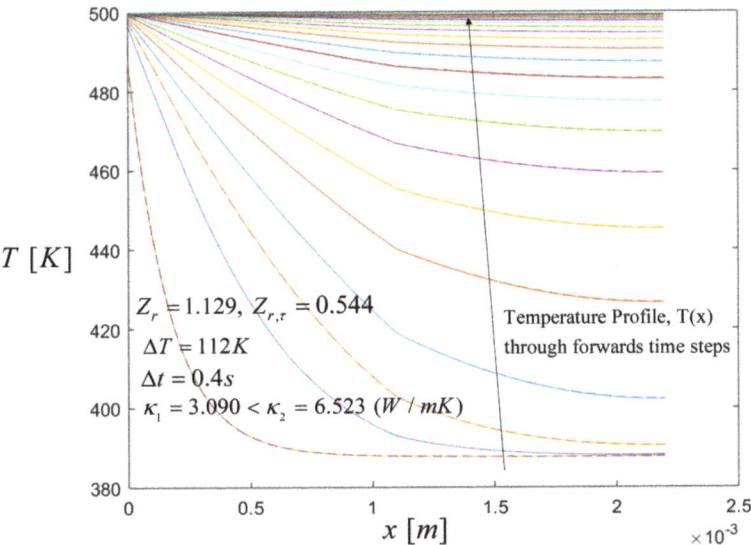

**Figure 4.** Spatial temperature profile during the heat pulse when $Z_r = 1.129$ and $Z_{r,\tau} = 0.544$, with Thomson and without Thomson coefficients, respectively. Comparison results with the Thomson effect (dashed lines) and without the Thomson effect (solid lines). The arrow direction point out the temperature distribution along the thermoelements through the time steps.

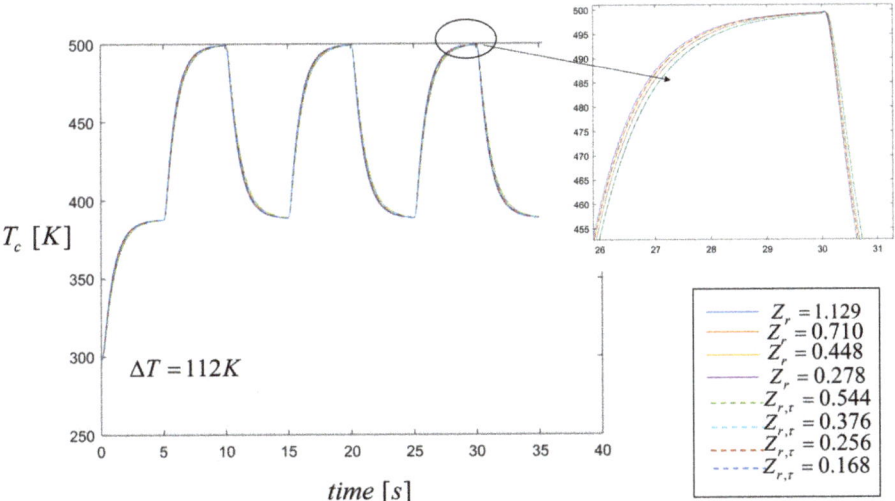

**Figure 5.** Temperature distribution on the cold side through time when two heat pulses are applied. Comparison results with the Thomson effect (dashed lines) and without the Thomson effect (solid lines).

## 5.2. Electrical Responses to Periodic Heat Fluxes

According to the law of heat transfer from the hot side to the cold side, the temperature of the cold side changes until thermal equilibrium is attained within a STEG. Temperature gradient values become smaller through time when the system approaches steady state and therefore when the generated voltage also decreases. Maintaining $T_c$ values constant in order to achieve higher temperature gradients is difficult which directly affects the electrical performance of the TEG. Higher voltage generation accuracy can be achieved by considering the heat input power as an efficient parameter to analyze its influence on the energy conversion of a STEG. As is well known from the literature, no previous study has focused on the effect of variable pulsed heat on the performance of a STEG as is done in this work. The duty cycle determines the ratio of the heating time to the period time in the simulation if the duty cycle value is very high, the TEG performance is reduced. In this case for the two heat pulses, the period of heating is 5 s and the period time is 10 s, then the duty cycle is set to 0.5 s, which is the value used in all the simulations. The heating effects on the maximum voltage generation are shown in Figure 6 for different $Z_r$ and $Z_{r,\tau}$ values. The output voltage follows Ohm's law, i.e., it is given by the current multiplied by the load resistance from Equation (19). It can be seen that during one period of pulse heating the voltage generation values, for all the different $Z_r$ and $Z_{r\tau}$, are very similar in all cases reaching a maximum value of $V_{max} = 55$ mV and $V_{max} = 59$ mV, without the Thomson effect and with the Thomson effect respectively. The difference in the energy generated by STEG under periodic heating is almost the same, with a difference of less than 0.5% which can be easily seen from Figure 6, indicating that working with two input pulses and changing the figure of merit of the second material, i.e., $Z_2$, does not significantly affect the voltage generation. These results indicate that the smaller the value of $Z_r$, the more output voltage in the system is generated, which is confirmed in the next section where changes in the figure of merit when working with three heat pulses, increase the voltage generated.

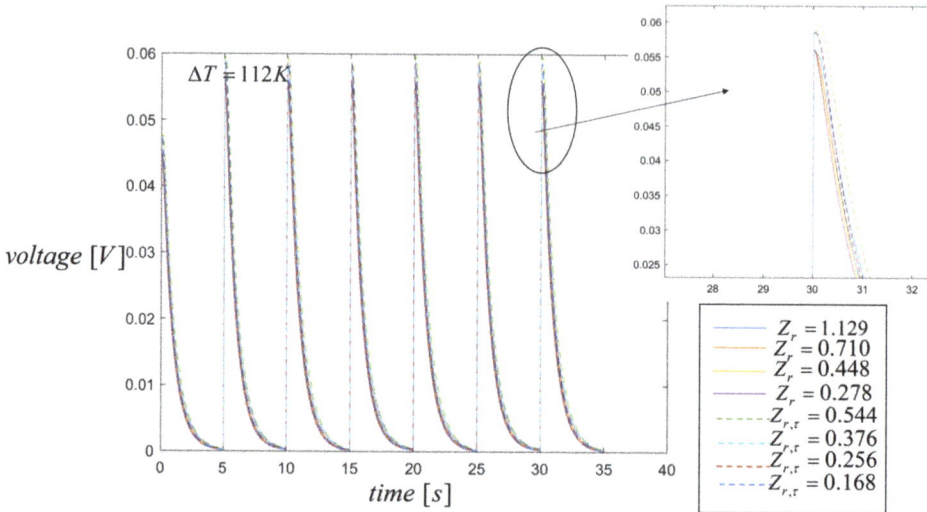

**Figure 6.** Voltage generation when two heat input pulses are applied with a duty cycle of 35 s. Comparison results with the Thomson effect (dashed lines) and without the Thomson effect (solid lines).

## 5.3. Three Heat Input Pulses: Thermal Behavior

In this section, we analyze the case in which three heat input pulses are applied, i.e., case (b) previously described, where the initial temperature for the STEG system is 298 K. The spatial temperature profiles were affected by the heat pulses applied and the properties of the materials.

Thereby, we calculated the temperature profiles through the time when $Z_r = 0.028$ and $Z_{r,\tau} = 0.019$. In the next section, we will show that this value improves the voltage output compared with other $Z_r$ and $Z_{r,\tau}$ values. Figure 7 shows the spatial temperature profile, evaluated when the heat pulse of 548 K is applied during 5 s until the system reaches a steady-state distribution. In this case, for three heat pulses, the time of heating is again 5 s and the period time is 15 s, then the duty cycle is set to 0.33 s. We should note that for the case when $\kappa_1 > \kappa_2$, the heat conduction is better in the first segment and from the interface forward to the second segment, the conduction of heat is reduced due to the lower thermal conductivity of the semiconductor material in the second segment. The higher temperature of the third pulse significantly changes the temperature profile, which can be appreciated from the interface when the Z values changes i.e., $Z_1 < Z_2$, compared to the case for two heat pulses.

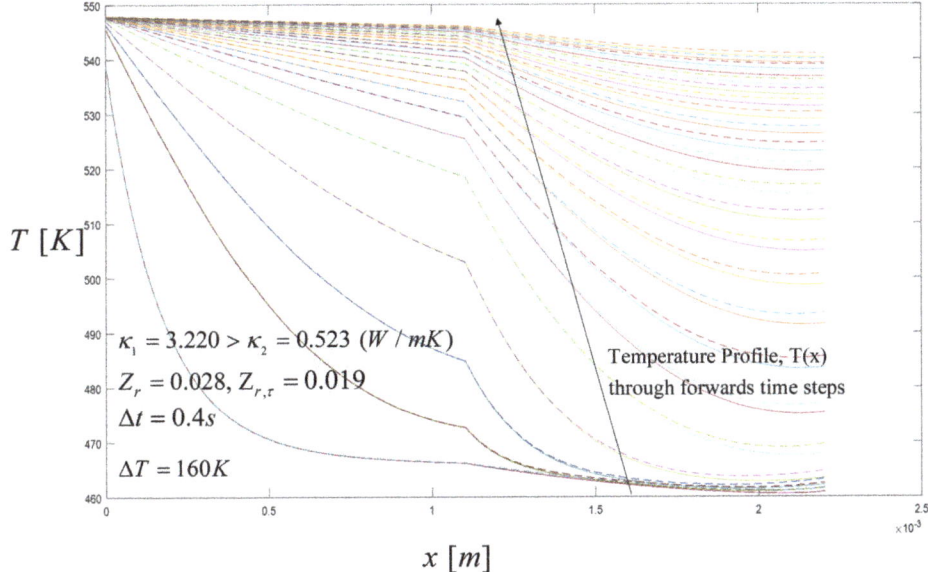

**Figure 7.** Spatial temperature profile during the pulse from 10.25 s to 15 s in time intervals of 0.25 s with $Z_r = 0.028$ and $Z_{r,\tau} = 0.019$, with Thomson and without Thomson coefficients, respectively. Comparison results with the Thomson effect (dashed lines) and without the Thomson effect (solid lines). The arrow direction point out the temperature distribution along the thermoelements through the time steps.

Figure 8 shows the temperature profiles on the cold side, $T_c$, for different $Z_r$ and $Z_{r,\tau}$. For the case when $Z_{r,\tau} = 0.019$, the temperature profile of the cold side showed a small increase in temperature at the moment when the source at 500 K was replaced by the source at 388 K. It is worthy to note that depending on the thermal conduction coefficient and considering the Thomson effect, which directly affects $Z_{r,\tau}$ values, the system reached its higher temperature values for $T_c$, due to the amount of heat flux across the interface for the case when $\kappa_1 > \kappa_2$. For all the other cases, the temperature profiles were close to each other as shown in Figure 8. From these results we can point that the spatial temperature profile played an essential role in power generation. In fact, there was a relation between $Z_r$ and $Z_{r,\tau}$ values with the output voltage and efficiency of the STEG system for the different temperature profiles. See the next section.

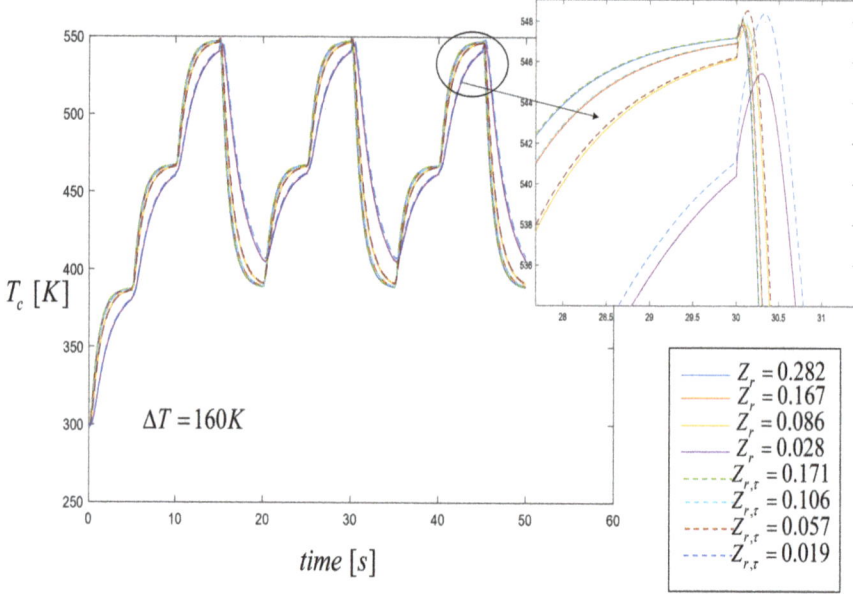

**Figure 8.** Temperature distribution on the cold side through time when three heat pulses are applied. Comparison results with the Thomson effect (dashed lines) and without the Thomson effect (solid lines).

### 5.4. Electrical and Efficiency Responses to Periodic Heat Fluxes

By using a third heat input pulse, higher values on the spatial temperature profiles were reached and consequently the output voltage increased, as shown in Figure 9 when $Z_r = 0.0028$ and $Z_{r,\tau} = 0.019$. The two important parameters in power generation are the figure of merit and temperature difference between the hot and cold sides of the STEG. Our results showed that a lower figure of merit value was required in the second segment to achieve a higher output voltage. In this paper, a small variation of the STEG output voltage was observed, since the value of the figure of merit changed within a small range in all other cases of $Z_r$ values. For this case, the difference in the energy generated by the STEG changed under three periodic heating compared with the case for two input pulses. Figure 9 shows that working with three input pulses and reducing the devalue of the second material's figure of merit directly affected the voltage generation. Maximum output voltages were $V_{max} = 88.7$ mV, without the Thomson effect, and $V_{max,\tau} = 108.3$, with the Thomson effect. As shown in the figure, the output voltage was significantly affected for the Thomson effect, with an 18.1% increase of voltage generated by the system considering the Thomson effect. Thus, the importance of including the Thomson effect was critical to the optimum characterization of STEG models in the transient state.

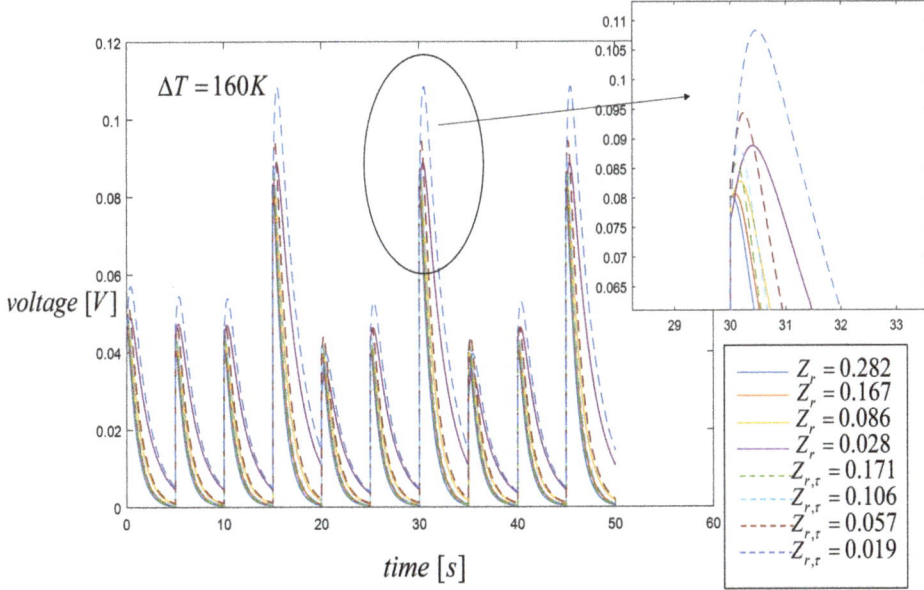

**Figure 9.** Voltage generation when three heat input pulses are applied with a duty cycle of 50 s. Comparison results with the Thomson effect (dashed lines) and without the Thomson effect (solid lines).

From Figure 10, for transient state efficiency, we can see that maximum efficiency could be obtained when there exists a maximum temperature difference following the voltage behavior. Results show that the highest efficiency values were 2.8% and 4%, for $Z_r = 0.028$ (orange line) and $Z_{r,\tau} = 0.019$ (blue line) values, respectively. Therefore, the lower value of the ratio of the figure of merit was needed for maximum efficiency. In both cases for the two first pulses, 388 K and 468 K, the voltage was nearly half of the peak voltage for the pulse of 548 K. These results show that the new design for thermoelectric generator incorporating the segmented materials under variable heat input pulse must take into account the variation of temperature differences through time. The efficiency increased to a maximum of 4%, but then decreased to reach a steady-state when the pulses were changed to consider the Thomson effect. The efficiency was highest in agreement with the output voltage results. The numerical simulation shows the voltage and efficiency values obtained for the system in transient state. In general, the energy generated was significantly affected by the ratio of the figure of merit $Z_r$. In a real power generation scenario, the heat input to the device was not always controlled to hold the temperature differential. Thus, we modeled a system behavior using variable heat input.

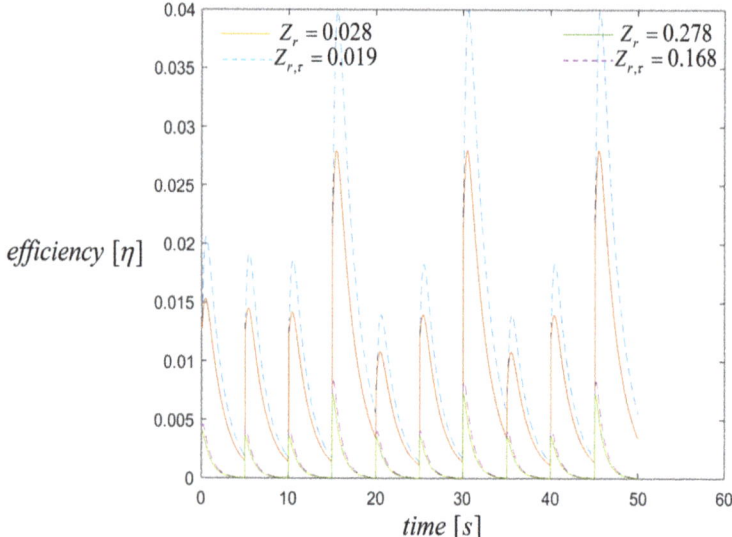

**Figure 10.** Efficiency when three heat input pulses are applied with a duty cycle of 50 s for the cases when $Z_r = 0.287$ and $Z_{r,\tau} = 0.168$, lower efficiency, and $Z_r = 0.028$ and $Z_{r,\tau} = 0.019$, maximum efficiency. Comparison results with the Thomson effect (dashed lines) and the without Thomson effect (solid lines).

## 6. Conclusions

In this paper, a numerical investigation of a STEG under variable pulsed heat input power was performed. The electrical performance, thermal behavior, and efficiency were obtained for different values of the ratio $Z_r$ and $Z_{r,\tau}$. Simulations were performed by varying the thermal conductivity and Seebeck coefficient of $Bi_2Te_3$. We gained different values of $Z_r$ and the results showed optimal $Z_r$ values for better system performance. The equations for the numerical method based on the finite-difference analysis were developed and solved in MATLAB. The most relevant conclusions and results of this study are the following.

In the segmented thermoelement, the value of the figure of merit for the first segmented, on which the temperature source was directly applied, must be lower than the figure of merit value of the second material in order to obtain better performance. In our case the materials used in the segmented thermoelement were $CoSb_3$ and $Bi_2Te_3$.

From spatial temperature profile, it was observed that when the figure of merit $Z_1 < Z_2$, the system required more time to reach steady-state. This is because the $Bi_2Te_3$ thermoelectric properties such as conductivity and the Seebeck coefficient directly affected the figure of merit. Our results proved that when the value of the figure of merit $Z_2$ increased, the voltage generated also increased, thus reaching its highest value.

Cold side temperature profiles through time showed that when $Z_{r,\tau} = 0.019$ the highest temperature was reached $T_c = 548.4$ K at the moment in which the temperature source of 548 K was replaced by the source of 388 K, therefore under these conditions a maximum $\Delta T$ between the hot side and the cold side was reached.

The optimum performance was obtained considering the Thomson effect when $Z_{r,\tau} = 0.019$, where the figure of merit of $CoSb_3$ is $Z_1 = 1.9 \times 10^{-3}$ and $Bi_2Te_3$ is $Z_2 = 93 \times 10^{-3}$ and the maximum difference of temperature was $\Delta T = 160.4$ K. The maximum voltage reached when considering the Thomson effect was $V_{max,\tau} = 108.3$ mV which was 18.1% higher in comparison to the voltage generated without the Thomson effect, $Vmax = 88.7$ mV.

**Author Contributions:** M.A.O.-R. designed the research and O.Y.E.-M.d.O. and P.E.R.-O. calculated the data. All authors wrote the paper. All authors have read and approved the final manuscript.

**Funding:** This research was funded by Research Grant 20190075 of Instituto Politecnico Nacional, CONACyT-Mexico (Grants CVU (Curriculum Vitae Unico) No. 490910) and CONACyT-Mexico (Grants CVU (Curriculum Vitae Unico) No. 900970).

**Conflicts of Interest:** The authors declare no conflict of interest.

## References

1. Kaltschmitt, M.; Streicher, W.; Wiese, A. *Renewable Energy: Technology, Economics and Environment*; Springer: New York, NY, USA, 2007.
2. Da Rosa, A.V. *Fundamentals of Renewable Energy Processes*; Elsevier: Amsterdam, The Netherlands, 2009.
3. Riffat, S.; Ma, X. Thermoelectrics: A review of present and potential applications. *Appl. Therm. Eng.* **2003**, *23*, 913–935. [CrossRef]
4. Champier, D. Thermoelectric generators: A review of applications. *Energy Convers. Manag.* **2017**, *140*, 167–181. [CrossRef]
5. Nolas, G.; Sharp, J.; Goldsmid, H. *Thermoelectrics: Basic Principles and New Materials Developments*; Springer: Berlin, Germany, 2001.
6. Ioffe, A.F. *Semiconductor Thermoelements and Thermoelectric Cooling*; Inforsearch limited London: London, UK, 1956.
7. Li, G.; Shittu, S.; Diallo, T.; Yu, M.; Zhao, X.; Ji, J. A review of solar photovoltaicthermoelectric hybrid system for electricity generation. *Energy* **2018**, *158*, 41–58. [CrossRef]
8. Nazri, N.; Fudholi, A.; Bakhtyar, B.; Yen, C.; Ibrahim, A.; Ruslan, M.; Mat, S.; Sopian, K. Energy economic analysis of photovoltaic–thermal-thermoelectric (PVTTE) air collectors. *Renew. Sustain. Energy Rev.* **2018**, *92*, 187–197. [CrossRef]
9. Kannan, N.; Vakeesan, D. Solar energy for future world: - A review. *Renew. Sustain. Energy Rev.* **2016**, *62*, 1092–1105. [CrossRef]
10. Gou, X.; Yang, S.; Xiao, H.; Ou, Q. A dynamic model for thermoelectric generator applied in waste heat recovery. *Energy* **2013**, *52*, 201–209. [CrossRef]
11. Yu, C.; Chau, K. Thermoelectric automotive waste heat energy recovery using maximum power point tracking. *Energy Convers. Manag.* **2009**, *50*, 1506–1512. [CrossRef]
12. Hsiao, Y.; Chang, W.; Chen, S. A mathematic model of thermoelectric module with applications on waste heat recovery from automobile engine. *Energy* **2010**, *35*, 1447–1454. [CrossRef]
13. Pei, Y.; Lalonde, A.; Iwanaga, S.; Snyder, G. High thermoelectric figure of merit in heavy hole dominated PbTe. *Energy Environ. Sci.* **2011**, *4*, 2085–2089. [CrossRef]
14. Ma, Q.; Fang, H.; Zhang, M. Theoretical analysis and design optimization of thermoelectric generator. *Appl. Therm. Eng.* **2017**, *127*, 758–764. [CrossRef]
15. Meng, F.; Chen, L.; Sun, F. Effects of thermocouples' physical size on the performance of the TEG–TEH system. *Int. J. Low-Carbon Technol.* **2016**, *1*, 375–382. [CrossRef]
16. Ge, Y.; Liu, Z.; Sun, H.; Liu, W. Optimal design of a segmented thermoelectric generator based on three-dimensional numerical simulation and multi-objective genetic algorithm. *Energy* **2018**, *147*, 1060–1069. [CrossRef]
17. Lundgaard, C.; Sigmund, O. Design of segmented off-diagonal thermoelectric generators with topology optimization. *Appl. Energy* **2019**, *236*, 950–960. [CrossRef]
18. Ming, T.; Wu, Y.; Peng, C.; Tao, Y. Thermal analysis on a segmented thermoelectric generator. *Energy* **2015**, *80*, 388–399. [CrossRef]
19. Shu, G.; Ma, X.; Tian, H.; Yang, H.; Chen, T.; Li, X. Configuration optimization of the segmented modules in an exhaust-based thermoelectric generator for engine waste heat recovery. *Energy* **2018**, *160*, 612–624. [CrossRef]
20. Paul, G. *Dynamic Behaviour of Thermoelectric Devices*; MIT Press Ltd.: Cambridge, MA, USA, 1960.
21. Crane, D.T. An introduction to system level steady-state and transient modeling and optimization of high power density thermoelectric generator devices made of segmented thermoelectric elements. *J. Electron. Mater.* **2011**, *40*, 561–569. [CrossRef]

22. Mahmoudinezhad, S.; Rezania, A.; Cotfas, P.; Cotfas, D.; Rosendahl, L. Transient behavior of concentrated solar oxide thermoelectric generator. *Energy* **2019**, *168*, 823–832. [CrossRef]
23. Samson, S.; Guiqiang, L.; Xudong, Z.; Xiaoli, M.; Yousef, G.; Emmanuel, A. Optimized high performance thermoelectric generator with combined segmented and asymmetrical legs under pulsed heat input power. *J. Power Sources* **2019**, *428*, 53–66.
24. Massaguer, A.; Massaguer, B.; Comamala, M.; Pujol, T.; Montoro, L.; Cardenas, M.; Carbonell, D.; Bueno, A. Transient behavior under a normalized driving cycle of an automotive thermoelectric generator. *Appl. Energy* **2017**, *206*, 1282–1296. [CrossRef]
25. Korzhuev, M.A.; Katin, I.V. On the Placement of Thermoelectric Generators in Automobiles. *J. Electron. Mater.* **2010**, *39*, 1390–1394. [CrossRef]
26. Mahmoudinezhad, S.; Rezania, A.; Rosendahl, L. Behavior of hybrid concentrated photovoltaic-thermoelectric generator under variable solar radiation. *Energy Convers. Manag.* **2018**, *164*, 443–452. [CrossRef]
27. Korzhuev, M.A; Avilov, E.S. Use of the Harman Technique for Figure of Merit Measurements of Cascade Thermoelectric Converters. *J. Electron. Mater.* **2010**, *39*, 1499–1503. [CrossRef]
28. Korzhuev, M.A. Symmetry Analysis of Thermoelectric Energy Converters with Inhomogeneous Legs. *J. Electron. Mater.* **2010**, *39*, 1381–1385. [CrossRef]
29. Bulat L.P.; Zakordonets V.S. The theoretical analysis of the thermoelectric semiconducting crystalline materials figure of merit. In Proceedings of the 15th International Conference on Thermoelectrics, Pasadena, CA, USA, 26–29 March 1996; pp. 197–200.
30. Li, G.; Zheng, Y.; Hu, J.; Guo, W. Experiments and a simplified theoretical model for a water-cooled, stove-powered thermoelectric generator. *Energy* **2019**, *185*, 437–448.
31. Li, G.; Zhang, S.; Zheng, Y.; Zhu, L.; Guo, W. Experimental study on a stove-powered thermoelectric generator (STEG) with self starting fan cooling. *Renew. Energy* **2018**, *121*, 502–512.
32. Swanson, B.; Somers, E.; Heikes, R. Optimization of a Sandwiched Thermoelectric Device. *J. Heat Transf.* **1961**, *83*, 77–82. [CrossRef]
33. Nguyen, N.; Pochiraju, K. Behavior of thermoelectric generators exposed to transient heat sources. *Appl. Therm. Eng.* **2013**, *51*, 1–9. [CrossRef]
34. Hadjistassou, C.; Kyriakides, E.; Georgiou, J. Designing high efficiency segmented thermoelectric generators. *Energy Convers. Manag.* **2013**, *66*, 165–172. [CrossRef]

© 2019 by the authors. Licensee MDPI, Basel, Switzerland. This article is an open access article distributed under the terms and conditions of the Creative Commons Attribution (CC BY) license (http://creativecommons.org/licenses/by/4.0/).

Article

# Power Conversion and Its Efficiency in Thermoelectric Materials

## Armin Feldhoff

Institute of Physical Chemistry and Electrochemistry, Leibniz University Hannover, Callinstraße 3A, D-30167 Hannover, Germany; armin.feldhoff@pci.uni-hannover.de; Tel.: +49-511-762-2940

Received: 26 June 2020; Accepted: 18 July 2020; Published: 22 July 2020

**Abstract:** The basic principles of thermoelectrics rely on the coupling of entropy and electric charge. However, the long-standing dispute of energetics versus entropy has long paralysed the field. Herein, it is shown that treating entropy and electric charge in a symmetric manner enables a simple transport equation to be obtained and the power conversion and its efficiency to be deduced for a single thermoelectric material apart from a device. The material's performance in both generator mode (thermo-electric) and entropy pump mode (electro-thermal) are discussed on a single voltage-electrical current curve, which is presented in a generalized manner by relating it to the electrically open-circuit voltage and the electrically closed-circuited electrical current. The electrical and thermal power in entropy pump mode are related to the maximum electrical power in generator mode, which depends on the material's power factor. Particular working points on the material's voltage-electrical current curve are deduced, namely, the electrical open circuit, electrical short circuit, maximum electrical power, maximum power conversion efficiency, and entropy conductivity inversion. Optimizing a thermoelectric material for different working points is discussed with respect to its figure-of-merit $zT$ and power factor. The importance of the results to state-of-the-art and emerging materials is emphasized.

**Keywords:** thermoelectrics; power conversion; efficiency; voltage-electrical current curve; working point; entropy pump mode; generator mode; power factor; figure of merit; Altenkirch-Ioffe model

## 1. Introduction

### 1.1. Controversial Points of View

Entropy is a central quantity in thermoelectrics, but seldom has it been addressed as such. The basic physical quantity that is known today as entropy is widely considered to be a derived quantity according to the approaches by Clausius [1–3] and Boltzmann [4–6] to quantify its value in certain situations. Both the perception of entropy as a derived quantity and the underestimation of its role in thermal processes are seen as residual outcomes of the Ostwald-Boltzmann battle, which is worth recalling and constitutes another chapter in the tragicomical history of thermodynamics [7]. In the frame of this work, entropy is considered to be a basic quantity. The benefits of this controversial point of view are made obvious on the example of thermoelectric materials.

### 1.2. Implications of Natural Philosophy

Clausius intended to borrow terms for important quantities from the ancient languages, so that they may be adopted unchanged in all modern languages. He proposed to call the quantity $S$, which had been introduced by him, the entropy of the body, from the Greek word τροπη (tropy), transformation [1–3]. Intentionally, he formed the word entropy to be as similar as possible to the word energy. In his opinion, the two quantities to be denoted by these words are so nearly allied in their physical meanings that a certain similarity in designation is desirable [1–3].

The importance of entropy was underlined by Gibbs in the very first words of his treatise on thermodynamics: "The comprehension of the laws which govern any material system is greatly facilitated by considering the energy and entropy of the system in the various states of which it is capable" [8,9]. However, the "Energeticist" [10] school in Germany, which rejected atomism and other matter theories, postulated energy as the primary substance in nature, and considered entropy as a superfluous derived concept [11–13]. The protagonist was Ostwald, cofounder of physical chemistry and its Nestor in Germany, and behind it was the natural philosophy of Mach [6,14,15]. Soon, the "Energeticist" school attracted much critical attention not only by the British pioneers [16] but also from a younger generation of German physicists [11]. The young Sommerfeld witnessed a memorable debate at the 1895 Assembly of the German Society of Scientists and Physicians in Lübeck, in which Boltzmann "like a bull defeated the torero [Helm as substitute to Ostwald] despite all his art of fencing [14]." In a follow-up critique, Boltzmann [17,18] condemned Ostwald's "Energetics" not only for perceived mathematical and physical error, but also for its false promise of easy rewards [11]. However, Ostwald never admitted that he had been defeated, and the object of the dispute has been kept alive to the present day [19,20]. Even though the personalities have changed over time, the battle has been newly inflamed in the controversy regarding the Karlsruhe Physics Course [21], which resulted in removing the entropy-treating educational course from German schools [22].

Today, the dissipation or "degradation" of energy is often treated without clear reference to entropy [19,20]. Preference is given to thermal energy ("heat") or enthalpy. Textbooks on classical thermodynamics take the approach of Clausius to quantify entropy in equilibrium conditions as the definition of entropy, which then is perceived as an energy-derived quantity. The success of Boltzmann's principle (called so by Einstein [6]) to quantify entropy in partitioned systems in equilibrium [23] renders it often to be a statistics-derived quantity [24]. However, the special cases considered herein do show only certain aspects of entropy, which should be considered in a wider context. By not considering entropy as a central basic quantity, clearness is lost, and uncertainty even creeps over authors who endeavor for accuracy and clarity when it comes to the description of thermal phenomena.

*1.3. Evolution of Thermodynamics*

The field of thermodynamics has evolved from the aim of understanding the thermodynamical engine (i.e., the steam engine) [11], which by principle operates under non-equilibrium conditions. However, for several reasons, thermodynamics has been limited to equilibrium conditions for a long time. For its suggestion to use entropy under non-equilibrium conditions, Planck's PhD thesis [25] was heavily criticized [19,20]. Planck was likely then intimidated and did not deepen this approach to entropy [19,20]. Alternately, the elegance and success of Gibbs' treatise on using equilibrium conditions did pave the way for thermodynamics under equilibrium conditions.

It took several decades until Callen [26,27] and de Groot [28] independently formulated a theory to describe thermodynamic systems in non-equilibrium conditions. This theory was helpful for quantitatively describing thermoelectric phenomena. However, the primary focus was the entropy production in irreversible processes and, thus, the excess entropy. No attention was given to entropy itself and its ability, which in older terms could be mentioned as the motive power of entropy, to drive a steam engine [29–31] or thermoelectric generator [32–34].

*1.4. Modern Thermodynamics*

Consistent with Falk [35], Fuchs [32], and Strunk [23,31], the author holds the view that entropy should be considered as a fundamental quantity. The characteristics of a fundamental quantity unfold from its relations with other fundamental quantities. Concise theories have been developed by Fuchs [32], Job & Rüffler [36,37], and Strunk [23,31,38].

In context of the development of physical concepts, it is worth noting that the basic physical quantity that is known today as entropy, was named quantity of heat by Joseph Black (1728–1799) [39–41] and calorique by Sadi Carnot (1796–1832) [29,30,40]. Indeed, calorique is the French word for quantity of heat. In his 1911 Presidential address to the Physical Society of London, Hugh Longbourne Callendar [29] outlined Carnot's calorique (i.e., entropy) as a quantity, that "any schoolboy could understand". Moreover, Callendar underlined that Carnot's calorique reappeared as a triple integral in Kelvin's 1852 paper, as the thermodynamic function of Rankine and as equivalence-value of a transformation in the 1854 paper of Clausius, and as entropy in the 1865 paper of Clausius [2] along with an abstract redefinition. No one at that time appears to have realized that entropy was merely calorique under another name. Callendar closed his remarks with the advice to distinguish a quantity of heat from a quantity of thermal energy.

Traditionally, thermal energy is called "heat". Concordant with Callendar [29] and Fuchs [32], in the author's opinion, heat is not energy, and entropy is the true measure of a quantity of heat as opposed to a quantity of thermal energy. Thus, the use this term for thermal energy should be avoided [42]. For clarity, the traditional term "heat" is put into quotation marks when it addresses the thermal energy. In this approach, entropy is a basic quantity. Thermoelectrics is an example par excellence to show the benefits of this philosophical perspective.

*1.5. Entropy in Thermoelectrics*

In the context of thermoelectrics, according to Boltzmann's principle, entropy is considered as a statistics-derived quantity when it is used to quantify the effect of spin and orbital degrees of freedom on the Seebeck coefficient in strongly correlated electron systems [43,44]. This, however, is a minor aspect. The approach by Clausius, to consider entropy as an energy-derived quantity does not play a significant role either.

In the so-called theory of thermodynamics of irreversible processes, as developed by Callen [26,27] and de Groot [28], it is rather the case that the thermal energy is derived from the entropy. Entropy is a fundamental quantity that is central to thermoelectrics. These texts can be read with great earning if entropy is considered as an indestructible substance-like quantity that is able to flow through the thermoelectric material and carries the thermal energy. The concept of energy carriers was developed by Falk et al. [45] and Herrmann [21].

However, the theory of thermodynamics of irreversible processes has the tendency to focus on the irreversibly produced excess entropy, but not on the entropy itself. Instead, energetic quantities are preferred. In §60 of his textbook, de Groot [28] presents an alternative presentation of thermoelectricity by the use of entropies of transfer, for which he has stated that the theory becomes somewhat more elegant compared to using energies of transfer. Unfortunately, he has not deepened this approach.

In a preceding paper [34], the author has shown that the rehabilitation of entropy into the theory by Callen [26,27] and de Groot [28] leads to a vivid description of thermoelectric devices. Like electrical charge carries the electrical energy, entropy carries the thermal energy. Thermal induction of an electrical current and electrical induction of a thermal current become understandable.

*1.6. Aim of This Work*

Like the preceding paper by the author [34], the present work aims to contribute to a better understanding of thermoelectrics by reconsidering it by treating entropy and electric charge as basic quantities of equal rank. This is semantically considered by naming the part of energy that flows together with entropy the thermal energy and part of energy flowing together with electrical charge the electrical energy. The energy flux through the thermoelectric material can thus be divided into thermal power and electrical power. Power conversion, which is in the focus of this article, implies that the system under consideration is not in equilibrium, but instead flown through by substance-like quantities. For the case of thermoelectric materials, these are entropy, electric charge, and energy.

By recalling the historical development of the perception of entropy, obstacles are identified, which have hindered the recognition of its important role in the field of thermoelectrics. The confused traditional approach and the use of model devices are avoided. Both power conversion and the efficiency of power conversion are accessed quantitatively for a thermoelectric material apart from a device. New physical insight into thermoelectrics is gained on the level of the thermoelectric material rather than on the device level. On the material's voltage–electrical current curve, distinct working points are identified (see Table 1), which not only allow for quantification of the material's properties and performance under specific operational conditions, but also relate generator mode (thermal-to-electrical power conversion) and entropy pump mode (electrical-to-thermal power conversion) of the same material to each other.

**Table 1.** Working points on the voltage–electrical current curve of a thermoelectric material in both operational modes, as addressed in this work.

| Abbreviation | Working Point | Operational Mode |
|---|---|---|
| MCEP | Maximum (power) conversion efficiency point | entropy pump mode |
| EICP | Entropy conductivity inversion point | entropy pump mode |
| OC | (electrical) open circuit | generator mode |
| MCEP | (see above) | generator mode |
| MEPP | Maximum (electrical) power point | generator mode |
| SC | (electrical) short circuit | generator mode |

The results are worked out in detail, and the outcome from the formalism is graphically illustrated and explained. The simplicity of thermoelectrics is clarified. The findings are linked to the outcome of the traditional approach to thermoelectrics and state-of-the-art thermoelectric materials.

## 2. Results

### 2.1. Categories

The results section is categorized, as follows.

- Section 2.2: Coupling currents of entropy and charge in thermoelectric materials
- Section 2.3: Material's voltage–electrical current and electrical power–electrical current characteristics
- Section 2.4: Material's thermal conductivity–electrical current characteristics
- Section 2.5: Thermoelectric material in generator mode
- Section 2.5.1: Working point for maximum electrical power
- Section 2.5.2: Thermal conductivity
- Section 2.5.3: Thermal power
- Section 2.5.4: Power conversion efficiency (thermal to electrical)
- Section 2.5.5: Working points for maximum conversion efficiency and maximum electrical power
- Section 2.6: Thermoelectric material in entropy pump mode
- Section 2.6.1: Power conversion efficiency (electrical to thermal)
- Section 2.6.2: Electrical and thermal power
- Section 2.7: Complete picture

### 2.2. Coupling Currents of Entropy and Charge in Thermoelectric Materials

When a thermoelectric material is simultaneously placed in a gradient of the electrochemical potential $\nabla \tilde{\mu}$ and a gradient of the temperature $\nabla T$, electrical flux density $j_q$, and entropy flux density $j_S$ are observed [34,46].

$$\begin{pmatrix} j_q \\ j_S \end{pmatrix} = \begin{pmatrix} \sigma & \sigma \cdot \alpha \\ \sigma \cdot \alpha & \sigma \cdot \alpha^2 + \Lambda_{OC} \end{pmatrix} \cdot \begin{pmatrix} -\nabla \tilde{\mu}/q \\ -\nabla T \end{pmatrix} \tag{1}$$

With the classical thermodynamic potential gradients $\nabla \tilde{\mu}$ (per electric charge $q$) and $\nabla T$ being employed, the basic transport Equation (1) has the following structure.

$$\text{flux densities} = \text{material tensor} \cdot \text{potential gradients} \tag{2}$$

The thermoelectric material tensor in Equation (1) is composed of only three quantities, which are the isothermal electrical conductivity $\sigma$, the Seebeck coefficient $\alpha$, and the entropy conductivity at electrical open circuit $\Lambda_{OC}$ (i.e., at vanishing electrical current). In principle, all three quantities are tensors themselves, but, for homogenous materials, they are often treated as scalars.

The entropy conductivity $\Lambda$ is related to the traditional "heat" conductivity $\lambda$ by the absolute temperature $T$ [32,34,37]. This, in principle, indicates that the traditional "heat" conduction is based on a more fundamental entropy conduction. The author proposes using the generic term thermal conductivity to address either the "heat" conductivity or the entropy conductivity [47,48].

$$\lambda = T \cdot \Lambda \tag{3}$$

It is emphasized that Equation (1) refers to a steady-state non-equilibrium situation. Instead of the quantities electric charge $q$ and entropy $S$, their local flux densities appear. According to Falk [35], considering local flux densities allows addressing local energy conversion or better to say local power conversion. Because flowing quantities are involved, preference should be given to local power density. Remember, power is the flux of energy. Equation (1) allows for locally varying quantities to be considered, which can be expressed with the positional vector $\mathbf{r}$: $\mathbf{j}_q = \mathbf{j}_q(\mathbf{r})$, $\mathbf{j}_S = \mathbf{j}_S(\mathbf{r})$, $\sigma = \sigma(\mathbf{r})$, $\alpha = \alpha(\mathbf{r})$, $\Lambda_{OC} = \Lambda_{OC}(\mathbf{r})$, $\nabla \tilde{\mu} = \nabla \tilde{\mu}(\mathbf{r})$, $\nabla T = \nabla T(\mathbf{r})$. Of course, the thermodynamic potentials are locally varying when gradients are present: $\tilde{\mu} = \tilde{\mu}(\mathbf{r})$, $T = T(\mathbf{r})$.

However, if the local variation of all quantities in Equation (1) is neglected, a simplified formulation of the transport equation can be observed [34,49,50]. If a further weak temperature dependence is assumed for the electron chemical potential $\mu$ (i.e., $\frac{\partial \mu}{\partial T} \approx 0$), the temperature dependence of the electrochemical potential $\tilde{\mu} = \mu + q \cdot \varphi$ is only in the electrical potential $\varphi$. With $\nabla \mu / q \approx 0$ follows $\nabla \tilde{\mu}/q = \nabla \mu/q + \nabla \varphi \approx \nabla \varphi$. The assumption of constant gradients (i.e., linear potential curves) allows for them to be substituted by the difference of the respective potential along the thermoelectric material of length $L$: $\nabla \varphi \to -\Delta \varphi / L$, $\nabla T \to -\Delta T / L$. Furthermore, for a thermoelectric material of cross-sectional area $A$, the local flux densities can be replaced by the integrative currents of electrical charge and entropy, respectively: $\mathbf{j}_q \to I_q / A$, $\mathbf{j}_S \to I_S / A$. Subsequently, the transport equation follows as:

$$\begin{pmatrix} I_q \\ I_S \end{pmatrix} = \frac{A}{L} \cdot \begin{pmatrix} \sigma & \sigma \cdot \alpha \\ \sigma \cdot \alpha & \sigma \cdot \alpha^2 + \Lambda_{OC} \end{pmatrix} \cdot \begin{pmatrix} \Delta \varphi \\ \Delta T \end{pmatrix} \tag{4}$$

Equation (4) describes the coupling of currents of electrical charge $I_q$ and entropy $I_S$ in the thermoelectric material, which causes the occurence of either an electrically-induced entropy current [51] (Peltier effect) or a thermally-induced electrical current [52,53] (Seebeck effect). Note that Equation (4) describes these effects in a thermoelectric material, which is schematically shown in Figure 1, apart from a device.

### 2.3. Material's Voltage—Electrical Current and Electrical Power—Electrical Current Characteristics

Different working conditions of the thermoelectric material in this article are discussed with reference to the voltage–electrical current curve, which is derived from Equation (4) as Equation (5). Remember that the voltage $\Delta \varphi$ is the electrical potential difference along the thermoelectric material.

$$\Delta \varphi = -\alpha \cdot \Delta T + \frac{I_q}{\frac{A}{L} \cdot \sigma} \tag{5}$$

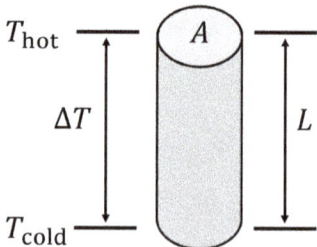

**Figure 1.** This paper discusses characteristics of a thermoelectric material of cross-sectional area $A$ and length $L$ when exposed to a temperature difference $\Delta T = T_{hot} - T_{cold}$ between a hot reservoir at $T_{hot}$ and a cold reservoir at $T_{cold}$.

According to Equation (5), the voltage–electrical current characteristics is a line, which has the material's electrical resistance $R = \frac{1}{\frac{A}{L} \cdot \sigma}$ as its slope. This line is only determined by the voltage $\Delta \varphi_{OC}$ under electrically open-circuited conditions (i.e., at zero electrical current) and the electrical current $I_{SC}$ at electrically short-circuited conditions (i.e., at zero voltage). The OC is of practical importance for the measurement of temperature using thermocouples.

$$\Delta \varphi_{OC} = -\alpha \cdot \Delta T \tag{6}$$

$$I_{q,SC} = \frac{A}{L} \cdot \alpha \cdot \sigma \cdot \Delta T \tag{7}$$

Obviously, the sign of the Seebeck coefficient $\alpha$ determines the sign of both the voltage $\Delta \varphi_{OC}$ under electrically short-circuited conditions and the electrical current $I_{q,SC}$ under electrically short-circuited conditions. Thus, the voltage–electrical current characteristics of p-type ($\alpha > 0$) or n-type ($\alpha < 0$) conductors differ from each other by principle (cf. Appendix A).

To discuss the materials independently of the sign of the Seebeck coefficient, the absolute of the voltage $|\Delta \varphi|$ is plotted in Figure 2 versus the absolute value of the electrical current $|I_q|$. In order to diminish Ohmic losses, the electrical resistance $R = \frac{1}{\frac{A}{L} \cdot \sigma}$ must be reduced, which, for the given geometry, requires the electrical conductivity $\sigma$ to be increased.

To make the discussion independent from even the material parameters and temperature difference $\Delta T$, the normalized electrical current $i$ and normalized voltage $u$, as normalized to electrically short-circuited and open-circuited conditions, respectively, are considered in subsequent sections.

$$i = \frac{I_q}{I_{q,SC}} = \frac{I_q}{\frac{A}{L} \cdot \alpha \cdot \sigma \cdot \Delta T} \tag{8}$$

$$u = \frac{\Delta \varphi}{\Delta \varphi_{OC}} = \frac{\Delta \varphi}{-\alpha \cdot \Delta T} = 1 - i \tag{9}$$

The electrical power $P_{el}$ is determined by the product of voltage and electrical current as given by Equation (10). It increases linearly with the electrical current, but it is parabolically damped at high electrical currents due to the limited electrical conductivity (Ohmic dissipation [54]).

$$\begin{aligned} P_{el} = \Delta \varphi \cdot I_q &= \left( -\alpha \cdot \Delta T + \frac{I_q}{\frac{A}{L} \cdot \sigma} \right) \cdot I_q \\ &= -\alpha \cdot \Delta T \cdot I_q + \frac{I_q^2}{\frac{A}{L} \cdot \sigma} \\ &= -\frac{A}{L} \cdot \sigma \cdot \alpha^2 \cdot (\Delta T)^2 \cdot (i - i^2) \end{aligned} \tag{10}$$

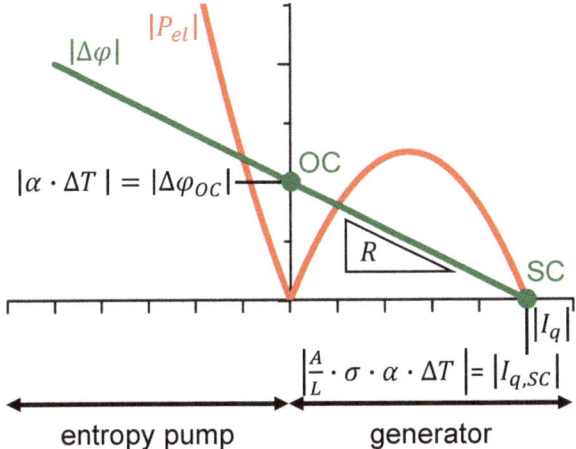

**Figure 2.** Absolute voltage $|\Delta\varphi|$ – electrical current $|I_q|$ curve (green), with slope given by the electrical resistance $R = \frac{1}{\frac{A}{L} \cdot \sigma}$, and the absolute electrical power $|P_{el}|$ – electrical current $|I_q|$ curve (red) for a thermoelectric material. Here, $\Delta T = \frac{T_{hot} - T_{cold}}{T_{hot}}$ is the temperature difference along the thermoelectric material of cross-sectional area $A$ and length $L$. These quantities, together with the (isothermal) electrical conductivity $\sigma$ and the Seebeck coefficient $\alpha$, determine the electrical current $I_{SC}$ under electrically short-circuited conditions. The voltage $\Delta\varphi_{OC}$ under electrically open-circuited conditions is determined by the Seebeck coefficient and the temperature difference. Generator mode refers to a positive sign and entropy pump mode to a negative sign of the electrical power (cf. Appendix A).

The absolute of the electrical power $|P_{el}|$ is plotted in Figure 2 versus the absolute value of the electrical current $|I_q|$ to discuss the thermoelectric materials independent of the sign of the Seebeck coefficient.

It is obvious from Figure 2 that the electrical power to be put into the material in entropy pump mode may distinctly exceed the electrical power that can be gained in generator mode if the material is applied to the same temperature difference.

### 2.4. Material's Thermal Conductivity—Electrical Current Characteristics

From Equation (4), the entropy current $I_S$ flowing through the material is obtained. It depends on not only the temperature difference $\Delta T$ but also the Peltier effect that is associated with the thermally induced electrical current $I_q$, which can be expressed by the normalized electrical current $i$ as given in Equation (8).

$$\begin{aligned} I_S &= \tfrac{A}{L} \cdot \Lambda_{OC} \cdot \Delta T + \alpha \cdot I_q \\ &= \tfrac{A}{L} \cdot \Lambda_{OC} \cdot \Delta T + \tfrac{A}{L} \cdot \sigma \alpha^2 \cdot i \cdot \Delta T \\ &= \tfrac{A}{L} \cdot (\Lambda_{OC} + \sigma \alpha^2 \cdot i) \Delta T \\ &= \tfrac{A}{L} \cdot \Lambda \cdot \Delta T \end{aligned} \qquad (11)$$

From Equation (11), it follows that the thermal conductivity, expressed here by the entropy conductivity $\Lambda$, is dependent on the electrical current $i$.

$$\Lambda = \Lambda(i) = \Lambda_{OC} + \sigma\alpha^2 \cdot i \qquad (12)$$

When compared to electrically open-circuited conditions, the power factor $\sigma\alpha^2$ gives an additional contribution to the entropy conductivity, which increases linearly with the electrical current. Under electrically short-circuited conditions (SC, i.e., $i = 1$), the entropy conductivity reaches its maximum value.

$$\Lambda_{SC} = \Lambda_{OC} + \sigma\alpha^2 \qquad (13)$$

Under electrically short-circuited conditions, the electrical potential is spatially constant (i.e., its gradient vanishes: $\nabla\varphi = 0$). Note that the entropy conductivity at electrical short circuit $\Lambda_{SC}$, as given by Equation (13), is identical to tensor element $M_{22}$ of the thermoelectric material tensor in the transport Equation (4).

To discuss the characteristics of the entropy conductivity in a general manner, it is normalized to its value under electrically open-circuited conditions:

$$\tilde{\Lambda} = \tilde{\Lambda}(i) = \frac{\Lambda}{\Lambda_{OC}} = 1 + \frac{\sigma\alpha^2}{\Lambda_{OC}} \cdot i = 1 + zT \cdot i \qquad (14)$$

In Equation (14), a figure-of-merit $zT$ has been identified, which only depends on the three material parameters $\sigma$, $\alpha$ and $\Lambda_{OC}$, which make up the material tensor of Equation (4).

$$zT = \frac{\sigma \cdot \alpha^2}{\Lambda_{OC}} \qquad (15)$$

Equation (14) is visualized in Figure 3 for some hypothetical thermoelectric materials with $zT = 0.1, 0.5, 1, 2, 4$ and $8$. Working points for electrically open-circuited (OC) conditions, maximum electrical power point (MEPP), and electrical short-circuited (SC) conditions are indicated on the voltage–electrical current curve. Note that the entropy conductivity inversion point (ECIP) is given by the negative reciprocal of the figure-of-merit $-1/zT$. Only for electrical currents being below the ECIP, effective entropy pump mode is reached with a negative entropy conductivity of the thermoelectric material. Only then, more entropy is pumped against the temperature difference than flows down it. Obviously, the measurements of the thermal conductivity of a thermoelectric material must refer to the working point on the voltage–electrical current curve.

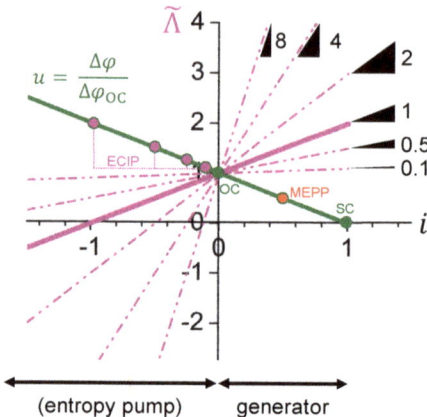

**Figure 3.** Normalized entropy conductivity $\tilde{\Lambda}$ as function of normalized electrical current $i$ for some hypothetical thermoelectric materials. Depending on the figure-of-merit $zT$, the curves pivot through the working point for electrically open-circuited (OC) conditions. The figure-of-merit $zT$ gives the slope of the curve and its negative reciprocal $-1/zT$ indicates the entropy conductivity inversion point (ECIP). For some thermoelectric materials, the respective ECIP is indicated as working point on the normalized voltage $u$–normalized electrical current $i$ curve. Note that the ECIP for materials with $zT = 0.1$ and $zT = 0.5$ is out of the applied scale. The term entropy pump mode is put into brackets because a net entropy current against the temperature difference will only occur if the magnitude of the electrical current is beyond the respective ECIP. For generator mode, the working points MEPP and SC are indicated.

## 2.5. Thermoelectric Material in Generator Mode

### 2.5.1. Working Point for Maximum Electrical Power

Remember, the characteristics of the thermoelectric material are all discussed for $\Delta T$ being different from zero, which implies non-isothermal conditions. It can be easily seen from Equation (10) that maximum electrical power output is obtained for half of the electrically short-circuited electrical current ($i_{MEPP} = \frac{1}{2}$, cf. Appendix B.1):

$$P_{el,\,max} = |\,P_{el}\,(i_{MEPP} = 0.5)\,| = \frac{1}{4} \cdot \frac{A}{L} \cdot \sigma \cdot \alpha^2 \cdot (\Delta T)^2 \qquad (16)$$

To make the discussion independent from material parameters and temperature difference, the normalized electrical power $p_{el}$, as normalized to the maximum electrical power in generator mode, is plotted in Figure 4.

$$p_{el} = \frac{|\,P_{el}\,|}{P_{el,\,max}} = 4 \cdot |\,i - i^2\,| \qquad (17)$$

The maximum electrical power point (MEPP) is indicated on the normalized voltage–electrical current curve in Figure 4. It is clearly seen that the MEPP ($i_{MEPP} = 0.5$, $u_{MEPP} = 0.5$) is at half of the open-circuited voltage as well as at half of the electrically short-circuited electrical current, which also follows from Equation (9).

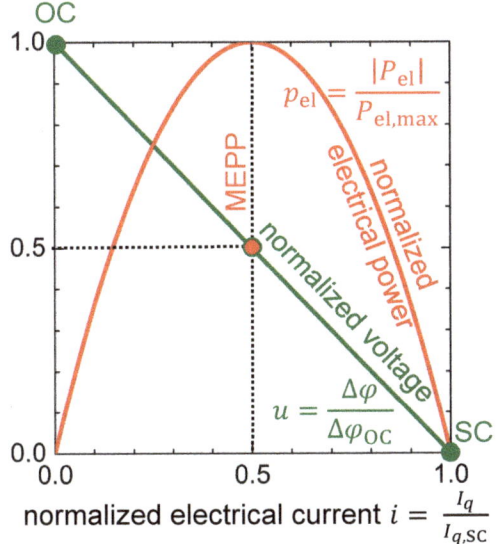

**Figure 4.** Normalized curves for both voltage $u$ – electrical current $i$ characteristics and electrical power $p_{el}$–electrical current $i$ characteristics of a thermoelectric material when it is operated in generator mode. The working points open-circuited (OC), maximum electrical power point (MEPP), and short-circuited (SC) are indicated.

### 2.5.2. Thermal Conductivity

For the thermoelectric material being operated in generator mode, Equation (12) is graphically expressed in Figure 5. The electrically open-circuited entropy conductivity $\Lambda_{OC}$ is purely dissipative, while the part of the entropy conductivity depending on the power factor $\sigma \cdot \alpha^2$ couples to the electrical current, and it fully contributes to the thermal-to-electric power conversion. Obviously, to maximize the electrical power at a given temperature difference, the power $\sigma \cdot \alpha^2$ must be maximized, which is in accordance with Equation (10).

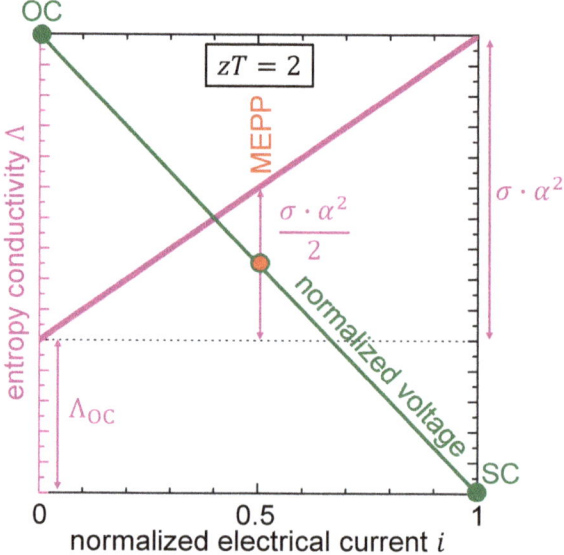

**Figure 5.** Entropy conductivity $\Lambda$ as function of the normalized electrical current $i$ for a thermoelectric material with $zT = 2$ in generator mode. The working points OC, MEPP, and SC are indicated on the normalized voltage–electrical current curve.

The thermally induced electrical current carries electrical energy, which, however, with increasing electrical current, is diminished by Ohmic losses due to the limited (isothermal) electrical conductivity $\sigma$ as discussed above. At maximum electrical power, the entropy conductivity is increased by half of the power factor as compared to electrically open-circuited conditions. Under electrically short-circuited conditions, the entropy conductivity reaches its maximum (see Equation (13)).

2.5.3. Thermal Power

The thermal input power and the thermal output power depend on the electrical current $i$. According to Fuchs [33], the available thermal power $P_{th}$ is determined by the fall of entropy down the temperature difference $\Delta T$ along the material.

$$P_{th} = I_S \cdot \Delta T = \Lambda \cdot (\Delta T)^2 = \frac{A}{L} \cdot \left( \Lambda_{OC} + \sigma \alpha^2 \cdot i \right) \cdot (\Delta T)^2 \tag{18}$$

Thus, the available thermal power, as given by Equation (18), depends on the electrical current in the same manner as the entropy conductivity in Figures 3 and 5.

## 2.5.4. Power Conversion Efficiency (Thermal to Electrical)

From Equations (10) and (18), the second-law power conversion efficiency for the thermoelectric material in generator mode is obtained:

$$\eta_{\text{II,gen}} = \left| \frac{P_{\text{el}}}{P_{\text{th,avail}}} \right| = \frac{\frac{A}{L} \cdot \sigma \cdot \alpha^2 \cdot (\Delta T)^2 \cdot (i - i^2)}{\frac{A}{L} \cdot (\Lambda_{\text{OC}} + \sigma \alpha^2 \cdot i) \cdot (\Delta T)^2}$$
$$= \frac{i - i^2}{i + \frac{\Lambda_{\text{OC}}}{\sigma \cdot \alpha^2}} \qquad (19)$$
$$= \frac{i - i^2}{i + \frac{1}{zT}}$$

Equation (19) is plotted in Figure 6 as solid blue curves for some hypothetical thermoelectric materials with different values of the figure-of-merit $zT$. Obviously, the figure-of-merit $zT$ must be maximized in order to maximize the thermal-to-electrical power conversion efficiency at a given (thermally induced) electrical current.

Equation (19) can be read as the coupled thermal power being converted into electrical power with the constraint; however, with increasing electrical current, Ohmic dissipation gains overhead. As a result, the optimum power conversion efficiency is obtained at lower electrical current than the optimum electrical power output, and the working points for one or other task differ from each other, which can be seen in Figure 6.

According to Fuchs [33], the second-law efficiency $\eta_{\text{II,gen}}$ is related to the first-law efficiency $\eta_{\text{I,gen}}$ by Carnot's efficiency $\eta_{\text{C}}$.

$$\eta_{\text{I,gen}} = \eta_{\text{C}} \cdot \eta_{\text{II,gen}} = \frac{T_{\text{hot}} - T_{\text{cold}}}{T_{\text{hot}}} \cdot \eta_{\text{II,gen}} \qquad (20)$$

Carnot's efficiency $\eta_{\text{C}}$ places a theoretical limit for the case in which the second-law efficiency $\eta_{\text{II,gen}} = 1$, which refers to the unrealistic case of vanishing dissipation. Nevertheless, the second-law efficiency $\eta_{\text{II,gen}}$ is the only material-dependent factor and has been used by Altenkirch [55] and Ioffe [56] in order to estimate the performance of thermoelectric materials by treating thermogenerators. It is worth noting that the entropy-based approach presented here allows for power conversion and its efficiency for a single thermoelectric material apart from a device to be discussed.

## 2.5.5. Working Points for Maximum Conversion Efficiency and Maximum Electrical Power

From the maximum of Equation (19), the maximum conversion efficiency point (MCEP) is obtained with the normalized electrical current $i_{\text{MCEP,gen}}$ being, as follows (cf. Appendix B.2):

$$i_{\text{MCEP,gen}} = \frac{1}{\sqrt{1 + zT} + 1} \qquad (21)$$

At the MCEP, the maximum power conversion efficiency of the thermoelectric material in generator mode is then obtained, as follows (cf. Appendix B.2):

$$\eta_{\text{II,gen,max}} = \eta_{\text{II,gen}}\left(i_{\text{MCEP,gen}}\right) = \frac{\sqrt{1 + zT} - 1}{\sqrt{1 + zT} + 1} \qquad (22)$$

Equation (23), which shows the variation of the MCEP with varying $i_{\text{MCEP,gen}}$ due to varying $zT$, is plotted in Figure 6 as dotted blue line.

$$\eta_{\text{II,gen,max}}\left(i_{\text{MCEP,gen}}\right) = 1 - 2 \cdot i_{\text{MCEP,gen}} \qquad (23)$$

Note that with increasing figure-of-merit $zT$, not only does the MCEP drift apart from the MEPP, but the electrical power output also decreases with respect to the MEPP (see Equation (16)), both of which can be seen in Figure 6 (cf. Appendix B.2).

$$P_{el,MCEP} = \frac{4 \cdot \sqrt{1+zT}}{\left(\sqrt{1+zT}+1\right)^2} \cdot P_{el,max} \qquad (24)$$

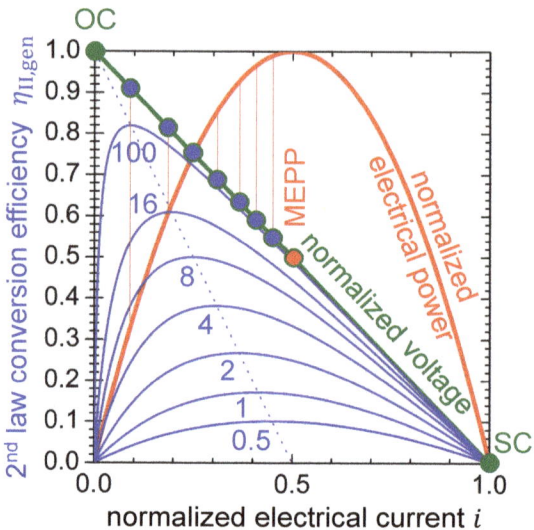

**Figure 6.** Thermal to electrical power conversion efficiency for some hypothetic materials with figure-of-merit $zT$ varying from 0.5 to 100. Respective working points MCEP (blue) are indicated on the voltage–electrical current curve as well as the MEPP (red). Vertical lines indicate the electrical power output at the MCEP for the example materials. Note that the MCEP drifts apart from the MEPP with increasing figure-of-merit $zT$. The dashed line indicates the dependence of the MCEP with varying $zT$.

Obviously, with increasing figure-of-merit $zT$, the electrical power at the MCEP converges to zero. Figure 7 shows that a notable difference in electrical power output between MCEP and MEPP can be expected for thermoelectric materials with $zT > 0.3$ only (red curves). A notable difference in the power conversion efficiency of the thermoelectric material being operated in the MCEP or the MEPP can only be expected when $zT > 2$. This is also obvious from Table 2, which, for some hypothetical values of the material's figure-of-merit $zT$, gives values of the second-law power conversion efficiency at the working points under discussion. The 2$^{nd}$ law power conversion efficiency at the MEPP is obtained as follows (cf. Appendix B.1).

$$\eta_{II,gen,MEPP} = \eta_{II,gen}(i_{MEPP} = 0.5) = \frac{1}{2} \cdot \frac{zT}{zT+2} \qquad (25)$$

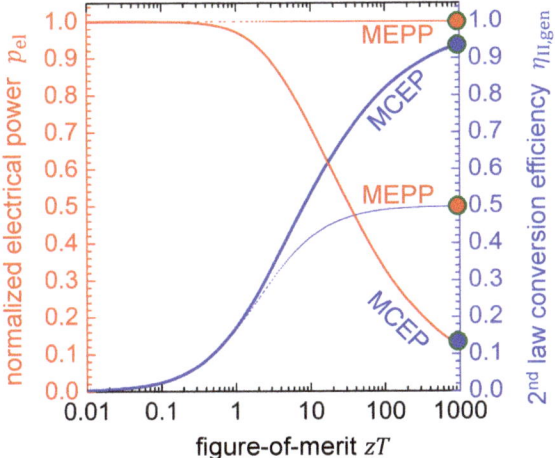

**Figure 7.** Electrical power output (red lines) and thermal-to-electrical power conversion efficiency (blue lines) for some hypothetic materials with figure-of-merit $zT$ varying from 0.01 to 1000 when operated in two distinct working points, respectively. Solid lines refer to the MCEP and dashed lines refer to the MEPP.

It is worth noting that, for a thermoelectric material with $zT < 2$, there is no benefit from operating it apart from the MEPP.

**Table 2.** Second-law power conversion efficiency of a thermoelectric material at the MCEP in either entropy pump mode or generator mode and at the MEPP in generator mode for some hypothetical values of the figure-of-merit $zT$.

| $zT$ | Maximum 2nd Law Efficiency | 2nd Law Efficiency at MEPP |
|---|---|---|
| 0.1 | 0.02 | 0.02 |
| 0.5 | 0.1 | 0.1 |
| 1 | 0.17 | 0.17 |
| 1.5 | 0.23 | 0.21 |
| 2 | 0.27 | 0.25 |
| 2.5 | 0.30 | 0.28 |
| 3 | 0.33 | 0.3 |
| 3.5 | 0.36 | 0.32 |
| 4 | 0.38 | 0.33 |
| 8 | 0.5 | 0.4 |
| 16 | 0.61 | 0.44 |
| 32 | 0.70 | 0.47 |
| 100 | 0.82 | 0.49 |

## 2.6. Thermoelectric Material in Entropy Pump Mode

### 2.6.1. Power Conversion Efficiency (Electrical to Thermal)

Traditional approaches consider a coefficient of performance when addressing the performance of a thermoelectric cooling or heating device [56,57]. Analogously, a coefficient of performance $COP$ of the thermoelectric material, when used in a cooler, can be considered. It is the thermal power removed from the cold side $T_{\text{cold}} \cdot I_S$ related to the electrical power (cf. Appendix C.1).

$$COP_{cooler} = \left|\frac{T_{cold} \cdot I_S}{P_{el}}\right| = \frac{T_{cold}}{\Delta T} \cdot \left|\frac{P_{th}}{P_{el}}\right|$$
$$= \frac{T_{cold}}{\Delta T} \cdot \eta_{II,ep} \qquad (26)$$

If instead of a cooler, the thermoelectric material is used in a heater (see Fuchs [32], p. 135ff), the thermal power released to the hot side $T_{hot} \cdot I_S$ becomes the reference parameter, and the $COP$ is then (cf. Appendix C.1):

$$COP_{heater} = \left|\frac{T_{hot} \cdot I_S}{P_{el}}\right| = \frac{T_{hot}}{\Delta T} \cdot \left|\frac{P_{th}}{P_{el}}\right|$$
$$= \frac{T_{hot}}{\Delta T} \cdot \eta_{II,ep} \qquad (27)$$
$$= \frac{1}{\eta_C} \cdot \eta_{II,ep}$$

In both cases, Equations (26) and (27), the $COP$ can be factorized into a temperature factor and the second-law efficiency for the thermoelectric material in entropy pump mode $\eta_{II,ep}$ (see Fuchs [32], p. 135ff). When the thermoelectric material is used in a heater (Equation (27)), the temperature factor is the inverse of Carnot's efficiency $\eta_C$ [32]. The second-law efficiency for the thermoelectric material in entropy pump mode $\eta_{II,ep}$ relates the thermal power $P_{th}$ that is needed to pump a certain entropy current from the cold side to the hot side to the electrical power $P_{el}$ (cf. Appendix C.1).

$$\eta_{II,ep} = \left|\frac{P_{th}}{P_{el}}\right| = \frac{i + \frac{1}{zT}}{-i^2 + i} \qquad (28)$$

The second-law efficiency for the thermoelectric material in entropy pump mode $\eta_{II,ep}$ only depends on the normalized electrical current $i$ (i.e., working point on the voltage–electrical current curve) and the material's figure-of-merit $zT$. It can be used to assess the performance of the thermoelectric material when it is used to pump entropy, regardless of whether the purpose is cooling or heating.

Note that the second-law efficiency for the thermoelectric material in entropy pump mode $\eta_{II,ep}$ (Equation (28)) is the inverse of the second-law efficiency for the thermoelectric material in generator mode (Equation (19)). Because a net entropy current from the cold side to the hot side will only be obtained for negative entropy conductivity (see Equation (14) and Figure 3), here $\eta_{II,ep}$ will make sense only for the normalized electrical current being $i \leq \frac{1}{zT}$. For this parameter range it is plotted in Figure 8 for some hypothetic thermoelectric materials with figure-of-merit $zT$ between 0.5 and 100.

The maximum 2$^{nd}$-law power conversion efficiency for a thermoelectric material operated in entropy pump mode is dependent on the material's figure-of-merit $zT$ (cf. Appendix C.2):

$$\eta_{II,ep,max} = \frac{\sqrt{1+zT}-1}{\sqrt{1+zT}+1} \qquad (29)$$

It is obtained at a normalized electrical current $i_{MCEP,ep}$, which corresponds to the thermoelectric material's maximum conversion efficiency point (MCEP) in entropy pump mode (cf. Appendix C.2). Respective working points for some hypothetic thermoelectric materials are indicated on the voltage–electrical current curve presented in Figure 8.

$$i_{MCEP,ep} = -\frac{1}{\sqrt{1+zT}-1} \qquad (30)$$

The dependence of the maximum second-law efficiency on the electrical current is shown in Figure 8 as a hyperbolic line (cf. Appendix C.2).

$$\eta_{II,ep,max}(i_{MCEP,ep}) = \frac{1}{1 - 2 \cdot i_{MCEP,ep}} \qquad (31)$$

Obviously, an ideal thermoelectric material would have an infinite $zT$, but the MCEP converges then to the OC working point at vanishing electrical current and, thus, zero electrical power. On the contrary, for the limit of vanishing $zT$, the maximum second-law efficiency converges to zero at infinite magnitude of the electrical current.

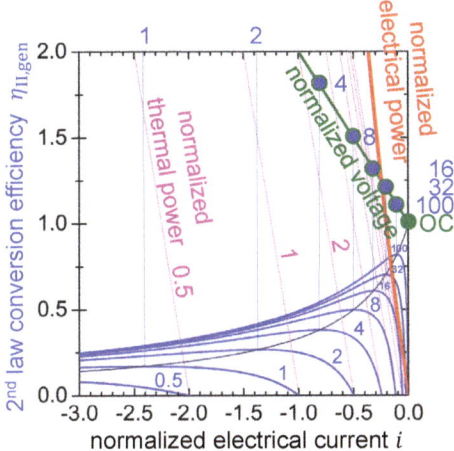

**Figure 8.** Electrical-to-thermal power conversion efficiency as a function of the reduced electrical current for some hypothetic materials with figure-of-merit $zT$ varying from 0.5 to 100. Respective working points MCEP (blue) are indicated on the voltage–electrical current curve for $zT = 100, 32, 18, 8$ and $4$. Further vertical lines (blue) indicate the MCEP for $zT = 2, 1$. The MCEP for $zT = 0.5$ is out of display. The hyperbolic curve indicates the dependence of the MCEP with varying $zT$. The red curve indicates electrical power–electrical current characteristics. The set of inclined parallel lines (magenta) indicate the thermal power–electrical current characteristics for the respective $zT$. All of the power curves are normalized to the MEPP in generator mode.

### 2.6.2. Electrical and Thermal Power

All of the power curves in Figure 8, for the thermoelectric material in entropy pump mode, are normalized to the MEPP in generator mode (see Figures 2 and 4) when the material is exposed to the same temperature difference $\Delta T$. According to Equations (16) and (18), the normalized thermal power $p_{th}$ in Figure 8 is given by a straight line that intersects the horizontal axis at $-\frac{1}{zT}$ and it has a slope of $-4$ (cf. Appendix C.3).

$$p_{th} = \frac{|P_{th}|}{P_{el,max}} = 4 \cdot \left| \frac{1}{zT} + i \right| \tag{32}$$

For different values of the figure-of-merit $zT$, a set of inclined parallel lines results. Only the lines for $zT = 0.5, 1$ and $2$ are labelled in Figure 8. With increasing figure-of-merit $zT$, the normalized thermal power curve approaches the normalized electrical power curve, which is in accordance with the increasing power conversion efficiency. However, when the thermoelectric material is operated in its MCEP, the thermal power will decrease with increasing figure-of-merit $zT$, which becomes obvious when Equation (30) is combined with Equation (32) (cf. Appendix C.2).

$$p_{th,MCEP} = p_{th}\left(i_{MCEP,ep}\right) = 4 \cdot \frac{\sqrt{1+zT}}{zT} \tag{33}$$

The normalized thermal power at MCEP would be steeply curved in Figure 8, with the data point out of scale for $zT < 8$, but has been skipped for clarity. Instead, relevant values for the MCEP are listed in Table 3, together with the normalized electrical power and the normalized electrical current.

**Table 3.** Values of normalized electrical current $i_{MCEP,ep}$, normalized thermal power $p_{th,MCEP}$, and normalized electrical power $p_{el,MCEP}$ at the MCEP in entropy pump mode for some hypothetical values of the figure-of-merit $zT$. Values of the second law power conversion efficiency can be read from Table 2.

| $zT$ | $i_{MCEP,ep}$ | $p_{th,MCEP}$ | $p_{el,MCEP}$ |
|---|---|---|---|
| 0.1 | −20.49 | 41.95 | 1761.32 |
| 0.5 | −4.45 | 9.80 | 97.01 |
| 1 | −2.41 | 5.66 | 32.87 |
| 1.5 | −1.72 | 4.22 | 19.67 |
| 2 | −1.36 | 3.46 | 12.83 |
| 2.5 | −1.48 | 2.99 | 10.77 |
| 3.0 | −1 | 2.68 | 8.93 |
| 3.5 | −0.89 | 2.42 | 7.56 |
| 4 | −0.80 | 2.2 | 5.76 |
| 8 | −0.50 | 1.5 | 3.00 |
| 16 | −0.32 | 1.03 | 1.69 |
| 32 | −0.21 | 0.71 | 1.02 |
| 100 | −0.11 | 0.40 | 0.49 |

## 2.7. Complete Picture

With the approach chosen here, working points on the voltage–electrical current curve relate the power conversion properties of the thermoelectric material in generator mode and entropy pump mode to each other. Figure 9 illustrates the concise result for a hypothetical thermoelectric material with figure-of-merit $zT = 3.5$.

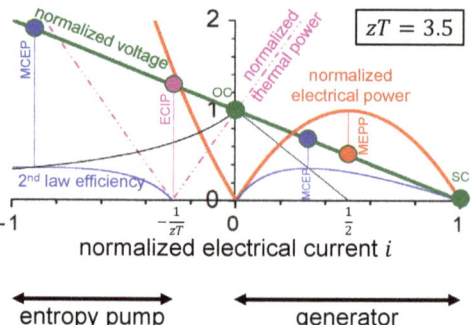

**Figure 9.** Related characteristics of a hypothetic thermoelectric material with figure-of-merit $zT = 3.5$ in entropy pump mode and generator mode: normalized voltage, normalized electrical power, normalized thermal power, and 2$^{nd}$-law conversion efficiency as a function of the normalized electrical current. Different working points are indicated on the voltage–electrical current curve. Note that, for current state-of-the-art materials, the MCEP in entropy pump mode would be out of display (see Table 3).

For a given figure-of-merit $zT$, according to Equations (22) and (29), the values of the maximum 2$^{nd}$-law conversion efficiency for both modes are identical. Some values are given in Table 2. In addition, values of the 2$^{nd}$-law conversion efficiency at the MEPP in generator mode are given (see Equation (25)). Remember, the obtained power requires consideration of the absolute value of the electrical power, as determined by the power factor (see Equation (16)).

## 3. Materials and Methods

Detailed calculations, as given in Appendixs B and C, were made using pencil and paper. The manuscript was prepared using Latex in MikTex distribution. Figures were drawn with the aid of Microcal's Origin and Microsoft's PowerPoint.

## 4. Discussion

### 4.1. Remarks on the Use of Working Points

Traditionally, a thermoelectric device is considered and, in generator mode, the operational conditions are set by an external load resistance. The approach of this work, which uses working points on the material's voltage–electrical voltage curve, gives consistent results, which is explicitly shown in Appendix B.3. However, consideration of working points comes with the advantage that the contribution of individual thermoelectric materials in a device can be easily understood [58]. Moreover, the material's voltage–electrical voltage curve directly relates generator mode and entropy pump mode.

### 4.2. Remarks on the Altenkirch-Ioffe Model

Due to the prominence of the Altenkirch-Ioffe model [55,56], it is worth comparing it to the model, which has been introduced in this work. A comparison of important quantities described by the model of this work and the Altenkirch-Ioffe model is shown in Figure 10.

Remember, Equation (4) has been derived for a thermoelectric material apart from a device. Furthermore, a constant temperature gradient has been assumed, which means a constant slope of the temperature profile, which then connects the hot side at $T_{hot}$ and the cold side at $T_{cold}$ by a straight line (solid line in Figure 10a). The further assumption of a temperature-independent entropy conductivity $\Lambda_{OC}$ at electrical open-circuit is plotted in Figure 10b as a solid line. As a consequence of these assumptions, at a given electrical current (including electrically open-circuited conditions), the entropy current will carry the highest energy current at the hot side of the thermoelectric material. When advancing through the thermoelectric material to lower temperatures, the entropy current cannot further carry all thermal energy ("heat"), which then needs to be dissipated. Following Walstrom's approach [59], thermal energy is assumed to be dissipated transversally together with instantaneously produced excess entropy as its carrier. It is important to emphasize that excess entropy leaves the thermoelectric material in directions transversal to the flow of the entropy inserted at the hot side. The ability to conduct thermal energy is decreased with decreasing temperature, which is reflected in a decreasing "heat" conductivity, as plotted in Figure 10c as a solid line.

Traced back to Altenkirch [55] and Ioffe [56], often a model is discussed that considers a two-leg thermogenerator and assumes constant "heat" conductivity. Concerning the thermoelectric material, the model is purely one-dimensional and does not allow for transversal dissipation of entropy and energy. All dissipation has to be considered parallel or antiparallel to the flow of entropy and thermal energy along the thermoelectric material. In fact, only the parallel option (i.e., down the temperature gradient) remains physically meaningful. Under electrically open-circuited conditions (i.e., vanishing electrical current), the temperature profile can still be linear. However, Heikes and Ure [60] have shown that, in the presence of a thermally-induced electrical current, the temperature profile is flattened at the hot side and steeply sloping at the cold side, which is shown in Figure 10a as a dashed line. As a consequence of the curved temperature profile and the constant "heat" conductivity (see dashed line in Figure 10c), the "heat" flux is diminished at the hot side (thermal energy input) and increased at the cold side (thermal energy output). The change in the temperature profile is such that, as compared to the zero electrical current situation, the thermal energy input is diminished by half of the Joule "heat" and the thermal energy release at the cold side is increased by half of the Joule "heat", as shown by Heikes and Ure [60,61]. This is to account for the dissipation of thermal energy being parallel to the flow of entropy and thermal energy. As a consequence, when compared to electrically open-circuited conditions, the thermoelectric material would be thermally less transparent when an electrical current flows.

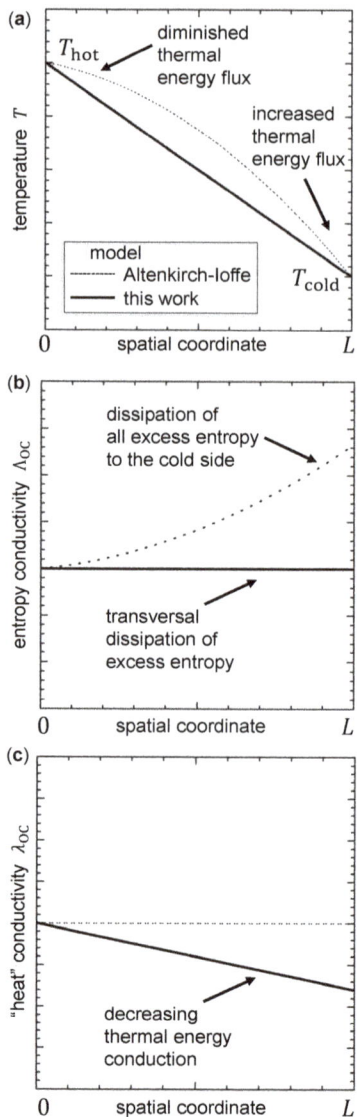

**Figure 10.** Comparison of the model of this work (constant entropy conductivity) to the Altenkirch-Ioffe model [33,55,56,60] (constant "heat" conductivity) with the schematic profiles of the following quantities over the thermoelectric material when the material is carrying a (thermally induced) electrical current: (**a**) temperature $T$; (**b**) electrically open-circuited entropy conductivity $\Lambda_{OC}$; and, (**c**) electrically open-circuited "heat" conductivity $\lambda_{OC}$. Note that profiles are not drawn to scale.

In contrast, the model of this work predicts the thermoelectric material to become thermally more transparent with increasing electrical current, which is reflected in the then reversible increased entropy conductivity $\Lambda(i)$ (see Equations (12) and (14)). In the author's opinion, this is an important characteristic of thermoelectric materials, which is fully embezzled in the traditional model.

In the Altenkirch-Ioffe model, all the excess entropy and excess thermal energy are dissipated to the cold side, which is reflected in an irreversible increase of the entropy conductivity along the thermoelectric material, as visualized in Figure 10b. The aforementioned assumption introduces a ratio of $T_{hot}/T_{cold}$ into the formula for the 2$^{nd}$-law efficiency at the MCEP (see Appendixs B.4 and B.5 for a device in generator mode; see Appendixs C.4 and C.5 for a device in entropy pump mode). Ioffe [56] has shown that the deviation from Equation (22) (generator mode) or Equation (29) (entropy pump mode), however, is only a few per cent when the efficiency itself is small. In other words, for a small temperature difference $\Delta T$, both of the models give nearly the same results.

It must be emphasized that both of the models rely on very special assumptions and, thus, cannot claim general validity [62]. In this sense, all of the results have to be considered semi-quantitatively when it comes to real thermoelectric materials and devices. More general considerations, as provided by Equation (1), need to consider the local variation of thermoelectric parameters but are beyond the scope of this work. Heikes and Ure [60] and Gryasnov et al. [63] have considered the local variation of thermoelectric parameters to some extent. However, the advantage of the model of this work is not only to consider the thermoelectric material apart from a device, but also to clearly separate the dissipation of entropy and thermal energy from the reversible thermoelectric coupling. The simplicity of thermoelectrics is manifested.

### 4.3. Remarks on Narducci's Model

Narducci has put the question "Do we really need high thermoelectric figures of merit?" and found in his calculations that, when considering constant $\Delta T$, the electrical power output of a two-leg thermogenerator device at the MEPP increases with increased thermal conductivity (see Narducci [64], Figure 2). The situation that is discussed by Narducci corresponds to a decreasing figure-of-merit (i.e., $zT \rightarrow 0$ limit) with the electrical power converging to what we have obtained here as $P_{el,\,max}$ (see Equation (16)). In light of this work, it becomes obvious that the MCEP and the MEPP of the thermoelectric material(s) then merge (see Figures 6 and 7).

### 4.4. Remarks on $\Lambda_{OC}$

In the model applied in this work, the electrically open-circuited entropy conductivity $\Lambda_{OC}$ originates only from non-charge transporting excitations of the solid (mostly phonons). Here, the contribution from electrons to the entropy conductivity solely originates from the power factor (see Figures 3 and 5). Subsequently, distinguishing contributions from electrons and phonons to the thermal conductivity is straightforward (see Ioffe [56], p. 44) and has been coined the "phonon glass–electron crystal" (PGEC) concept by Slack [65]. In this case, $\Lambda_{OC}$ is identical to the phonon contribution to the entropy conductivity.

However, as mentioned by Ioffe (see Ioffe [56], p. 46), in materials with charge carriers of both signs (electrons and holes from multiple bands), the situation is more intricate. Subsequently, important electronic contribution to the thermal conductivity can be expected for vanishing net flux of charge. In other words, the electrically open-circuited entropy conductivity $\Lambda_{OC}$ has contributions from both phonons and electrons. The application of the empirical Wiedemann-Franz law to describe the relationship between thermal and electrical conductivity is questionable for these materials [48,56]. In practice, this is the case for many semiconductors and metals. To improve the thermoelectric properties of these materials, it is not sufficient to reduce the phonon contribution by the PGEC concept. In addition, electronic band engineering is required in order to diminish the electron contribution to $\Lambda_{OC}$. The theory in this work can be easily extended to treat this case by introducing a second type of charge carrier into Equations (1) and (4).

### 4.5. Remarks on Figure-of-Merit $zT$

In this work, the figure-of-merit has been introduced in context with the entropy conductivity (cf. Equation (14)) to underline that it is the dimensionless ratio of two entropy conductivities. Initially, the thermoelectric figure-of-merit was introduced by Ioffe [56] as a parameter $z = \frac{\sigma \cdot \alpha^2}{\lambda_{OC}}$ in the treatment of a thermogenerator referring to the "heat" conductivity. In subsequent treatment, Ioffe has taken into account the medium temperature $\bar{T}$ of the device and elucidated the thermoelectric material's figure-of-merit to be $z\bar{T} = \frac{\sigma \cdot \alpha^2}{\lambda_{OC}} \cdot \bar{T}$, which has subsequently been widely used as $zT$. With this formulation of the figure-of-merit, researchers often have been confused by the intensive variable temperature $\bar{T}$ showing up explicitly besides material parameters [66]. It is seen as a persistent residual outcome of the historical dispute between Ostwald and Boltzmann (see Section 1.2) that it has not been realized that the use of entropy conductivity $\Lambda$ instead of the "heat" conductivity $\lambda$ makes the figure-of merit depend on three material parameters only, which all implicitly depend on temperature (see Equations (3) and (15)).

The author has used $zT$ to be consistent with the conventional nomenclature of the thermoelectric community. All of the formulas in this article, which contain the figure-of-merit, however, would look more straightforward if $zT$ were to be substituted by a single letter, for instance, $f$ as used by Zener [67].

$$f = \frac{\sigma \cdot \alpha^2}{\Lambda_{OC}} = \frac{\sigma \cdot \alpha^2}{\lambda_{OC}} \cdot T = zT \tag{34}$$

### 4.6. Remarks on State-of-the-Art and Emerging Thermoelectric Materials

It is worth noting that, for a thermoelectric material with $zT < 2$, there is no benefit from operating it apart from the MEPP (see Figure 6, Figure 7 and Table 2). In this context, it is important to perceive that current state-of-the-art materials hardly exceed a $zT$ value of 2. The values listed in Table 4 are peak values. Among the materials of Table 4, PbTe$_{0.7}$S$_{0.3}$-2.5%K has a peak $zT$ of 2.2 at 923 K and a record high average $zT$ of 1.56 in the temperature interval of 300–900 K [68]. Conclusively, the tracking of the MEPP [69], but not of the MCEP, is reported for thermogenerators. However, for the application of emerging thermoelectric materials with further improved figure-of-merit, and thus more distant working points, tracking of the MCEP might become relevant.

**Table 4.** Maximum figure-of-merit $zT_{max}$ and corresponding power factor $\sigma \cdot \alpha^2$ of some state-of-the-art and emerging thermoelectric materials at temperature $T$ with indication of conduction type.

| Material | Type | $zT_{max}$ | $\sigma \cdot \alpha^2$ [$\mu$Wcm$^{-1}$K$^{-2}$] | $T$ [K] | Ref. |
|---|---|---|---|---|---|
| (Bi$_{0.25}$Sb$_{0.75}$)$_2$Te$_3$ | p | 1.05 | 43 | 323 | [70] |
| FeNb$_{0.8}$Ti$_{0.2}$Sb | p | 1.10 | 53 | 973 | [48,71] |
| Hf$_{0.6}$Zr$_{0.4}$Hf$_{0.25}$NiSn$_{0.995}$Sb$_{0.005}$ | n | 1.20 | 47 | 900 | [48,72] |
| Bi$_2$(Te$_{0.94}$Se$_{0.06}$)$_3$ (0.017 wt.% Te, 0.068 wt.% I) | n | 1.25 | 57 | 298 | [73] |
| (Bi$_{0.25}$Sb$_{0.75}$)$_2$Te$_3$ (8wt.% Te) | p | 1.27 | 58 | 298 | [73] |
| nano (Bi$_{0.25}$Sb$_{0.75}$)$_2$Te$_3$ | p | 1.4 | 38 | 373 | [70] |
| ZrCoBi$_{0.65}$Sb$_{0.15}$Sn$_{0.20}$ | p | 1.42 | 38 | 973 | [48,74] |
| FeNb$_{0.88}$Hf$_{0.12}$Sb | p | 1.45 | 51 | 1200 | [48,75] |
| Bi$_{0.88}$Ca$_{0.06}$Pb$_{0.06}$CuSeO | p | 1.5 | 8 | 873 | [48,76] |
| β-Cu$_{2-x}$Se | p | 1.5 | 12 | 1000 | [77] |
| Ti$_{0.5}$Zr$_{0.25}$Hf$_{0.25}$NiSn$_{0.998}$Sb$_{0.002}$Se | n | 1.5 | 62 | 700 | [48,78] |
| Mg$_3$Sb$_{1.48}$Bi$_{0.4}$Te$_{0.04}$ | n | 1.65 | 13 | 725 | [79] |
| Ba$_{0.08}$La$_{0.05}$Yb$_{0.04}$Co$_4$Sb$_{12}$ | n | 1.7 | 51 | 850 | [80] |
| Mg$_{3.175}$Mn$_{0.025}$Sb$_{1.5}$Bi$_{0.49}$Te$_{0.01}$ | n | 1.71 | 20 | 700 | [48,81] |
| B-doped Si$_{80}$Ge$_{20}$ + YSi$_2$ | p | 1.81 | 39 | 1073 | [48,82] |
| Cu$_{2-y}$S$_{1/3}$Se$_{1/3}$Te$_{1/3}$ | p | 1.9 | 8 | 1000 | [83] |
| AgPb$_m$SbTe$_{2+m}$ | n | 2.2 | 11 | 800 | [84] |
| PbTe$_{0.7}$S$_{0.3}$-2.5%K | p | 2.2 | 14 | 923 | [68] |
| PbTe-4%SrTe-2%Na | p | 2.2 | 24 | 915 | [85] |
| Ge$_{0.89}$Sb$_{0.1}$In$_{0.01}$Te | p | 2.3 | 37 | 650 | [86] |
| PbTe-8%SrTe | p | 2.5 | 30 | 923 | [87] |
| SnSe single crystal's b-axis | p | 2.6 | 10 | 923 | [88] |
| β-Cu$_2$Se/CuInSe$_2$ (1% In) | p | 2.6 | 12.5 | 850 | [89] |
| SnSe$_{0.97}$Br$_{0.03}$ single crystal's a-axis | n | 2.8 | 9 | 773 | [90] |

The benefit of an increased figure-of-merit $zT$ will be an increased power conversion efficiency at the MEPP anyway. Figure 6, Figure 7, and Table 2 indicate that the material's second-law power conversion efficiency at the MEPP will not exceed the value of 0.5 (see also Equation (25)). Interestingly, this value corresponds to the lower limit of the Curzon-Ahlborn efficiency of a Carnot engine operated at its MEPP [91,92]. At the MEPP, a real thermoelectric material will always be operated at less than half of the Carnot efficiency.

### 4.7. Remarks on the Importance of the Power Factor and Choice of Materials for Thermogenerators

Because normalized curves are discussed in this work, one might lose sight of the fact that the power factor $\sigma \cdot \alpha^2$ is at least as important as the figure-of-merit $zT$. According to Equation (16), it rules over the maximum achievable absolute electrical power when the thermoelectric material is operated in generator mode at MCEP. For a material with high $zT$ (e.g., 100), the electrical power is much lower at the MCEP compared to the MEPP (Figures 6 and 7). This is because, at the low electrical current of the MCEP, the thermoelectric material is less permeable to entropy when compared to the MEPP (see Figure 5). Thus, less thermal power is available to be thermoelectrically converted into electrical power. The amount of useful thermal power depends on the power factor and the electrical current (see the second summand in Equation (18)).

The open-circuited entropy conductivity $\Lambda_{OC}$ causes a thermoelectrically-inactive bypass, which eventually leads the temperature difference $\Delta T$, which squared determines the maximum electrical power in Equation (16), to drop. To provide large $\Delta T$, the open-circuited entropy conductivity $\Lambda_{OC}$ should be kept small. Here, in addition to a high power factor $\sigma \alpha^2$, the figure-of-merit $zT$ comes into play, which relates the aforementioned contributions to the entropy conductivity (see Equation (12) and Equation (15)). The materials that are listed in Table 4 represent those with the highest values of the figure-of-merit reported thus far. In the author's opinion, the most interesting materials are those that also have a high power factor of at least 30 $\mu\text{Wcm}^{-1}\text{K}^{-2}$.

A high electrical conductivity $\sigma$ is also advantageous, as already mentioned in Section 2.3. The choice of materials can easily be made with the help of type-1 Ioffe plots [56] ($\sigma \alpha^2 - \sigma$) and type-2 Ioffe plots ($\Lambda_{OC} - \sigma$) [56,93], which have been recently revitalized on the example of current thermoelectric materials [47,48,94]. The reader is referred to Fuchs [32] (p. 135ff) for further details.

### 4.8. Remarks on the Second-Law Power Conversion Efficiency vs. Coefficient of Performance for Entropy Pumps

While the upper limit of the coefficient of performance will depend on temperature conditions, as involved in the Carnot efficiency $\eta_C$ (Equation (27)) or the temperature factor $\frac{T_{cold}}{\Delta T}$ (Equation (26)), the upper limit for the second-law efficiency is fixed to unity (i.e., $\eta_{II,ep} \leq 1$). The unity value of the second-law efficiency refers to an ideal material. While the coefficient of performance is related to a floating scale, the second-law efficiency allows for the estimation of how far from ideal a thermoelectric material is. Another advantage is that the second-law efficiency in Equation (28) only depends on the figure-of-merit and the electrical current and, thus, allows for evaluation of the performance of the thermoelectric material apart from specific temperature conditions, as well as independent from use in a cooler or a heater.

Note that, according to Equations (29) and (22), the maximum second-law efficiency of a thermoelectric material is identical in entropy pump mode and generator mode:

$$\eta_{II,ep,max} = \eta_{II,ep}(i_{MCEP,ep}) = \eta_{II,gen}(i_{MCEP,gen}) = \eta_{II,gen,max} \qquad (35)$$

This is also apparent from Figure 9.

### 4.9. Remarks on the Choice of Materials for Entropy Pumps

Remember, electrical and thermal power in Figure 8 are normalized to the MEPP in generator mode (see Equations (17) and (32)). Thus, the absolute thermal power in entropy pump mode is determined by the material's power factor $\sigma \cdot \alpha^2$ (see Equation (16)). A low open-circuited entropy conductivity $\Lambda_{OC}$ is desired to prevent the thermoelectrically inactive fall of entropy along the temperature difference $\Delta T$, which would make it difficult to maintain the $\Delta T$. Thus, in addition to a high power factor $\sigma \cdot \alpha^2$, a high figure-of-merit $zT$ is favourable, which relates the aforementioned quantities (see Equation (15)).

Operating the thermoelectric material in entropy pump mode requires good performance at ambient temperature and below (e.g., for cooling 150–300 K) or above (e.g., for heating 300–400 K). Among the materials listed in Table 4, only bismuth telluride-based materials fulfil all requirements; and, they are the current materials of choice for the mentioned applications and are conclusively found in commercial devices.

According to Figure 8, emerging materials with improved figure-of-merit at a power factor comparable to bismuth telluride-based materials would have the benefit that comparable thermal power could be pumped from the cold to hot side at a lower electrical current and electrical power.

## 5. Conclusions

Treating entropy and electrical charge as basic quantities allows for a concise description of thermoelectric transport phenomena (entropy, charge, thermal energy, and electrical energy) and it is the key to comprehensibility. The basic transport equation involves classical thermodynamic potentials (temperature and electrical potential) and enables the identification of a thermoelectric material tensor. On the material's voltage–electrical current cure, distinct working points can be identified, which allow for consideration of the power conversion and its efficiency of the thermoelectric material apart from a device. The power depends on the power factor, and the conversion efficiency depends on the figure-of-merit $zT$. A clear physical meaning is given to the power factor as the part of the entropy conductivity that couples to the electrical current. The thermal conductivity, expressed here as entropy conductivity, depends on the electrical current and becomes negative when the thermoelectric material is operated in entropy pump mode. The dimensionless figure-of-merit $zT$ is the ratio of two entropy conductivities, the one under electrically open-circuited conditions and the one that couples to the electrical current. The performance of the thermoelectric material in generator mode and entropy pump mode are related to each other and they can be considered on a single voltage–electrical current curve apart from a device.

**Funding:** This research was funded by the Deutsche Forschungsgemeinschaft (DFG, German Research Foundation) – project number 325156807. The publication of this article was funded by the Open Access Fund of the Leibniz Universität Hannover.

**Acknowledgments:** The author is grateful to Jürgen Caro for his continuous encouragement and sustainable support. The author is grateful to Mario Wolf and Richard Hinterding for critical reading of the manuscript and their suggestions.

**Conflicts of Interest:** The author declares no conflicts of interest.

## Abbreviations

The following abbreviations are used in this manuscript:

| | |
|---|---|
| ECIP | Entropy Conductivity Inversion Point |
| MCEP | Maximum Conversion Efficiency Point (either in generator mode or entropy pump mode) |
| MEPP | Maximum Electrical Power Point (in generator mode) |
| OC | (Electrical) Open Circuit |
| SC | (Electrical) Short Circuit |

## Symbols

The following symbols are used in this manuscript:

## Geometry

| | |
|---|---|
| $A$ | cross-sectional area of thermoelectric material |
| $L$ | length of thermoelectric material |

## Material properties

| | |
|---|---|
| $\alpha$ | Seebeck coefficient |
| $f$ | figure-of-merit (as proposed by Zener [67]) |
| $\lambda$ | "heat" conductivity |
| $\lambda_{OC}$ | "heat" conductivity under electrically open-circuited (OC) conditions |
| $\Lambda$ | entropy conductivity |
| $\Lambda_{OC}$ | entropy conductivity under electrically open-circuited (OC) conditions |
| $\Lambda_{SC}$ | entropy conductivity under electrically open-circuited (SC) conditions |
| $\tilde{\Lambda}$ | normalized entropy conductivity |
| $M_{22}$ | tensor element (of the thermoelectric material tensor) |
| $R$ | electrical resistance (of thermoelectric material) |
| $\sigma$ | isothermal electrical conductivity |
| $z$ | thermoelectric factor (as introduced by Ioffe [56]) |
| $zT$ | figure-of-merit (as introduced by Ioffe [56]) |
| $zT_{max}$ | maximum figure-of-merit |

## Thermodynamic potentials

| | |
|---|---|
| $\mu$ | chemical potential |
| $\tilde{\mu}$ | electrochemical potential ($\tilde{\mu} = \mu + q \cdot \varphi$) |
| $\nabla \tilde{\mu}$ | gradient of the electrochemical potential |
| $\nabla \tilde{\mu}/q$ | gradient of the electrochemical potential per electric charge ($\nabla \tilde{\mu}/q = \nabla \mu/q + \nabla \varphi$) |
| $\varphi$ | electrical potential |
| $\nabla \varphi$ | gradient of the electrical potential |
| $\Delta \varphi$ | difference of electrical potential (along the thermoelectric material) |
| $\Delta \varphi_{OC}$ | voltage under electrically open-circuited (OC) conditions |
| $T$ | absolute temperature |
| $T_{cold}$ | temperature of the thermoelectric material at its cold side |
| $T_{hot}$ | temperature of the thermoelectric material at its hot side |
| $\nabla T$ | gradient of the temperature |
| $\Delta T$ | difference of temperature (along the thermoelectric material) |
| $u$ | normalized voltage |
| $u_{MEPP}$ | normalized voltage at the maximum electrical power point (MEPP) |

## Fluxes

| | |
|---|---|
| $A$ | cross-sectional area of thermoelectric material |
| $L$ | length of thermoelectric material |
| $i$ | normalized electrical current |
| $i_{MCEP,ep}$ | normalized electrical current at the maximum conversion efficiency point (MCEP) in entropy pump mode |
| $i_{MCEP,gen}$ | normalized electrical current at the maximum conversion efficiency point (MCEP) in generator mode |
| $i_{MEPP}$ | normalized electrical current at the maximum electrical power point (MEPP) |
| $I_q$ | electrical current |
| $I_{q,SC}$ | electrical current at electrically short-circuited (SC) conditions |
| $I_S$ | entropy current |
| $j_q$ | electrical flux density |
| $j_S$ | entropy flux density |
| $q$ | electric charge |
| $S$ | entropy |

## Performance

| | |
|---|---|
| $COP_{cooler}$ | coefficient of performance of the thermoelectric material when used in a cooler |
| $COP_{heater}$ | coefficient of performance of the thermoelectric material when used in a heater |
| $\eta_{I,gen}$ | first-law power conversion efficiency of the thermoelectric material in generator mode |
| $\eta_{II,gen}$ | second-law power conversion efficiency of the thermoelectric material in generator mode |
| $\eta_{II,gen,max}$ | maximum second-law power conversion efficiency of the thermoelectric material in generator mode |
| $\eta_{II,ep}$ | second-law power conversion efficiency of the thermoelectric material in entropy pump mode |
| $\eta_{II,ep,max}$ | maximum second-law power conversion efficiency of the thermoelectric material in entropy pump mode |
| $\eta_C$ | Carnot's efficiency |
| $p_{el}$ | normalized electrical power |
| $P_{el}$ | electrical power, needed for lifting electrical charge (generator mode) or made available by the fall of electric charge (entropy pump mode); simplified called output (generator mode) or input (entropy pump mode), when the electrical potential on one side of the thermoelectric material is set to zero |
| $P_{el,\,max}$ | maximum electrical power output of the thermoelectric material in generator mode (at the MEPP) |
| $P_{el,MCEP}$ | electrical power output, of the thermoelectric material in generator mode, at the MCEP |
| $P_{th}$ | thermal power, made available by the fall of entropy (generator mode) or needed for lifting entropy (entropy pump mode) |

## Appendix A. Voltage–Electrical Current and Electrical Power–Electrical Current Characteristics: p- and n-Type Materials

The voltage–electrical current characteristics (green curve) and the electrical power–electrical current characteristics (red curve) of a thermoelectric material with either p-type or n-type conduction are given in Figure A1.

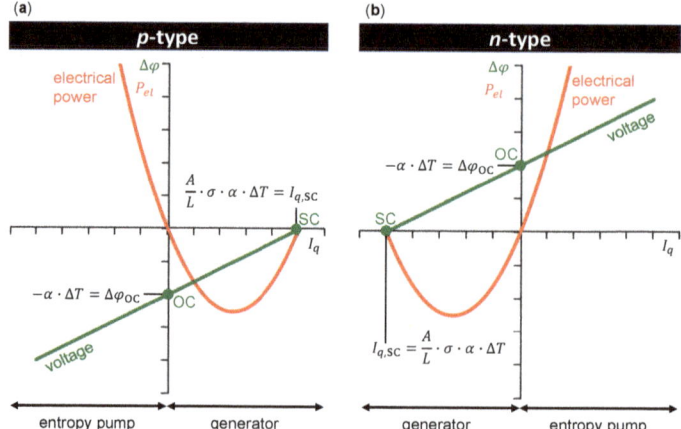

**Figure A1.** Voltage $\Delta\varphi$ – electrical current $I_q$ characteristics (green curves) and electrical power $P_{el}$ – electrical current characteristics $I_q$ (red curves) for materials with: (a) Seebeck coefficient $\alpha$ being positive, which refers to p-type conduction and (b) Seebeck coefficient $\alpha$ being negative, which refers to n-type conduction. Here, $\Delta T = \frac{T_{hot}-T_{cold}}{T_{hot}}$ is the temperature difference along a thermoelectric material of length $L$ and cross-sectional area $A$. These quantities, together with the (isothermal) electrical conductivity $\sigma$ and the Seebeck coefficient, determine the electrical current $I_{SC}$ under electrical short-circuited conditions. The voltage $\Delta\varphi_{OC}$ under electrical short-circuited conditions is determined by the Seebeck coefficient and the temperature difference. When the electrical power $P_{el}$ is negative (electrical power output), the material is in generator mode (thermal-to-electrical power conversion). When the electrical power $P_{el}$ is positive (electrical power input), the material is in entropy pump mode (electrical-to-thermal power conversion).

## Appendix B. Thermal-to-Electrical Power Conversion: Calculations and Established Models

*Appendix B.1. Maximum Electrical Power Point (MEPP): Material in Generator Mode*

The MEPP is found by looking for the vanishing first derivative of the electrical power.

$$0 = \frac{\partial P_{el}}{\partial i} = -\frac{A}{L} \cdot \sigma \cdot \alpha^2 \cdot (\Delta T)^2 \cdot \frac{\partial}{\partial i}(i - i^2) \quad (A1)$$
$$= -\frac{A}{L} \cdot \sigma \cdot \alpha^2 \cdot (\Delta T)^2 \cdot (1 - 2 \cdot i)$$

The derivative vanishes if the term in the brackets vanishes, and the normalized current at the MEPP is as follows.

$$i_{MEPP} = \frac{1}{2} \quad (A2)$$

At the MEPP, the maximum electrical power is obtained as follows.

$$P_{el,max} = P_{el}(i_{MEPP}) = -\frac{A}{L} \cdot \sigma \cdot \alpha^2 \cdot (\Delta T)^2 \cdot (i_{MEPP} - i_{MEPP}^2)$$
$$= -\frac{A}{L} \cdot \sigma \cdot \alpha^2 \cdot (\Delta T)^2 \cdot \left(\frac{1}{2} - \left(\frac{1}{2}\right)^2\right) \quad (A3)$$
$$= -\frac{A}{L} \cdot \sigma \cdot \alpha^2 \cdot (\Delta T)^2 \cdot \left(\frac{1}{2} - \frac{1}{4}\right)$$
$$= -\frac{1}{4} \cdot \frac{A}{L} \cdot \sigma \cdot \alpha^2 \cdot (\Delta T)^2$$

The 2$^{nd}$-law power conversion efficiency at the MEPP is then obtained as follows.

$$\eta_{II,gen,MEPP} = \eta_{II,gen}(i_{MEPP}) = \frac{i_{MEPP} - i_{MEPP}^2}{i_{MEPP} + \frac{1}{zT}}$$
$$= \frac{\frac{1}{2} - \left(\frac{1}{2}\right)^2}{\frac{1}{2} + \frac{1}{zT}}$$
$$= \frac{\frac{1}{2} - \frac{1}{4}}{\frac{1}{2} + \frac{1}{zT}} \quad (A4)$$
$$= \frac{1}{4} \cdot \frac{1}{\frac{1}{2} + \frac{1}{zT}}$$
$$= \frac{zT}{4} \cdot \frac{1}{\frac{zT}{2} + 1}$$
$$= \frac{1}{2} \cdot \frac{zT}{zT + 2}$$

*Appendix B.2. Maximum Conversion Efficiency Point (MCEP): Material in Generator Mode*

The 2$^{nd}$-law power conversion efficiency for a thermoelectric material operated in generator mode is obtained as follows.

$$\eta_{II,gen} = \left|\frac{P_{el}}{P_{th}}\right| = \frac{\alpha \cdot I_q \cdot \Delta T - \frac{I_q^2}{\frac{A}{L} \cdot \sigma}}{\alpha \cdot I_q \cdot \Delta T + \frac{A}{L} \cdot \Lambda_{OC} \cdot (\Delta T)^2} = \frac{\frac{I_q}{I_{q,SC}} - \left(\frac{I_q}{I_{q,SC}}\right)^2}{\frac{I_q}{I_{q,SC}} + \frac{\Lambda_{OC}}{\sigma \cdot \alpha^2}} \quad (A5)$$

Substituting in Equation (A5) the dimensionless normalized electrical current $i = \frac{|I_q|}{|I_{q,SC}|}$ and the figure-of-merit $zT = \frac{\sigma \cdot \alpha^2}{\Lambda_{OC}}$, the 2nd law power conversion efficiency can be written as follows.

$$\eta_{II,gen} = \frac{i - i^2}{i + \frac{1}{zT}} \tag{A6}$$

The maximum power conversion efficiency point (MCEP) can then be found by the first derivative to vanish.

$$\begin{aligned} 0 &= \frac{\partial \eta_{II,gen}}{\partial i} \\ &= \frac{\partial}{\partial i}\left(\frac{i-i^2}{i+\frac{1}{zT}}\right) \\ &= \frac{(1-2 \cdot i) \cdot \left(i+\frac{1}{zT}\right) - \left(i-i^2\right) \cdot 1}{\left(i+\frac{1}{zT}\right)^2} \\ &= \frac{i+\frac{1}{zT} - 2 \cdot i^2 - 2 \cdot \frac{1}{zT} \cdot i - i + i^2}{\left(i+\frac{1}{zT}\right)^2} \\ &= \frac{-i^2 - 2 \cdot \frac{1}{zT} \cdot i + \frac{1}{zT}}{\left(i+\frac{1}{zT}\right)^2} \end{aligned} \tag{A7}$$

The derivative vanishes when the numerator vanishes.

$$i^2 + 2 \cdot \frac{1}{zT} \cdot i - \frac{1}{zT} = 0 \tag{A8}$$

This quadratic equation has two solutions, from which only one gives a positive-definite normalized current $i$ at the maximum conversion efficiency point (MCEP).

$$\begin{aligned} i_{MCEP,gen} &= \frac{\sqrt{1+zT}-1}{zT} \\ &= \frac{\left(\sqrt{1+zT}-1\right) \cdot \left(\sqrt{1+zT}+1\right)}{zT \cdot \left(\sqrt{1+zT}+1\right)} \\ &= \frac{1+zT+\sqrt{1+zT}-\sqrt{1+zT}-1}{zT \cdot \left(\sqrt{1+zT}+1\right)} \\ &= \frac{zT}{zT \cdot \left(\sqrt{1+zT}+1\right)} \\ &= \frac{1}{\sqrt{1+zT}+1} \end{aligned} \tag{A9}$$

At the MCEP, the maximum power conversion efficiency of the thermoelectric material in generator mode is then obtained as follows.

$$\eta_{\text{II,gen,max}} = \eta_{\text{II,gen}}\left(i_{\text{MCEP,gen}}\right) = \frac{i_{\text{MCEP,gen}} - i^2_{\text{MCEP,gen}}}{i_{\text{MCEP,gen}} + \frac{1}{z\overline{T}}}$$

$$= i_{\text{MCEP,gen}} \cdot \frac{1 - i_{\text{MCEP,gen}}}{i_{\text{MCEP,gen}} + \frac{1}{z\overline{T}}}$$

$$= i_{\text{MCEP,gen}} \cdot \frac{\frac{1}{i_{\text{MCEP,gen}}} - 1}{1 + \frac{1}{i_{\text{MCEP,gen}} \cdot z\overline{T}}}$$

$$= \frac{1}{\sqrt{1+z\overline{T}}+1} \cdot \frac{\frac{1}{\frac{1}{\sqrt{1+z\overline{T}}+1}} - 1}{1 + \frac{1}{\frac{1}{\sqrt{1+z\overline{T}}+1} \cdot z\overline{T}}}$$

$$= \frac{1}{\sqrt{1+z\overline{T}}+1} \cdot \frac{\sqrt{1+z\overline{T}}+1-1}{1 + \frac{\sqrt{1+z\overline{T}}+1}{z\overline{T}}}$$

$$= \frac{1}{\sqrt{1+z\overline{T}}+1} \cdot \frac{z\overline{T} \cdot \sqrt{1+z\overline{T}}}{z\overline{T} + \sqrt{1+z\overline{T}}+1} \quad (A10)$$

$$= \frac{1}{\sqrt{1+z\overline{T}}+1} \cdot \frac{z\overline{T} \cdot \sqrt{1+z\overline{T}}}{1+z\overline{T} + \sqrt{1+z\overline{T}}}$$

$$= \frac{1}{\sqrt{1+z\overline{T}}+1} \cdot \frac{z\overline{T}}{\sqrt{1+z\overline{T}}+1}$$

$$= \frac{z\overline{T}}{\left(\sqrt{1+z\overline{T}}+1\right)^2}$$

$$= \frac{1+z\overline{T}-1}{\left(1+\sqrt{1+z\overline{T}}\right)^2}$$

$$= \frac{\left(\sqrt{1+z\overline{T}}+1\right) \cdot \left(\sqrt{1+z\overline{T}}-1\right)}{\left(1+\sqrt{1+z\overline{T}}\right)^2}$$

$$= \frac{\sqrt{1+z\overline{T}}-1}{\sqrt{1+z\overline{T}}+1}$$

By combining Equation (A9) and Equation (A10), the dependence of the maximum second-law efficiency on the electrical current can be shown to be linear.

$$\eta_{\text{II,gen,max}}\left(i_{\text{MCEP,gen}}\right) = \frac{\sqrt{1+z\overline{T}}-1}{\sqrt{1+z\overline{T}}+1}$$

$$= \frac{\sqrt{1+z\overline{T}}+1-2}{\sqrt{1+z\overline{T}}+1} \quad (A11)$$

$$= \frac{\frac{1}{i_{\text{MCEP,gen}}} - 2}{\frac{1}{i_{\text{MCEP,gen}}}}$$

$$= 1 - 2 \cdot i_{\text{MCEP,gen}}$$

The electrical power at the MCEP is as follows.

$$P_{el,MCEP} = P_{el}(i_{MCEP,gen}) = -\frac{A}{L} \cdot \sigma \cdot \alpha^2 \cdot (\Delta T)^2 \cdot \left(i_{MCEP,gen} - i^2_{MCEP,gen}\right)$$

$$= -\frac{A}{L} \cdot \sigma \cdot \alpha^2 \cdot (\Delta T)^2 \cdot \left(\frac{1}{\sqrt{1+zT}+1} - \frac{1}{\left(\sqrt{1+zT}+1\right)^2}\right)$$

$$= -\frac{A}{L} \cdot \sigma \cdot \alpha^2 \cdot (\Delta T)^2 \cdot \left(\frac{\sqrt{1+zT}+1}{\left(\sqrt{1+zT}+1\right)^2} - \frac{1}{\left(\sqrt{1+zT}+1\right)^2}\right)$$

$$= -\frac{A}{L} \cdot \sigma \cdot \alpha^2 \cdot (\Delta T)^2 \cdot \frac{\sqrt{1+zT}}{\left(\sqrt{1+zT}+1\right)^2} \quad \text{(A12)}$$

$$= -\frac{1}{4} \cdot \frac{A}{L} \cdot \sigma \cdot \alpha^2 \cdot (\Delta T)^2 \cdot \frac{4\cdot\sqrt{1+zT}}{\left(\sqrt{1+zT}+1\right)^2}$$

$$= P_{el,max} \cdot \frac{4\cdot\sqrt{1+zT}}{\left(\sqrt{1+zT}+1\right)^2}$$

*Appendix B.3. Comparison to Power Conversion Efficiency after Fuchs: Thermogenerator Device*

By accepting temperature and entropy as primitive quantities, Fuchs [33] has created aggregate dynamical models of a Peltier device. Suggesting the Peltier device to function analogously to a battery, he has derived linear voltage-electrical current characteristics and identified the only two dissipative processes, which are the diffusion of electric charge and the diffusion of entropy. For the case of the device being operated as a thermogenerator, Fuchs [33] has derived its 2nd-law efficiency by the ratio of useful to available power and expressed the efficiency with respect to the internal resistance of the device $R_{TEG}$ and an external load resistance $R_{ext}$.

$$\eta_{II,TEG} = \frac{R_{ext}}{R_{TEG} + (R_{TEG} + R_{ext}) \cdot \frac{1}{zT}} \cdot \frac{R_{TEG}}{R_{TEG} + R_{ext}} \quad \text{(A13)}$$

For a given figure-of-merit $zT$, the 2nd-law efficiency of the device has its maximum at.

$$R_{ext} = \sqrt{1+zT} \cdot R_{TEG} \quad \text{(A14)}$$

Thus, the maximum 2nd-law power conversion efficiency is as follows.

$$\eta_{II,TEG,max} = \frac{\sqrt{1+zT} \cdot R_{TEG}}{R_{TEG} + (R_{TEG} + \sqrt{1+zT} \cdot R_{TEG}) \cdot \frac{1}{zT}} \cdot \frac{R_{TEG}}{R_{TEG} + \sqrt{1+zT} \cdot R_{TEG}}$$

$$= \frac{\sqrt{1+zT}}{1+(1+\sqrt{1+zT}) \cdot \frac{1}{zT}} \cdot \frac{1}{1+\sqrt{1+zT}}$$

$$= \frac{\sqrt{1+zT}}{zT+(1+\sqrt{1+zT})} \cdot \frac{zT}{1+\sqrt{1+zT}}$$

$$= \frac{\sqrt{1+zT}}{1+zT+\sqrt{1+zT}} \cdot \frac{zT}{1+\sqrt{1+zT}}$$

$$= \frac{1}{\sqrt{1+zT}+1} \cdot \frac{zT}{1+\sqrt{1+zT}} \quad (A15)$$

$$= \frac{zT}{(1+\sqrt{1+zT})^2}$$

$$= \frac{1+zT-1}{(1+\sqrt{1+zT})^2}$$

$$= \frac{(\sqrt{1+zT}+1) \cdot (\sqrt{1+zT}-1)}{(1+\sqrt{1+zT})^2}$$

$$= \frac{\sqrt{1+zT}-1}{\sqrt{1+zT}+1}$$

Of note, Fuchs has neglected the Joule "heat", which would only have a small impact when the device is operated in generator mode. Note that Equation (A15) is equivalent to what has been obtained in this work for a thermoelectric material apart from a device (cf. Equation (A10)).

*Appendix B.4. Comparison to Power Conversion Efficiency after Altenkirch: Thermogenerator Device*

Altenkirch [55] has estimated the power conversion efficiency for a thermogenerator (called thermopile at that time), which has been assumed to be made of two legs of dissimilar materials. For a small temperature difference along the device, which will cause only a small thermally-induced electrical current and allows neglect the Joule heating as well as the Thomson effect, he has derived his Equation (4) for the 1st-law power conversion efficiency. Altenkirch [55] has factorized the 1st law power conversion efficiency into the Carnot efficiency and what we call here the 2nd-law power conversion efficiency $\eta_{II}$. The latter has been of the following form.

$$\eta_{II,TEG} = \frac{zT \cdot \frac{R_{ext}}{R_{TEG}}}{zT \cdot (1+\frac{R_{ext}}{R_{TEG}}) + (1+\frac{R_{ext}}{R_{TEG}})^2}$$

$$= \frac{zT \cdot R_{ext} \cdot R_{TEG}}{zT \cdot (R_{TEG}^2 + R_{TEG} \cdot R_{ext}) + (R_{TEG} + R_{ext})^2} \quad (A16)$$

$$= \frac{R_{ext} \cdot R_{TEG}}{R_{TEG} \cdot (R_{TEG} + R_{ext}) + (R_{TEG} + R_{ext})^2 \cdot \frac{1}{zT}}$$

$$= \frac{R_{ext}}{R_{TEG} + (R_{TEG} + R_{ext}) \cdot \frac{1}{zT}} \cdot \frac{R_{TEG}}{R_{TEG} + R_{ext}}$$

Here, Altenkirchs's nomenclature has been substituted by $\frac{R_{ext}}{R_{TEG}} = x$ and $zT = 10^7 \cdot \eta'$. In his treatment, the factor $10^7$ appeared due to the use of the calorie as the energy units, and "$\eta'$" was called the effective thermopower of the device, which however contained the Seebeck coefficient multiplied with the square root of the ratio of specific thermal and specific electrical conductivities of the thermoelectric materials involved. Equation (A16) is equivalent to the result observed by Fuchs (cf. Equation (A13)).

Subsequently, Altenkirch derived the efficiency to be maximized for the following.

$$x = \frac{R_{ext}}{R_{TEG}} = \sqrt{1+z\overline{T}} \tag{A17}$$

Note that Equation (A17) is equivalent to the result obtained by Fuchs (cf. Equation (A14)).

For the thermoelectric generator (TEG), Altenkirch derived the maximum 2$^{nd}$-law power conversion efficiency $\eta_{II,TEG,max}$ to be (see Altenkirch [55], Equation (5)) as follows.

$$\eta_{II,TEG,max} = \frac{\sqrt{1+z\overline{T}}-1}{\sqrt{1+z\overline{T}}+1} \tag{A18}$$

Note that Equation (A18) is equivalent to the result obtained by Fuchs (cf. Equation (A15)).

Even though Altenkirch did not use the term figure-of-merit (compare Altenkirch [55], Figure 3), he plotted the maximum 2$^{nd}$ law power conversion efficiency $\eta_{II,TEG,max}$ as a function of $x = \frac{R_{ext}}{R_{TEG}}$ for different values of his "$\eta'$", which despite a dimensionless factor has been identified with $z\overline{T}$. In the plot, he indicated the shift of the MCEP with varied figure-of-merit.

Altenkirch extended his approach by considering the impact of the Thomson effect on the power conversion efficiency. Moreover, he added remarks on the rate of thermal power exchange of the device with a hot reservoir and cold reservoir and its impact on the effective temperature difference along the device.

*Appendix B.5. Comparison to Power Conversion Efficiency after Ioffe: Thermogenerator Device*

Ioffe [56] has considered a thermocouple in which legs of materials 1 and 2 of equal length are joined by a metallic bridge. The Seebeck coefficient of the device has been estimated from those of the two legs: $\alpha = |\alpha_1| + |\alpha_2|$. From equal length and the individual values of the electrical resistivities ($\rho_1, \rho_2$), "heat" conductivities ($\lambda_{OC,1}, \lambda_{OC,2}$) and cross-sectional area, he has calculated the total electrical resistance $R_{TEG}$ and thermal conductance of the device $K_{TEG}$ (see Ioffe [56], p. 36). To calculate the efficiency of thermal-to-electrical power conversion of the device, he has neglected the Thomson "heat". Furthermore, he made an assumption regarding the Joule "heat" (see Ioffe [56], p. 38): "Of the total Joule 'heat' $I_q^2 \cdot R_{TEG}$ generated in the thermoelement, half passes to the hot junction, returning the power $\frac{1}{2} \cdot I_q^2 \cdot R_{TEG}$ and the rest is transferred to the cold junction." As a result, the temperatures of the hot $T_{hot}$ and cold junction $T_{cold}$ appear in the maximum second-law efficiency.

$$\eta_{II,TEG,max} = \frac{\sqrt{1+z\overline{T}}-1}{\sqrt{1+z\overline{T}}+\frac{T_{hot}}{T_{cold}}} \tag{A19}$$

The aforementioned argument, which was probably inspired by Altenkirch's [55] article, is based on misunderstanding the dissipation, which in the author's opinion is thermal energy to leave the system together with produced entropy. The entropy, and thus the thermal energy, will not have driving force to flow to higher temperature. Anyway, following the argument, the thermal input power is diminished by half of the dissipated Joule "heat". In this work, it has been outlined that the effect of Joule "heat" would be a diminished thermal power supply due to a changed temperature profile (cf. Section 4.2 in the the main text).

Neglecting the Joule "heat", Ioffe has derived the following equation (see Ioffe [56], p. 40).

$$\eta_{II,TEG,max} = \frac{\sqrt{1+z\overline{T}}-1}{\sqrt{1+z\overline{T}}+1} \tag{A20}$$

Note that this is equivalent to what has been obtained in this work for a thermoelectric material apart from a device.

In the factor $z$, which Ioffe deduced (see Ioffe [56], p. 39), the cross-sectional areas $A_1$ and $A_2$ and length $L$ cancel out, so it depends only on the thermoelectric properties of both materials but not their dimensions.

$$z = \frac{\alpha^2}{K_{TEG} \cdot R_{TEG}} = \frac{\alpha^2}{\left(\sqrt{\lambda_{OC,1} \cdot \rho_1} + \sqrt{\lambda_{OC,2} \cdot \rho_2}\right)^2} = \frac{\alpha^2}{\left(\sqrt{\frac{\lambda_{OC,1}}{\sigma_1}} + \sqrt{\frac{\lambda_{OC,2}}{\sigma_2}}\right)^2} \tag{A21}$$

In the case that the electrical conductivities ($\sigma = \sigma_n = \sigma_p$) and "heat" conductivities ($\lambda_{OC} = \lambda_{OC,1} = \lambda_{OC,2}$) are equal in both legs of the device, respectively, Equation (A21) becomes the following.

$$z = \frac{\sigma \cdot \alpha^2}{\lambda_{OC}} \tag{A22}$$

Ioffe used Equation (A22) when discussing a thermoelectric cooler (see Ioffe [56], p. 100) but derived an equivalent expression – using the thermal conductance instead of the thermal conductivity – when discussing the thermogenerator (see Ioffe [56], p. 38ff.). Anyway, in Equations (A19) and (A20) for the maximum power conversion efficiency, there appears not the factor $z$ but this factor multiplied with the average temperature $\overline{T}$.

$$z\overline{T} = z \cdot \overline{T} = z \cdot \frac{T_{hot} + T_{cold}}{2} \tag{A23}$$

Because of Ioffe's Equations (A21)–(A23), the figure-of-merit of a thermoelectric material is currently termed $z\overline{T}$ or $zT$.

## Appendix C. Electrical-to-Thermal Power Conversion: Calculations and Established Models

### Appendix C.1. Power Conversion Efficiency

When the thermoelectric material is used in a cooler, the coefficient of performance $COP$ is the ratio of the thermal power removed from the cold side $T_{cold} \cdot I_S$ related to the electrical power $P_{el}$.

$$COP_{cooler} = \left|\frac{T_{cold} \cdot I_S}{P_{el}}\right| = \left|\frac{T_{cold} \cdot I_S}{P_{th}}\right| \cdot \left|\frac{P_{th}}{P_{el}}\right| = \frac{T_{cold}}{\Delta T} \cdot \left|\frac{P_{th}}{P_{el}}\right|$$

$$= \frac{T_{cold}}{\Delta T} \cdot \frac{\frac{A}{L} \cdot (\lambda_{OC} + \sigma\alpha^2 \cdot i) \cdot (\Delta T)^2}{\frac{A}{L} \cdot \sigma \cdot \alpha^2 \cdot (\Delta T)^2 \cdot (i - i^2)}$$

$$= \frac{T_{cold}}{\Delta T} \cdot \frac{\frac{\lambda_{OC}}{\sigma\alpha^2} + i}{i - i^2} \tag{A24}$$

$$= \frac{T_{cold}}{\Delta T} \cdot \frac{i + \frac{1}{zT}}{-i^2 + i}$$

$$= \frac{T_{cold}}{\Delta T} \cdot \eta_{II,ep}$$

When the thermoelectric material is used in a heater, the coefficient of performance $COP$ is the ratio of the thermal power released to the hot side $T_{hot} \cdot I_S$ related to the electrical power $P_{el}$.

$$COP_{heater} = \left| \frac{T_{hot} \cdot I_S}{P_{el}} \right| = \left| \frac{T_{hot} \cdot I_S}{P_{th}} \right| \cdot \left| \frac{P_{th}}{P_{el}} \right| = \frac{T_{hot}}{\Delta T} \cdot \left| \frac{P_{th}}{P_{el}} \right|$$

$$= \frac{T_{hot}}{\Delta T} \cdot \frac{i + \frac{1}{zT}}{-i^2 + i}$$ 

(A25)

$$= \frac{T_{hot}}{\Delta T} \cdot \eta_{II,ep}$$

$$= \frac{\eta_{II,ep}}{\eta_C}$$

The second-law efficiency for the thermoelectric material in entropy pump mode $\eta_{II,ep}$ is as follows.

$$\eta_{II,ep} = \left| \frac{P_{th}}{P_{el}} \right| = \frac{\frac{A}{L} \cdot (\Lambda_{OC} + \sigma \alpha^2 \cdot i) \cdot (\Delta T)^2}{\frac{A}{L} \cdot \sigma \cdot \alpha^2 \cdot (\Delta T)^2 \cdot (i - i^2)}$$

$$= \frac{\frac{\Lambda_{OC}}{\sigma \alpha^2} + i}{i - i^2}$$ 

(A26)

$$= \frac{i + \frac{1}{zT}}{-i^2 + i}$$

The electrical power $P_{el}$ used in Equations (A24)–(A26) is available by the fall of electric charge along the electrical potential difference $\Delta \varphi$. It drives the pumping of entropy from the material's cold side to its hot side. The thermal power $P_{th} = \Delta T \cdot I_S = T_{hot} \cdot I_S - T_{cold} \cdot I_S$ is needed for lifting entropy along the temperature difference $\Delta T$. Some illustration is given in Figure A2.

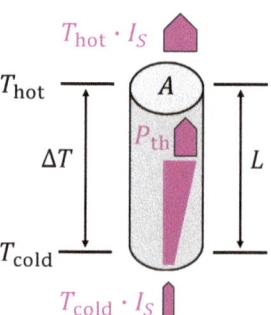

**Figure A2.** When the thermoelectric material is operated in entropy pump mode, electrical power $P_{el}$, which is available by the fall of electric charge along $\Delta \varphi$, drives the pumping of entropy from the cold side to hot side. The thermal power $P_{th} = \Delta T \cdot I_S = T_{hot} \cdot I_S - T_{cold} \cdot I_S$ for lifting entropy along the temperature difference $\Delta T$ adds to the thermal power removed from the cold side $T_{cold} \cdot I_S$ to give the thermal power released to the hot side $T_{hot} \cdot I_S$. Different width of arrows refers to different magnitudes of thermal power at the opposite sides of the material, which is due to thermoelectric power conversion.

## Appendix C.2. Maximum Conversion Efficiency Point (MCEP): Material in Entropy Pump Mode

The maximum power conversion efficiency point (MCEP) follows when the first derivative of the 2nd-law power conversion efficiency, as given by Equation (A26), vanishes.

$$
\begin{aligned}
0 &= \frac{\partial \eta_{II,ep}}{\partial i} \\
&= \frac{\partial}{\partial i}\left(\frac{i+\frac{1}{zT}}{-i^2+i}\right) \\
&= \frac{1 \cdot \left(-i^2+i\right) - \left(i+\frac{1}{zT}\right)\cdot(-2\cdot i+1)}{\left(-i^2+i\right)^2} \\
&= \frac{-i^2+i+2\cdot i^2+\frac{2}{zT}\cdot i-i-\frac{1}{zT}}{\left(-i^2+i\right)^2} \\
&= \frac{i^2+\frac{2}{zT}\cdot i-\frac{1}{zT}}{\left(-i^2+i\right)^2}
\end{aligned}
\tag{A27}
$$

The derivative vanishes when the numerator vanishes.

$$
i^2 + \frac{2}{zT}\cdot i - \frac{1}{zT} = 0 \tag{A28}
$$

The quadratic Equation (A28) has two solutions.

$$
\begin{aligned}
i_{1,2} &= -\frac{1}{zT} \pm \sqrt{\left(\frac{1}{zT}\right)^2 + \frac{1}{zT}} \\
&= -\frac{1}{zT} \pm \frac{1}{zT}\cdot\sqrt{1+zT}
\end{aligned}
\tag{A29}
$$

From the two solutions shown in Equation (A29) only one fulfils the requirement $i \leq -\frac{1}{zT}$ for the material's maximum conversion efficiency point (MCEP) in entropy pump mode. Thus, the normalized electrical current at the maximum conversion efficiency point (MCEP) is obtained as follows.

$$
\begin{aligned}
i_{MCEP,ep} &= -\frac{1}{zT} - \frac{1}{zT}\cdot\sqrt{1+zT} \\
&= -\frac{1+\sqrt{1+zT}}{zT} \\
&= -\frac{\sqrt{1+zT}+1}{zT} \quad ! \\
&= -\frac{\sqrt{1+zT}+1}{1+zT-1} \\
&= -\frac{\sqrt{1+zT}+1}{\left(\sqrt{1+zT}+1\right)\cdot\left(\sqrt{1+zT}-1\right)} \\
&= -\frac{1}{\sqrt{1+zT}-1}
\end{aligned}
\tag{A30}
$$

The maximum 2nd-law power conversion efficiency for a thermoelectric material operated in entropy pump mode is then as follows.

$$\eta_{\text{II,ep,max}} = \eta_{\text{II,ep,max}}\left(i_{\text{MCEP,ep}}\right) = \frac{i_{\text{MCEP,ep}} + \frac{1}{zT}}{-i_{\text{MCEP,ep}}^2 + i_{\text{MCEP,ep}}}$$

$$= \frac{-\frac{\sqrt{1+zT}+1}{zT} + \frac{1}{zT}}{-\left(\frac{\sqrt{1+zT}+1}{zT}\right)^2 - \frac{\sqrt{1+zT}+1}{zT}}$$

$$= \frac{zT}{zT} \cdot \frac{-\sqrt{1+zT}-1+1}{-\frac{1}{zT}\cdot\left(\sqrt{1+zT}+1\right)^2 - \left(\sqrt{1+zT}+1\right)}$$

$$= \frac{1}{-\left(\sqrt{1+zT}+1\right)} \cdot \frac{-\sqrt{1+zT}}{\frac{1}{zT}\cdot\left(\sqrt{1+zT}+1\right)+1}$$

$$= \frac{zT}{\sqrt{1+zT}+1} \cdot \frac{\sqrt{1+zT}}{\left(\sqrt{1+zT}+1\right)+zT} \qquad \text{(A31)}$$

$$= \frac{zT}{\sqrt{1+zT}+1} \cdot \frac{\sqrt{1+zT}}{1+zT+\sqrt{1+zT}}$$

$$= \frac{zT}{\sqrt{1+zT}+1} \cdot \frac{1}{\sqrt{1+zT}+1}$$

$$= \frac{1+zT-1}{\sqrt{1+zT}+1} \cdot \frac{1}{\sqrt{1+zT}+1}$$

$$= \frac{\left(\sqrt{1+zT}+1\right)\cdot\left(\sqrt{1+zT}-1\right)}{\sqrt{1+zT}+1} \cdot \frac{1}{\sqrt{1+zT}+1}$$

$$= \frac{\sqrt{1+zT}-1}{\sqrt{1+zT}+1}$$

By combining Equations (A30) and (A31), the dependence of the maximum second-law efficiency on the electrical current can be shown to be hyperbolic.

$$\eta_{\text{II,ep,max}}\left(i_{\text{MCEP,ep}}\right) = \frac{\sqrt{1+zT}-1}{\sqrt{1+zT}+1}$$

$$= \frac{\sqrt{1+zT}-1}{\sqrt{1+zT}-1+2}$$

$$= \frac{\frac{-1}{i_{\text{MCEP,ep}}}}{\frac{-1}{i_{\text{MCEP,ep}}}+2} \qquad \text{(A32)}$$

$$= \frac{1}{1-2\cdot i_{\text{MCEP,ep}}}$$

The normalized thermal power (cf. Appendix C.3) at the MCEP is obtained by combining Equations (A35) and (A30).

$$p_{th,MCEP} = p_{th}(i_{MCEP,ep}) = 4 \cdot \left| \frac{1}{zT} - \frac{1}{\sqrt{1+zT}-1} \right|$$

$$= \frac{4}{zT} \cdot \left| 1 - \frac{zT}{\sqrt{1+zT}-1} \right|$$

$$= \frac{4}{zT} \cdot \left| 1 - \frac{1+zT-1}{\sqrt{1+zT}-1} \right|$$

$$= \frac{4}{zT} \cdot \left| 1 - \frac{(\sqrt{1+zT}+1) \cdot (\sqrt{1+zT}-1)}{\sqrt{1+zT}-1} \right|$$

$$= \frac{4}{zT} \cdot \left| 1 - (\sqrt{1+zT}+1) \right| \quad (A33)$$

$$= \frac{4}{zT} \cdot \left| 1 - \sqrt{1+zT} - 1 \right|$$

$$= \frac{4}{zT} \cdot \left| -\sqrt{1+zT} \right|$$

$$= 4 \cdot \left| -\frac{\sqrt{1+zT}}{zT} \right|$$

$$= 4 \cdot \frac{\sqrt{1+zT}}{zT}$$

The absolute thermal power at the MCEP in entropy pump mode, which is related to the MEPP in generator mode, is thus the following:

$$|P_{th,MCEP}| = p_{th,MCEP} \cdot P_{el,max}$$

$$= 4 \cdot \frac{\sqrt{1+zT}}{zT} \cdot P_{el,max} \quad (A34)$$

*Appendix C.3. Normalized Thermal Power*

The normalized thermal power $p_{th}$ is obtained as follows.

$$p_{th} = \frac{|P_{th}|}{P_{el,max}}$$

$$= \frac{\left| \frac{A}{L} \cdot (\Lambda_{OC} + \sigma \alpha^2 \cdot i) \cdot (\Delta T)^2 \right|}{\frac{1}{4} \cdot \frac{A}{L} \cdot \sigma \cdot \alpha^2 \cdot (\Delta T)^2} \quad (A35)$$

$$= 4 \cdot \left| \frac{\Lambda_{OC}}{\sigma \alpha^2} + i \right|$$

$$= 4 \cdot \left| \frac{1}{zT} + i \right|$$

*Appendix C.4. Comparison to Power Conversion Efficiency after Altenkirch: Thermoelectric Cooler Device*

For a thermoelectric cooler made of two legs of dissimilar thermoelectric materials (called a thermopile) in steady-state condition, Altenkirch [57] has derived an expression for the minimum electrical power input related to a given cooling power (see Altenkirch [57], Equation (12)), which factorizes into a Carnot-type factor $\frac{T_{hot}-T_{cold}}{T_{cold}}$ and the reciprocal of what he called the dissipation factor for the electro-thermal device. It must be emphasized that the Carnot-type factor introduced by Altenkirch is different from Carnot's efficiency because it relates the temperature difference $T_{hot} - T_{cold}$

to the temperature of the cold side $T_{\text{cold}}$ instead of the hot side $T_{\text{hot}}$. This is due to the thermal energy current removed from the cold side being related to the electrical power input.

When Altenkirch's nomenclature is substituted by $zT = 10^7 \cdot \eta'$, his dissipation factor (see Altenkirch [57], Equation (13)) for the thermoelectric cooler (TEC), which is the device-related analogue of what we here call the maximum 2$^{\text{nd}}$-law power conversion efficiency for a thermoelectric material operated in entropy pump mode $\eta_{\text{II,ep,max}}$, becomes as follows.

$$\eta_{\text{II,TEC,max}} = \frac{\sqrt{1+zT} - \frac{T_{\text{hot}}}{T_{\text{cold}}}}{\sqrt{1+zT}+1} \tag{A36}$$

Altenkirch [57] states that, for small temperature differences (i.e., $\frac{T_{\text{hot}}}{T_{\text{cold}}} \approx 1$), the maximum 2$^{\text{nd}}$-law power conversion efficiency for thermoelectric cooler $\eta_{\text{II,TEC,max}}$ becomes the following.

$$\eta_{\text{II,TEC,max}} = \frac{\sqrt{1+zT}-1}{\sqrt{1+zT}+1} \tag{A37}$$

Altenkirch's result of Equation (A37) for a device is identical to the maximum 2$^{\text{nd}}$-law power conversion efficiency for a thermoelectric material operated in entropy pump mode $\eta_{\text{II,ep,max}}$ as obtained in this work (see Equation (A31)).

*Appendix C.5. Comparison to Power Conversion Efficiency after Ioffe: Thermoelectric Cooler Device*

For a thermoelectric cooler made of two legs of dissimilar thermoelectric materials, Ioffe [56] (see Ioffe [56], p. 99) has derived a maximum coefficient of performance *COP*, which he factorized into the inverse of a Carnot-type factor $\frac{T_{\text{cold}}}{T_{\text{hot}}-T_{\text{cold}}}$ and what we here call the maximum 2$^{\text{nd}}$-law efficiency $\eta_{\text{II,ep,max}}$. After Ioffe [56], the device-related analogue of the latter has been as follows.

$$\eta_{\text{II,TEC,max}} = \frac{\sqrt{1+\frac{1}{2}\cdot z\cdot(T_{\text{hot}}+T_{\text{cold}})} - \frac{T_{\text{hot}}}{T_{\text{cold}}}}{\sqrt{1+\frac{1}{2}\cdot z\cdot(T_{\text{hot}}+T_{\text{cold}})}+1} \tag{A38}$$

In the case of small temperature difference (i.e., $\frac{T_{\text{cold}}}{T_{\text{hot}}} \approx 1$) and when identifying the average temperature $T = \frac{1}{2} \cdot (T_{\text{hot}} + T_{\text{cold}})$, it becomes identical to the result of this work for a thermoelectric material (see Equation (A31)).

## References and Notes

1. Clausius, R. *Abhandlungen über die mechanische Wärmetheorie*; Friedrich Vieweg und Sohn: Braunschweig, Germany, 1864.
2. Clausius, R. Ueber verschiedene für die Anwendung bequeme Formen der Hauptgleichungen der mechanischen Wärmetheorie. *Poggendorffs Ann. Phys. Chem.* **1865**, *125*, 353–400. [CrossRef]
3. Clausius, R. *Mechanical Theory of Heat*; John van Voorst: London, UK, 1867.
4. Boltzmann, L. Über die Beziehung zwischen dem zweiten Hauptsatze der mechanischen Wärmetheorie und der Wahrscheinlichkeitsrechnung. *Wiener Berichte* **1877**, *76*, 373–435.
5. Boltzmann, L. *Wissenschaftliche Abhandlungen, Band 2*; J. A. Barth: Leipzig, Germany, 1909.
6. Flamm, D. Ludwig Boltzmann and his influence on science. *Stud. Hist. Phil. Sci.* **1983**, *14*, 225–278. [CrossRef]
7. Truesdell, C.A. *The Tragicomical History of Thermodynamics 1822-1854*; Springer: New York, NY, USA, 1980. [CrossRef]
8. Gibbs, J.W. On the equilibrium of heterogeneous substances. *Trans. Conn. Acad.* **1875**, *3*, 108–248. [CrossRef]
9. Gibbs, J.W. *The Collected Works of J. Willard Gibbs, Volume 1, Thermodynamics*; Longmans, Green and Co.: New York, NY, USA, 1928.

10. The monism of the "Energeticist" school should not be confused with "energetics" of the British pioneers in thermodynamics [11].
11. Smith, C. *The Science of Energy: A Cultural History of Energy Physics in Victorian Britain*; The University of Chicago Press: Chicago, IL, USA, 1998.
12. Ostwald, W. Studien zur Energetik: 2. Grundlinien der allgemeinen Energetik. *Berichte über die Verhandlungen der Königlich-Sächsischen Gesellschaft der Wissenschaften zu Leipzig, Mathematisch-Physische Klasse* **1892**, *44*, 211–237.
13. Helm, G. *Die Lehre von der Energie*; Arthur Felix: Leipzig, Germany, 1887.
14. Sommerfeld, A. Ludwig Boltzmann zum Gedächtnis. *Wien.-Chem.-Ztg.* **1944**, *3-4*, 25–28.
15. The aim here is not to discredit Wilhelm Ostwald or Ernst Mach who both have made remarkable contributions to science, but instead to give an understanding of how the actual perception of entropy in the scientific community has developed. Readers who are interested in more background information, including the impact of Mach's natural philosophy on the development of quantum mechanics, are referred to Flamm [6].
16. The late William Thomson [Lord Kelvin] wrote in 1906: "Young persons who have grown up in scientific work within the last fifteen years seem to have forgotten that energy is not an absolute existence. Even the Germans laugh on the 'Energetikers' [11]".
17. Boltzmann, L. Ein Wort der Mathematik an die Energetik. *Ann. Phys.* **1896**, *39*, 39–71. [CrossRef]
18. Boltzmann, L. Zur Energetik. *Ann. Phys.* **1896**, *39*, 595–598. [CrossRef]
19. Müller, I. Max Planck – a life for thermodynamics. *Ann. Phys.* **2008**, *17*, 73–87. [CrossRef]
20. Müller, I. Ein Leben für die Thermodynamik. *Physik Journal* **2008**, *7*, 39–45.
21. Herrmann, F. The Karlsruhe Physics Course. *Eur. J. Phys.* **2000**, *21*, 49–58. [CrossRef]
22. Jorda, S. Kontroverse um Karlsruher Physikkurs [Controverse about the Karlsruhe Physiscs Course]. *Phys. J.* **2013**, *12*, 6–7.
23. Strunk, C. *Moderne Thermodynamik—Band 2: Quantenstatistik aus experimenteller Sicht*, 2 ed.; De Gruyter: Berlin, Germany, 2018.
24. It is worth noting that there was another dispute. Max Planck, together with his scholar Ernst Zermelo, also agitated heavily against Ludwig Boltzmann's atomistic-statistical principle, which has been fought also in the pages of the Annalen der Physik [19;20] Only later did Planck became an aglow follower of Boltzmann, and Zermelo translated Gibb's book on statistical mechanics [19,20]. Planck used Boltzmann's principle to estimate the entropy of the electromagnetic field in his formulation of the spectrum of black body radiation [19,20].
25. Planck, M. *Über den zweiten Hauptsatz der mechanischen Wärmetheorie*; Theodor Ackermann: München, Germany, 1879. [CrossRef]
26. Callen, H. The application of Onsager's reciprocal relations to thermoelectric, thermomagnetic, and galvanomagnetic effects. *Phys. Rev.* **1948**, *489*, 414–418. [CrossRef]
27. Callen, H.B. *Thermodynamics—An Introduction to the Physical Theories of Equilibrium Thermostatics and Irreversible Thermodynamics*; John Wiley and Son: New York, NY, USA, 1960.
28. de Groot, G. *Thermodynamics of Irreversible Processes*, 1 ed.; North-Holland Publishing Company: Amsterdam, The Netherlands, 1951.
29. Callendar, H. The caloric theory of heat and Carnot's principle. *Proc. Phys. Soc. London* **1911**, *23*, 153–189. [CrossRef]
30. Carnot, S. *Réflexions sur la puissance motrice du feu*; Bachelier: Paris, France, 1824.
31. Strunk, C. *Moderne Thermodynamik – Band 1: Physikalische Systeme und ihre Beschreibung*, 2 ed.; De Gruyter: Berlin, Germany, 2018.
32. Fuchs, H.U. *The Dynamics of Heat—A Unified Approach to Thermodynamics and Heat Transfer*, 2 ed.; Springer: New York, NY, USA, 2010.
33. Fuchs, H.U. A direct entropic approach to uniform and spatially continuous dynamical models of thermoelectric devices. *Energy Harvest. Syst.* **2014**, *1*, 253–265. [CrossRef]
34. Feldhoff, A. Thermoelectric material tensor derived from the Onsager – de Groot – Callen model. *Energy Harvest. Syst.* **2015**, *2*, 5–13. [CrossRef]
35. Falk, G. *Physik—Zahl und Realität*, 1 ed.; Birkhäuser: Basel, Switzerland, 1990.

36. Job, G. *Neudarstellung der Wärmelehre—Die Entropie als Wärme*; Akademische Verlagsgesellschaft: Frankfurt, Germany, 1972.
37. Job, G.; Rüffler, R. *Physical Chemistry from a Different Angle*, 1 ed.; Springer: Berlin/Heidelberger, Germany, 2016.
38. Strunk [23,31] has shaped the conceptional approach by Falk [35] to put thermodynamics first and assign the statistical behaviour not to an ensemble, but to the individual quantum state itself. Strunk's approach solves the paradox of doubled statistics, which has been inherent to the traditional approach, and it overcomes attempts to interpret quantum statistics as a modified variant of Newtonian mechanics-based kinetic gas theory. Strunk [23,31] suggests to consider heat as to involve entropy and energy. In his approach, entropy is a basic quantity.
39. Neave, E. Joseph Black's lectures on the elements of chemistry. *Isis* **1936**, *25*, 372–390. [CrossRef]
40. Falk, G. Entropy, a resurrection of caloric—A look at the history of thermodynamics. *Eur. J. Phys.* **1985**, *6*, 108–115. [CrossRef]
41. Robison, J. *Lectures on the Elements of Chemistry—Delivered in the University of Edinburgh by the Late Joseph Black*; William Creech Edinburgh: Edinburgh, UK, 1803; Volume 1.
42. To not offend his readers, Strunk has chosen a slightly different point of view by stating that heat comprises entropy and energy and that its use should be avoided when it addresses thermal energy solely.
43. Koshibae, W.; Maekawa, S. Effects of spin and orbital degeneracy on the thermopower of strongly correlated systems. *Phys. Rev. Lett.* **2001**, *87*, 236603–1–236603–4. [CrossRef]
44. Wang, Y.; Rogado, N.S.; Cava, R.; Ong, O. Spin entropy as the likely source of enhanced thermopower in $Na_xCo_2O_4$. *Phys. Rev. Lett.* **2003**, *423*, 425–428. [CrossRef]
45. Falk, G.; Herrmann, F.; Schmid, G. Energy forms or energy carriers? *Am. J. Phys.* **1983**, *51*, 1074–1077. [CrossRef]
46. Treating a device, Fuchs has derived a corresponding equation [32,33].
47. Wolf, M.; Menekse, K.; Mundstock, A.; Hinterding, R.; Nietschke, F.; Oeckler, O.; Feldhoff, A. Low thermal conductivity in thermoelectric oxide-based multiphase composites. *J. Electron. Mater.* **2019**, *48*, 7551–7561. [CrossRef]
48. Wolf, M.; Hinterding, R.; Feldhoff, A. High power factor vs. high $zT$—A review of thermoelectric materials for high-temperature application. *Entropy* **2019**, *21*, 1058. [CrossRef]
49. Geppert, B.; Brittner, A.; Helmich, L.; Bittner, M.; Feldhoff, A. Enhanced flexible thermoelectric generators based on oxide-metal composite materials. *J. Electron. Mater.* **2017**, *46*, 2356–2365. [CrossRef]
50. A constant Seebeck coefficient $\alpha$ of the thermoelectric material being in a temperature gradient implies that the Thomson coefficient $\tau = T \cdot \frac{\partial \alpha}{\partial T} \approx 0$ is negligible.
51. Peltier, J.C.A. Nouvelles expériences sur la caloricité des courants électrique. *Ann. Chim. Phys.* **1834**, *56*, 371–386.
52. Seebeck, T.J. Magnetische Polarisation der Metalle und Erze durch Temperatur-Differenz. *Physicalische und medicinische Abhandlungen der königlichen Academie der Wissenschaften zu Berlin* **1822**, *1820–21*, 265–373.
53. Velmre, E. Thomas Johann Seebeck (1770–1831). *Proc. Estonian Acad. Sci. Eng.* **2007**, *13*, 276–282.
54. Ohmic losses are often referred to as Joule heating.
55. Altenkirch, E. Über den Nutzeffekt der Thermosäule. *Physikalische Zeitschrift* **1909**, *10*, 560–568.
56. Ioffe, A.F. *Semiconductor Thermoelements and Thermoelectric Cooling*, 1 ed.; Infosearch Ltd.: London, UK, 1957.
57. Altenkirch, E. Elektrothermische Kälteerzeugung und reversible elektrische Heizung. *Physikalische Zeitschrift* **1911**, *12*, 920–924.
58. Wolf, M.; Rybakov, A.; Feldhoff, A. Understanding and improving thermoelectric generators via optimized material working points. *Entropy* in preparation.
59. Walstrom, P. Satial dependence of thermoelectric voltages and reversible heats. *Am. J. Phys.* **1988**, *56*, 890–894. [CrossRef]
60. Heikes, R.R.; Ure, R.W. *Thermoelectricity: Science and Engineering*; Interscience Publishers: New York, NY, USA, 1961.
61. This has confused Ioffe [56], who misinterpreted the situation as an uphill "heat" flow: "Of the total Joule 'heat' $I_q^2 \cdot R_{TEG}$ generated in the thermoelement, half passes to the hot junction, returning the power $\frac{1}{2} \cdot I_q^2 \cdot R_{TEG}$ and the rest is transferred to the cold junction."
62. Fuchs, H.U. Personal communication, 8 December 2018.

63. Gryasnov, O.; Moizhes, B.; Nemchinskii, V. Generalized thermoelectric effectivness. *J. Tech. Phys.* **1978**, *48*, 1720–1728.
64. Narducci, D. Do we really need high thermoelectric figures of merit? A critical appraisal to the power conversion efficiency of thermoelectric materials. *Appl. Phys. Lett.* **2011**, *99*, 102104:1–102104:3. [CrossRef]
65. Slack, A. New Materials and Performance Limits for Thermoelectric Cooling. In *CRC Handbook of Thermoelectrics*; Rowe, D., Ed.; CRC Press: New York, NY, USA, 1994.
66. Goupil, C.; Seifert, W.; Zabrocki, K.; Müller, E.; Snyder, G.J. Thermodynamics of thermoelectric phenomena and applications. *Entropy* **2011**, *13*, 1481–1517. [CrossRef]
67. Zener, C. Putting electrons to work. *Trans. ASM* **1961**, *53*, 1052–1068.
68. Wu, H.J.; Zhao, L.D.; Zheng, F.S.; Wu, D.; Pei, Y.L.; Tong, X.; Kanatzidis, M.G. Broad temperature plateau for thermoelectric figure of merit $zT > 2$ in phase-separated $PbTe_{0.7}S_{0.3}$. *Nat. Commun.* **2014**, *5*, 4515. [CrossRef]
69. Risseh, A.E.; Nee, H.P.; Goupil, C. Electrical power conditioning system for thermoelectric waste heat recovery in commercial vehicles. *IEEE Trans. Transp. Electrif.* **2018**, *4*, 548–562. [CrossRef]
70. Poudel, B.; Hao, Q.; Ma, Y.; Lan, Y.; Minnich, A.; Yu, B.; Yan, X.; Wang, D.; Muto, A.; Vashaee, D.; et al. High-thermoelectric performance of nanostructured bismuth antimony telluride bulk alloys. *Science* **2008**, *320*, 634–638. [CrossRef]
71. He, R.; Kraemer, D.; Mao, J.; Zeng, L.; Jie, Q.; Lan, Y.; Li, C.; Shuai, J.; Kim, H.S.; Liu, Y.; et al. Power factor and output power density in p-type half-Heuslers $Nb_{1-x}Ti_xFeSb$. *Proc. Natl. Acad. Sci. USA* **2016**, *113*, 13576–13581. [CrossRef]
72. Chen, L.; Gao, S.; Zeng, X.; Mehdizadeh Dehkordi, A.; Tritt, T.; Poon, S. Uncovering high thermoelectric figure of merit in (Hf,Zr)NiSn half-Heusler alloys. *Appl. Phys. Lett.* **2015**, *107*, 041902. [CrossRef]
73. Yamashita, O.; Ochi, T.; Odahara, H. Effect of the cooling rate on the thermoelectric properties of p-type $(Bi_{0.25}Sb_{0.75})_2Te_3$ and n-type $Bi_2(Te_{0.94}Se_{0.06})_3$ after melting in the bismuth-telluride system. *Mater. Res. Bull.* **2009**, *44*, 1352–1359. [CrossRef]
74. Zhu, H.; He, R.; Mao, J.; Zhu, Q.; Li, C.; Sun, J.; Ren, W.; Wang, Y.; Liu, Z.; Tang, Z.; et al. Discovery of ZrCoBi based half-Heuslers with high thermoelectric conversion efficiency. *Nat. Commun.* **2018**, *9*, 1–9. [CrossRef] [PubMed]
75. Pei, Y.; Shi, X.; Lalonde, A.; Wang, H.; Chen, L.; Snyder, G. Convergence of electronic bands for high performance bulk thermoelectrics. *Nature* **2011**, *473*, 66–69. [CrossRef]
76. Liu, Y.; Zhao, L.D.; Zhu, Y.; Liu, Y.; Li, F.; Yu, M.; Liu, D.B.; Xu, W.; Lin, Y.H.; Nan, C.W. Synergistically optimizing electrical and thermal transport properties of BiCuSeO via a dual-doping approach. *Adv. Energy Mater.* **2016**, *6*, 1502423. [CrossRef]
77. Liu, H.; Shi, X.; Xu, F.; Zhang, L.; Zhang, W.; Chen, L.; Li, Q.; Uher, C.; Day, T.; Snyder, G.J. Copper ion liquid-like thermoelectrics. *Nat. Mater.* **2012**, *11*, 422–425. [CrossRef] [PubMed]
78. Shutoh, N.; Sakurada, S. Thermoelectric properties of the $Ti_x(Zr_{0.5}Hf_{0.5})_{1-x}NiSn$ half-Heusler compounds. *J. Alloy. Compd.* **2005**, *389*, 204–208. [CrossRef]
79. Zhang, J.; Song, L.; Pedersen, S.H.; Yin, H.; Hung, L.T.; Iversen, B.B. Discovery of high-performance low-cost n-type $Mg_3Sb_2$-based thermoelectric materials with multi-valley conduction bands. *Nat. Commun.* **2017**, *8*, 13901. [CrossRef]
80. Shi, X.; Yang, J.; Salvador, J.R.; Chi, M.; Cho, J.Y.; Wang, H.; Bai, S.; Yang, J.; Zhang, W.; Chen, L. Multiple-filled skutterudites: High thermoelectric figure of merit through separately optimizing electrical and thermal transports. *J. Am. Chem. Soc.* **2011**, *133*, 7837–7846. [CrossRef]
81. Chen, X.; Wu, H.; Cui, J.; Xiao, Y.; Zhang, Y.; He, J.; Chen, Y.; Cao, J.; Cai, W.; Pennycook, S.J.; et al. Extraordinary thermoelectric performance in n-type manganese doped $Mg_3Sb_2$ Zintl: High band degeneracy, tuned carrier scattering mechanism and hierarchical microstructure. *Nano Energy* **2018**, *52*, 246–255. [CrossRef]
82. Ahmad, S.; Singh, A.; Bohra, A.; Basu, R.; Bhattacharya, S.; Bhatt, R.; Meshram, K.N.; Roy, M.; Sarkar, S.K.; Hayakawa, Y.; et al. Boosting thermoelectric performance of p-type SiGe alloys through in-situ metallic $YSi_2$ nanoinclusions. *Nano Energy* **2016**, *527*, 282–297. [CrossRef]
83. Zhao, K.; Zhu, C.; Qiu, P.; Qiu, P.; Blichfeld, A.B.; Eikeland, E.; Ren, D.; Iversen, B.B.; Xu, F.; Shi, X.; et al. High thermoelectric performance and low thermal conductivity in $Cu_{2-y}S_{1/3}Se_{1/3}Te_{1/3}$ liquid-like materials with nanoscale mosaic structures. *Nano Energy* **2017**, *42*, 43–50. [CrossRef]

84. Hsu, K.F.; Loo, S.; Guo, F.; Chen, W.; Dyck, J.S.; Uher, C.; Hogan, T.; Polychroniadis, E.K.; Kanatzidis, M.G. Cubic AgPb$_m$SbTe$_{2+m}$: Bulk thermoelectric materials with high figure of merit. *Science* **2014**, *303*, 818–821. [CrossRef] [PubMed]
85. Biswas, K.; He, J.; Blum, I.D.; Wu, C.I.; Hogan, T.P.; Seidman, D.N.; Dravid, V.P.; Kanatzidis, M.G. High-performance bulk thermoelectrics with all-scale hierarchical architectures. *Nature* **2012**, *489*, 414–418. [CrossRef] [PubMed]
86. Hong, M.; Chen, Z.G.; Yang, L.; Zou, Y.C.; Dargusch, M.S.; Wang, H.; Zou, J. Realizing $zT$ of 2.3 in Ge$_{1-x-y}$Sb$_x$In$_y$Te via Reducing the phase-transition temperature and introducing resonant energy doping. *Adv. Mater.* **2018**, *30*, 1705942. [CrossRef]
87. Tan, G.; Shi, F.; Hao, S.; Zhao, L.D.; Chi, H.; Zhang, X.; Uher, C.; Wolverton, C.; Dravid, V.P.; Kanatzidis, M.G. Non-equilibrium processing leads to record high thermoelectric figure of merit in PbTe-SrTe. *Nat. Commun.* **2016**, *7*, 12167. [CrossRef] [PubMed]
88. Zhao, L.D.; Lo, S.H.; Zhang, Y.; Sun, H.; Tan, G.; Uher, C.; Wolverton, C.; Dravid, V.P.; Kanatzidis, M.G. Ultralow thermal conductivity and high thermoelectric figure of merit in SnSe crystals. *Nature* **2014**, *508*, 373. [CrossRef]
89. Olvera, A.A.; Moroz, N.A.; Sahoo, P.; Ren, P.; Bailey, T.P.; Page, A.A.; Uher, C.; Poudeu, P.F.P. Partial indium solubility induces chemical stability and colossal thermoelectric figure of merit in Cu$_2$Se. *Energy Environ. Sci.* **2017**, *10*, 1668–1676. [CrossRef]
90. Chang, C.; Wu, M.; He, D.; Pei, Y.; Wu, C.F.; Wu, X.; Yu, H.; Zhu, F.; Wang, K.; Chen, Y. 3D charge and 2D phonon transports leading to high out-of-plane $zT$ in n-type SnSe crystals. *Science* **2018**, *360*, 778–782. [CrossRef]
91. Curzon, F.; Ahlborn, B. Efficiency of a Carnot engine at maximum power output. *Am. J. Phys.* **1975**, *43*, 22–24. [CrossRef]
92. Leff, H.S. Thermal efficiency at maximum work ouptut: New results for old heat engines. *Am. J. Phys.* **1987**, *55*, 602–610. [CrossRef]
93. Ioffe (see [56], p. 45) has introduced it as "heat" conductivity–electrical conductivity plot ($\lambda_{OC} - \sigma$).
94. Bittner, M.; Kanas, N.; Hinterding, R.; Steinbach, F.; Räthel, J.; Schrade, M.; Wiik, K.; Einarsrud, M.A.; Feldhoff, A. A comprehensive study on improved power materials for high-temperature thermoelectric generators. *J. Power Sources* **2019**, *410–411*, 143–151. [CrossRef]

© 2020 by the author. Licensee MDPI, Basel, Switzerland. This article is an open access article distributed under the terms and conditions of the Creative Commons Attribution (CC BY) license (http://creativecommons.org/licenses/by/4.0/).

Article

# Discrepancy between Constant Properties Model and Temperature-Dependent Material Properties for Performance Estimation of Thermoelectric Generators

**Prasanna Ponnusamy [1,\*], Johannes de Boor [1] and Eckhard Müller [1,2,\*]**

[1] Institute of Materials Research, German Aerospace Center (DLR), D-51170 Köln, Germany; Johannes.deboor@dlr.de
[2] Institute of Inorganic and Analytical Chemistry, Justus Liebig University Gießen, D-35392 Gießen, Germany
\* Correspondence: prasanna.ponnusamy@dlr.de (P.P.); Eckhard.mueller@dlr.de (E.M.);
Tel.: +49-2203-601-3386 (P.P.); +49-2203-601-3556 (E.M.)

Received: 6 July 2020; Accepted: 30 September 2020; Published: 4 October 2020

**Abstract:** The efficiency of a thermoelectric (TE) generator for the conversion of thermal energy into electrical energy can be easily but roughly estimated using a constant properties model (CPM) developed by Ioffe. However, material properties are, in general, temperature ($T$)-dependent and the CPM yields meaningful estimates only if physically appropriate averages, i.e., spatial averages for thermal and electrical resistivities and the temperature average (TAv) for the Seebeck coefficient ($\alpha$), are used. Even though the use of $\alpha_{TAv}$ compensates for the absence of Thomson heat in the CPM in the overall heat balance, we find that the CPM still overestimates performance (e.g., by up to 6% for PbTe) for many materials. The deviation originates from an asymmetric distribution of internally released Joule heat to either side of the TE leg and the distribution of internally released Thomson heat between the hot and cold side. The Thomson heat distribution differs from a complete compensation of the corresponding Peltier heat balance in the CPM. Both effects are estimated quantitatively here, showing that both may reach the same order of magnitude, but which one dominates varies from case to case, depending on the specific temperature characteristics of the thermoelectric properties. The role of the Thomson heat distribution is illustrated by a discussion of the transport entropy flow based on the $\alpha(T)$ plot. The changes in the lateral distribution of the internal heat lead to a difference in the heat input, the optimum current and thus of the efficiency of the CPM compared to the real case, while the estimate of generated power at maximum efficiency remains less affected as it is bound to the deviation of the optimum current, which is mostly <1%. This deviation can be corrected to a large extent by estimating the lateral Thomson heat distribution and the asymmetry of the Joule heat distribution. A simple guiding rule for the former is found.

**Keywords:** TEG performance; device modeling; temperature profile; constant properties model; Fourier heat; Thomson heat; Joule heat

## 1. Introduction

Thermoelectric generator (TEG) materials convert a certain fraction of the heat passed through them into useful electrical power, as the charge carriers (holes/electrons) absorb the thermal energy and move from the hot side to the cold side, carrying entropy [1,2]. The transport entropy flux related to the convective heat transport is given by $\alpha j$, with the Seebeck coefficient $\alpha(T)$ and current density $j$. Typically, a thermoelectric (TE) module consists of a series of pn leg pairs (thermocouples), electrically connected in series and thermally in parallel [3]. In steady-state conditions, the exact performance of the TEG is obtained by solving the thermoelectric heat balance equation [4] for the temperature profile $T(x)$. In 1D, it reads

$$\kappa(T)\frac{\partial^2 T}{\partial x^2} + \frac{d\kappa}{dT}\left(\frac{\partial T}{\partial x}\right)^2 - jT\frac{d\alpha}{dT}\frac{\partial T}{\partial x} = -\rho(T)j^2 \qquad (1)$$

where the thermal conductivity $\kappa$, the electrical resistivity $\rho$ and $\alpha$ are the three main temperature-dependent thermoelectric properties. Here, $\frac{\partial}{\partial x}\cdot\{\kappa(T)\frac{\partial T}{\partial x}\} = \kappa(T)\frac{\partial^2 T}{\partial x^2} + \frac{d\kappa}{dT}\left(\frac{\partial T}{\partial x}\right)^2$ corresponds to the (negative) divergence of the Fourier heat flux, i.e., its local change; $jT\frac{d\alpha}{dT}\frac{\partial T}{\partial x}$ corresponds to the local Thomson heat absorption driven by the change of the convective entropy flux $\alpha j$ related to the temperature dependence of $\alpha(T)$, and $\rho(T)j^2$ corresponds to the local Joule heat dissipation. With a typical TE material, with $\kappa(T)$ falling with $T$ and the amount of $\alpha(T)$ rising with $T$, Thomson heat will be released and Fourier heat flow will grow from the hot to the cold side along a TE leg in TEG operation, where the current flow is driven by the thermo-voltage generated by the leg. Equation (1) is a second-order non-linear partial differential equation, which can be solved using numerical methods like finite element methods (FEMs) [5,6], finite volume methods (FVMs) [7–9] or finite difference methods (FDMs) [6]. However, these solution methods are costly and time-consuming.

On the other hand, when assuming constant properties of the TE properties, an approximate solution can be found analytically, as suggested by Ioffe [1]. This solution by the constant property model (CPM) involves a discrepancy from the exact results due to the underlying simplification. Moreover, the choice of the averaged constant properties to be obtained from the actual temperature-dependent data is not straightforward. As can be seen from Equation (1), the Thomson heat vanishes when the Seebeck coefficient (and with that the convective entropy flux $\alpha j$) remains constant. Various models corrected the CPM to compensate for this "missing Thomson heat" [10–16] have been proposed. Meanwhile, Sandoz et al. [17] attempted to explain the use of the $T$-averaged Seebeck coefficient in predicting exact power in the CPM mathematically, but did not recognize the importance of the asymmetry in heat distribution for the prediction of efficiency.

In a previous study [18], on the physically appropriate choice of averages in the CPM, we highlighted that spatial averages (SpAv) for resistivities (electrical and thermal) and temperature averaging (TAv) for the Seebeck coefficient are essential for a meaningful CPM estimate. However, there is still a remaining deviation due to unconsidered local redistribution of internal heat release or absorption and of thermal conduction in the CPM, which is linked to a change in the $T$ profile $T(x)$ [1,12,18,19]. Here, we will analyze the individual heat contributions exemplarily for six representative thermoelectric materials that we considered previously [18] plus PbTe [20], as this is one of the best TE materials in practice and shows an especially large deviation between CPM and exact results.

Initially, the effect of the $T$ dependence of each of the TE properties, $\alpha$, $\rho$ and $\kappa$, leading to locally shifted heat release and transport over the TE element, for performance estimation, is studied separately. Calculated maximum efficiency in the full temperature-dependent case, $\eta_{max}$, is compared with tailored model materials, in order to separate and quantify the individual contributions. Model materials are defined by setting one or two of the three TE properties constant at its respective average while keeping the other properties $T$ dependent. Next, we explain the physical origin of a relevant part of the discrepancy between CPM results and the real situation using a schematic plot of the convective entropy flux derived from an $\alpha(T)$ graph, alongside showing that the net Peltier/Thomson heat is correctly considered by the CPM when appropriate temperature averaging is used for $\alpha(T)$. Marked areas in the entropy flux diagram quantify the exchange of Peltier and Thomson heat, and with that, a correction for the related deviation in CPM efficiency estimation, $d\eta_{max} = \frac{\eta_{max} - \eta_{max}^{CPM}}{\eta_{max}}$, is suggested and demonstrated.

## 2. Methods, Results and Discussion

### 2.1. Role of the T Dependence of Material Properties in Performance Estimation

Since a generalized temperature dependence study for all types of $T$ dependence is quite elaborate, a comparative study based on seven well-known and representative TE materials [20–24] was conducted. To understand the role of the $T$ dependence of each of $\alpha$, $\rho$ and $\kappa$ in performance estimation, the calculated maximum efficiencies when all properties are considered as $T$ dependent (referred to as "real case" or "exact" from now on) were compared with the calculated efficiencies of model materials. These model materials have the same $T$ dependence as the real materials for one or two of the three thermoelectric transport properties, while the remaining properties are kept constant; these materials are denoted as two temperature-dependent property (2TD) materials and 1TD materials, respectively. The constants used to define the model materials were obtained using the spatial averages (SpAv; for electrical and thermal resistivity) at a current density corresponding to the maximum efficiency of the real material and the temperature average (TAv; for the Seebeck coefficient). The SPAv and TAv of a $T$-dependent quantity $p$ for a hot side temperature $T_h$ and a cold side temperature $T_c$ are given by [1,12,18,25]

$$p_{TAv} = \bar{p} = \frac{1}{\Delta T} \int_{T_c}^{T_h} p(T) dT \qquad (2)$$

$$p_{SpAv} = \langle p \rangle = \frac{1}{L} \int_0^L p(T(x)) dx \qquad (3)$$

where $\Delta T = T_h - T_c$ and $L$ are the length of the TE leg. The exact efficiency using $T$-dependent properties was obtained using the 1D solution algorithm developed in [18] by calculating

$$P = V \cdot I,$$
$$\text{where } V = V_o - R_{in} I, V_o = \alpha \Delta T \text{ and} \qquad (4)$$

$$\eta = P/Q_{in} \qquad (5)$$

Here, $P$ is the output power, $V$ is the net output voltage which is given by the Seebeck voltage generated, $V_o = \int_{T_c}^{T_h} \alpha(T) dT$, minus the voltage drop due to internal resistance $R_i = \frac{\rho_{SpAv} L}{A}$, where $A$ is the area of the TE leg and $\rho_{SpAv} = \frac{1}{L} \int_0^L \rho(T(x)) dx$. $I = jA$ is the current passing through the TE material due to the generated voltage. The efficiency ($\eta$) is given by the ratio of output power to the input heat flow ($Q_{in}$) as in Equation (5), where $Q_{in}$ is given by

$$Q_{in} = -\kappa_h \cdot A \cdot \frac{dT}{dx}_h + I \cdot \alpha_h \cdot T_h \qquad (6)$$

$Q_{in}$ consists of the Fourier heat flow $-\kappa_h \cdot A \cdot \frac{dT}{dx}_h$ (including the fraction of Joule and Thomson heat contributions released in the leg which is flowing to the hot side) plus the Peltier heat ($I \cdot \alpha_h \cdot T_h$) absorbed at the hot side. The suffix h indicates the hot side values, i.e., $\kappa_h = \kappa(T_h)$ and $\alpha_h = \alpha(T_h)$. As the spatial averages depend on $T(x)$, which in turn varies with current, they were formed pre-assuming the optimum current of the real materials. For brevity, the efficiency was also calculated at the optimum current of the real material. The optimum current in the numerical calculation was obtained by finding the current where $\frac{d\eta}{dI}$ becomes zero.

The relative deviation (RD) of the calculated maximum efficiency between the 2TD model materials and the real materials, $\delta\eta_{max}^{model} = \frac{\eta_{max} - \eta_{max}^{model}}{\eta_{max}}$, is shown in Figure 1a. Here, and in the following, for brevity, we will use $\delta$ and $d$ to denote a relative and absolute deviation, respectively. The comparison shows how strongly each of the contributing $T$ dependences alone would shift efficiency. Obviously, the $T$ dependence of $\rho$ will affect the calculated efficiency to a lower extent than

$\alpha(T)$ and $\kappa(T)$ will do for some materials (middle section of Figure 1a); the asymmetry of Joule heat generation mostly plays a minor role. However, this does not hold for all materials and it does not mean that the RD between the CPM and a real material due to asymmetric distribution of Joule heat, $\delta\eta_{\text{maxJ}} = \frac{d\eta_{\text{max}}}{d\dot{Q}_J^h}\delta\dot{Q}_J^h$, would be insignificant, as all of the three identified effects will act simultaneously when comparing the CPM and the real case. Although the effects of the $T$ dependence of $\alpha(T)$ and $\kappa(T)$ are much larger for some materials, they often partly cancel each other. A comparison of the real Joule heat partial $T$ profiles in Figure 1b shows a considerable asymmetry, in correlation to the deviations in the $\rho(T) = const.$ case for SnSe and PbTe (Figure 1a, mid); however, the RD contribution related to the profiles in Figure 1b is larger as they contain an asymmetry due to the asymmetry of axial heat conduction linked to $\kappa(T)$, in addition to the asymmetry of Joule heat generation which alone is represented by Figure 1a. Calculation of the partial $T$ profiles is explained in Appendix A.2. It should be noted that unlike for $\alpha(T)$, where the absence of the $T$ dependence means an absence of Thomson heat, the absence of the $T$ dependence of $\rho$ just means that there is no local asymmetry in Joule heat generation, whereas the amount of Joule heat that appears remains unchanged. Both symmetrically or asymmetrically released Joule heat will contribute, together with Thomson heat, to the effect of a $T$ dependence of $\kappa(T)$ that consists in shifting the distribution of the inner reversible and irreversible heat towards the hot and cold sides. Accordingly, the magnitude of the effect of a $T$ dependence of $\kappa(T)$ will scale with the total amount of inner heat.

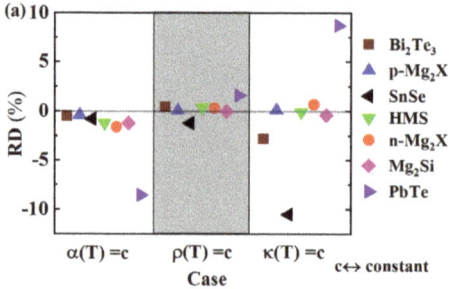
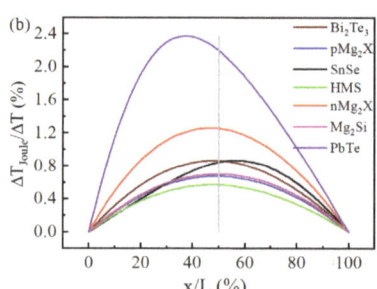

**Figure 1.** (a) Comparison of the relative deviation of the calculated maximum efficiency of 2TD (two temperature-dependent property) model materials (one of the thermoelectric properties kept constant,) to their real counterpart for the example materials, (b) $T$ profile bending caused by Joule heat for example materials. Distinct asymmetry is observed particularly for PbTe and SnSe, correlated to maximum offset values in the middle part of Figure (a).

When $\alpha$ or $\kappa$ is kept constant, there can be large discrepancies, as seen from the scatter in the left and right section of Figure 1a. Switching off Thomson heat results in a change from non-constant to constant convective entropy flux linked to a different partition of reversible (Peltier + Thomson-bound) heat to both sides of the leg. When setting $\kappa(T) = const.$, net Fourier heat transmitted does not change as the thermal resistance of the leg is fixed by the definition of the SpAv. Rather, the observed differences are merely due to a changed lateral distribution of Thomson and Joule heat. Comparing this to Figure 2a reveals that a large RD for $\kappa(T) = const.$ correlates to strongly non-linear $T$ profiles linked to $\kappa(T)$ (see $T$ profiles for $j = 0$); see also Appendix A.2, Figure A1a, where SnSe, Bi$_2$Te$_3$ and PbTe have significantly different $\kappa_h$ and $\kappa_c$ and Figure A2a, showing that the weight of Joule and Thomson heat to $Q_{\text{in}}$ is comparably large for these materials.

The dominating effect of the $T$ dependence of $\kappa$ and $\alpha$ on the estimated performance is also seen by comparing the $T$ profiles of the model cases with the real temperature profile of n-type Mg$_2$(Si,Sn) (referred to as n-Mg$_2$X), Figure 2b. All profiles are calculated for the optimum current for maximum efficiency of the real material. Here, in addition to the 2TD materials, 1TD materials were also involved.

$\alpha(T)$ and $\kappa(T)$ play a dominating role in the shaping of the temperature profile, which is reflected by the closeness of the $\alpha(T) \neq const.$, $\kappa(T) \neq const.$ case to the real material.

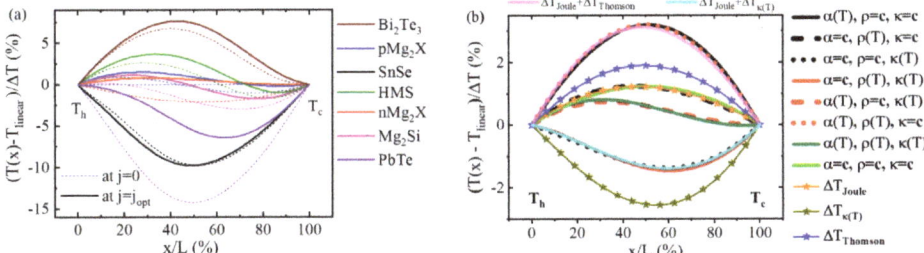

**Figure 2.** (a) Bending of $T$ profiles for the real materials at $j = 0$ (dotted lines) and $j = j_{opt}$ (solid lines), normalized to $\Delta T$, (b) $T$ profile bending for the 1TD and 2TD model materials in comparison to the full $T$-dependent case and the constant properties case, along with the individual contributions to the fully $T$-dependent profile for an n-Mg$_2$(Si,Sn) TE leg with $T_h = 723$ K and $T_c = 383$ K.

The effects of the 2TD cases on the overall inflowing Fourier heat and thus on the efficiency of n-Mg$_2$X from Figure 1a (red dots) can be discussed in terms of the hot side slopes of the corresponding temperature profiles (red lines) in Figure 2b when comparing between cases with the same $\kappa(T)$. The downward $\frac{dT}{dx}\big|_h$ for the 2TD material with $\alpha(T) = const.$(red solid line) indicates an increase in the inflowing Fourier due to missing Thomson heat, compared to the actual case (dark green line). Simultaneously, but only partly compensated in the $Q_{in}$ balance by missing Thomson heat, less Peltier heat is absorbed at the hot side and therefore the efficiency is overestimated (Figure 1a left side, red dot). The 2TD $\kappa(T) = const.$(red dotted line) deforms the $T$ profile considerably but hardly increases the heat input (Equation (6)) compared to the real material, as the SpAv of $\kappa(T)$ maintains an unchanged thermal resistance of the TE leg. We can conclude that replacing the $T$ dependence of $\alpha(T)$ and $\kappa(T)$ by adequate constants will, although significantly changing the $T$ profile, influence the inflowing heat and thus efficiency to a much lower extent due to compensating effects. The RD of CPM efficiency in effect arises mainly from a redistribution of internal Joule and Thomson heat due to considerable deformation of the $T$ profile by neglecting the $T$ dependence of $\kappa(T)$ and $\alpha(T)$ and local redistribution of reversible heat generation as a consequence of neglect of the $T$ dependence of the convective entropy flux.

When comparing the 1TD and 2TD model materials, additionally a shift of the SpAv values of $\rho$ and $\kappa$ as a consequence of different $T$ profiles, as well as coupling effects among the individual contributions, play a role, but only to a very minor extent, as proven by the close coincidence of their profiles to combinations of the individual partial $T$ profiles of the real material, see Figure 2b (pink and cyan lines). The latter represent the physical contributions to the real temperature profile, $\Delta T_{Joule}$, $\Delta T_{Thomson}$ and $\Delta T_{\kappa(T)}$ and are plotted by symbols and lines in Figure 2b. They sum up, together with the linear part, $T_{lin}(x) = T_h - x\frac{\Delta T}{L}$, to the total temperature profile

$$T(x) = T_{lin}(x) + \Delta T_{Joule}(x) + \Delta T_{Thomson}(x) + \Delta T_{\kappa(T)}(x) \tag{7}$$

The procedure to calculate the partial profiles is described in Appendix A.3.

From the close coincidence of combinations of the real partial $T$ profiles to the $T$ profiles of the 1TD and 2TD model materials, as evident from Figure 2b, we can conclude that the contributions from each of the effects (Thomson heat, Joule heat, $T$ dependence of $\kappa$) to the total $T(x)$ behave in good approximation and are independent and additive (a small note on this is given in the Appendix A.1.). The reason for the overall weak cross-coupling between the contributing effects is the small amplitude of the partial $T$ profiles $\Delta T_{Joule}$, $\Delta T_{Thomson}$, $\Delta T_{\kappa(T)}$ compared to the overall $\Delta T$ but also the fact that $\Delta T_{Thomson}$ and $\Delta T_{\kappa(T)}$ often partially compensate. Therefore, the $T$ profiles of a real material and the

CPM may also be quite close to each other for some materials. It is evident that the shape of $\alpha(T)$ and $\kappa(T)$ affects the temperature profile much more than that of $\rho(T)$ but this does not mean that the asymmetry of Joule heat distribution between the hot and cold side would contribute insignificantly to the difference of the inflowing heat between the CPM case and a real material. The redistribution of Joule heat affects the maximum efficiency to a relevant extent along with the redistribution of Thomson heat. Thus, we can split the RD of the maximum efficiency according to the physical origin—redistribution of Peltier–Thomson heat and Joule heat—as $\delta\eta_{max} = \frac{\eta_{max} - \eta_{max}^{CPM}}{\eta_{max}^{CPM}} = \delta\eta_{max\pi\tau} + \delta\eta_{maxJ}$.

Depending on the slope ratio of $\kappa(T)$ and $\alpha(T)$, the efficiency discrepancy due to Joule heat asymmetry, $\delta\eta_{maxJ}$, will vary considerably between different materials and may change sign from case to case, as observed in [18].

Now, let us proceed to understand in more detail how the absence of Thomson heat in the CPM will affect the efficiency calculation. We will see that it is partially and usually not entirely compensated by the difference in Peltier heat between a real material and its CPM approximation.

### 2.2. Peltier–Thomson Heat Balance and the Resulting Uncertainty in CPM Efficiency

Consider a TE material with constant $\kappa$ and a linearly increasing $\alpha(T)$ curve (which is typical for a TE material below the peak $zT$ temperature), as schematically shown in Figure 3. In a TE material under current flow, the convective entropy flux is given by $\dot{s}(T) = j\alpha(T)$. Hence, in a TE leg with a current flow $I$, the convective entropy flow $\dot{S}(T) = I\alpha(T)$ is directly linked to the temperature dependence of the Seebeck coefficient.

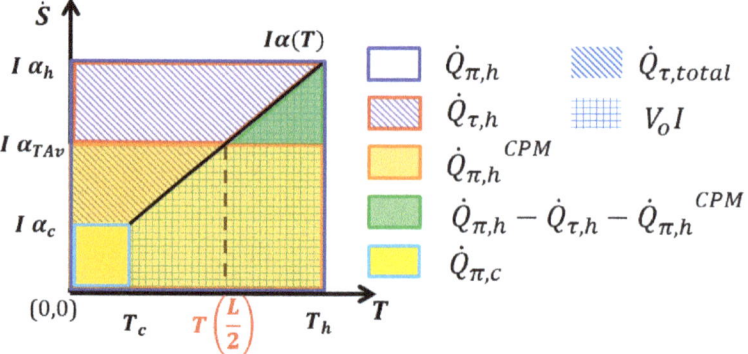

**Figure 3.** Schematic representation of reversible heat exchange in a TE leg for a linear $\alpha(T)$ curve (black line) in a plot of a convective 1D entropy flow with a constant current $I$. According to the relation $\dot{Q}_\pi = I\alpha T$, areas in the $\dot{S}(T)$ diagram represent certain amounts of (flowing or exchanged) Peltier (including Thomson) heat. The dark blue and light blue rectangles—in- and outflowing Peltier heat; trapezium above the $\dot{S}(T)$ curve—Thomson heat (marked with slant lines); trapezium below the $\dot{S}(T)$ curve (marked in checked lines) —gross electrical power generated ($V_0 I$); red trapezium—Thomson heat flowing to the hot side; orange rectangle—hot side Peltier heat (CPM). The green triangle indicates part of the difference in the amount of absorbed Peltier heat at the hot side between the actual and the CPM cases that is not compensated in the real material by backflowing Thomson heat $\dot{Q}_{\tau,h}$.

Peltier heat absorbed at the hot side ($T_h$) in the real case is given by $\dot{Q}_{\pi,h} = I\alpha_h T_h$, while at the cold side, it is $\dot{Q}_{\pi,c} = I\alpha_c T_c$. Areas in the diagram of Figure 3 represent certain amounts of Peltier and Thomson heat but also generated electric power. This allows a schematic comparison of reversible heat exchange in a $T$-dependent material to its CPM approximation. The difference in the Peltier heat balance, $I(\alpha_h T_h - \alpha_c T_c)$, is given by the difference of the light and dark blue line-marked areas. It is composed of the area below the $\dot{S}(T)$ curve (marked in checked lines) given by $P_0 = IV_0 = I \int_{T_c}^{T_h} \alpha(T) dT$, which is the gross produced electrical power (which includes Joule heat). The area to the left from the $\dot{S}(T)$ curve (indicated by slant lines) is

$$\int_{I\alpha_c}^{I\alpha_h} T d\dot{S} = I \int_{T_c}^{T_h} T \frac{d\alpha}{dT} dT = I \int_{T_c}^{T_h} \tau dT = \dot{Q}_\tau \qquad (8)$$

where $\tau = T\frac{d\alpha}{dT}$ is the Thomson coefficient. This area represents the net Thomson heat generated in the TE leg, $\dot{Q}_\tau$, which is directly linked to the variation of the convective entropy flow over the leg. The reversible heat balance

$$\dot{Q}_{\pi,h} - \dot{Q}_{\pi,c} = \dot{Q}_\tau + P_0 \qquad (9)$$

shows that the loss of Peltier heat in the sample equals released Thomson heat plus produced gross electrical power. $\dot{Q}_\tau$ and $P_0$ are counted here as positive when going out of the system. Part of the Thomson heat will flow back, as a contribution to the overall Fourier heat flow, to the hot side. For simplification we assume that Thomson heat that is released at any point in the leg will flow out to the closer side. This is physically not strict but sufficient to qualitatively illustrate the relevant effect of undercompensation of the difference in Peltier heat exchanged at the hot side in a real material compared to the CPM by Thomson heat flowing back to the hot side, i.e., compensation of $d\dot{Q}_{\pi,h} = \dot{Q}_{\pi,h} - \dot{Q}_{\pi,c}{}^{CPM} = IT_h(\alpha_h - \bar{\alpha})$ by $\dot{Q}_{\tau,h} = I \int_{\alpha_{\tau,ex}}^{\alpha_h} Td\alpha$. The relevant question on the Seebeck value $\alpha_{\tau,ex}$, from which the integration gives the correct amount of $\dot{Q}_{\tau,h}$ (and its corresponding temperature $T_{\tau,ex}$ with $\alpha_{\tau,ex} = \alpha(T_{\tau,ex})$), will be touched on below.

In the CPM, the Peltier heat at the hot side is given by $I\bar{\alpha}T_h$, while at the cold side it is $I\bar{\alpha}T_c$, where $\bar{\alpha} = \alpha_{TAv}$ is the temperature average of $\alpha(T)$ (see Equation (2)). Therefore, the following equation holds:

$$\dot{Q}_{\pi,h}{}^{CPM} - \dot{Q}_{\pi,c}{}^{CPM} = I\bar{\alpha}(T_h - T_c) = I\int_{T_c}^{T_h} \alpha(T)dT \qquad (10)$$

i.e., Peltier heat is completely balanced by electrical production.

From Equations (9) and (10), it is obvious that globally the explicit absence of Thomson heat in the CPM is taken care of correctly by the use of temperature averaged $\bar{\alpha}$ in the CPM, i.e.,

$$\dot{Q}_{\pi,h} - \dot{Q}_{\pi,c} - \dot{Q}_\tau = \dot{Q}_{\pi,h}{}^{CPM} - \dot{Q}_{\pi,c}{}^{CPM} = I\bar{\alpha}\Delta T = P_0 \qquad (11)$$

With this choice of $\bar{\alpha}$ as the CPM value, the gross power generated is exactly the same in the CPM as in the real material, at the same current. On the other hand, it implies that, typically, considerably less Peltier heat is absorbed at the hot side in the CPM case than in reality, whereas back-flowing Thomson heat partly compensates the actually higher Peltier heat intake. Figure 3 visualizes with the green triangle that this compensation is incomplete, i.e., $d\dot{Q}_{\pi\tau,h} = d\dot{Q}_{\pi,h} - \dot{Q}_{\tau,h} > 0$. Accordingly, more Thomson heat is leaving at the cold side. It is evident that this holds not only for a linear but also for a left- or right-hand bowed Seebeck curve.

In a less typical case with strongly asymmetric heat conduction, i.e., $\kappa(T)$ strongly increasing with $T$, or if $\alpha(T)$ forms a significant maximum, this typical tendency could reverse, but mostly it leads to underestimation of the inflowing heat in the CPM case $Q_{in}{}^{CPM}$ and hence to overestimation of the efficiency by the CPM. With p-Mg$_2$X, a particular example is given in Appendix A.3.2 (Figure A2c) where, with $\alpha(T)$ weakly changing between $T_c$ and $T_h$ but peaking inside, this compensation can also be almost perfect, or, as for SnSe (Figures 4, 5b and 6), overcompensation may even occur.

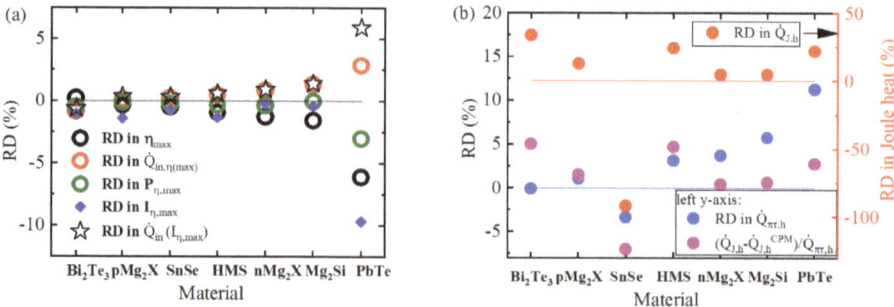

**Figure 4.** Calculated relative deviation (RD) of (**a**) the maximum efficiency, $\delta\eta_{max}$, heat input, $\delta\dot{Q}_{in}$, power at maximum efficiency, $\delta P_{\eta_{max}}$, and optimum current, $\delta I_{opt,\eta}$; additionally, $\delta\dot{Q}_{in}$ when neglecting $\delta I_{opt,\eta}$ (black stars), (**b**) Joule heat, $\delta\dot{Q}_J^h$, reversible heat, $\delta\dot{Q}_{\pi\tau}^h$, (see Equation (12)) and, for direct comparison, also $d\dot{Q}_J^h / \dot{Q}_{\pi\tau}^h$.

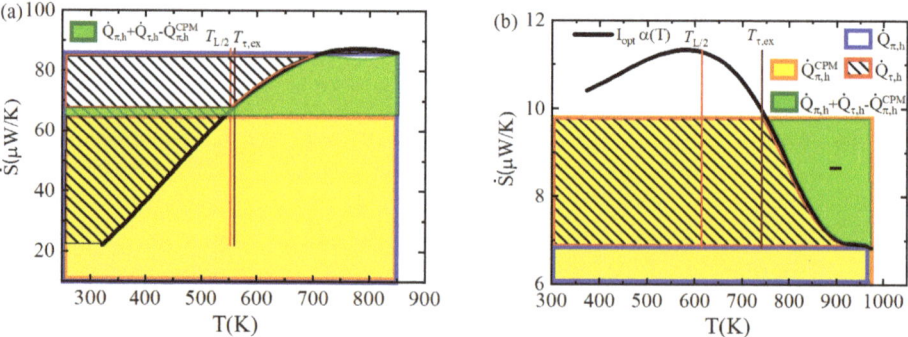

**Figure 5.** Plot of the convective 1D entropy flow at constant current $I$ for (**a**) PbTe and (**b**) SnSe. Relevant areas are marked to determine the uncompensated Peltier–Thomson heat $d\dot{Q}_{\pi\tau}^h$ (green area). Note that the $\frac{l}{2}$ temperature and the temperature $T_{\tau,ex}$ according to the extremum of $\Delta T_{Thomson}(x)$ may be located quite far apart (**b**) whereas $T_{\tau,ex}$ is very close to the crossing point of $\alpha(T)$ to $\bar{\alpha}$.

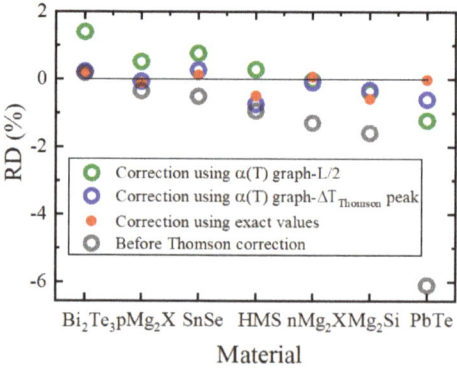

**Figure 6.** RD in maximum efficiency, $\delta\eta_{max}^{corr}$, corrected with respect to $d\dot{Q}_{\pi\tau}^{h, I=const}$ ($T_{\tau,ex}$ according to the peak of the exact Thomson profile; blue), $d\dot{Q}_{\pi\tau}^h$ (exact numerical calculation; red; compare also Equation (12)) and a first guess by the $\frac{l}{2}$ position.

Overall, the efficiency deviation between the real and CPM cases would be negligible if $Q_{in} = Q_{in}{}^{CPM}$. For a rising $\alpha(T)$ curve, which is the typical case applied for most of the established TE materials, the Peltier–Thomson part, $\dot{Q}_{\pi\tau}^{h,CPM}$, of $\dot{Q}_{in}$ will remain lower than the real $\dot{Q}_{\pi\tau}^{h}$. Thus, the efficiency is often overestimated by the CPM. Furthermore, a shift in $I_{\eta,opt}{}^{CPM}$ against the true $I_{\eta,opt}$ has to be taken into consideration due to a change in the current-dependent contributions to $\dot{Q}_{in}$. The usually higher intake of reversible heat at the hot side in the real case, $\dot{Q}_{\pi\tau}^{h}$, compared to the CPM ($\delta \dot{Q}_{\pi\tau,h} > 0$) results in a steeper curve $\dot{Q}_{in}(I)$ than $\dot{Q}_{in}{}^{CPM}(I)$. Efficiency, as defined by $\eta(I) = P/\dot{Q}_{in}$, will accordingly have a lower slope in reality than for the CPM, equivalent to a lower maximum position $I_{opt,\eta}$. Thus, usually, the CPM will overestimate the optimum current, $\delta I_{opt,\eta} = \frac{I_{opt,\eta} - I_{opt,\eta}^{CPM}}{I_{opt,\eta}^{CPM}} < 0$, and hence will overestimate output power at maximum efficiency ($\delta P_{\eta max} < 0$), which adds to the overestimate of maximum efficiency: $\delta \eta_{max} = \delta P_{\eta max} - \delta \dot{Q}_{in}$, amplifying the effect of $\delta \dot{Q}_{in}$ (see Figure 4a). Hence, for a quantitative analysis, we have to consider three contributions to the (absolute) deviation of $\dot{Q}_{in}$

$$d\dot{Q}_{in} = d\dot{Q}_{\pi\tau}^{h} - d\dot{Q}_{J}^{h} = d\dot{Q}_{\pi\tau}^{h,I=const} - d\dot{Q}_{J}^{h,I=const} + \left.\frac{\partial \dot{Q}_{in}}{\partial I}\right|_{I_{opt,\eta}} dI_{opt,\eta} \qquad (12)$$

where, similar to the outflowing Thomson heat, outflowing Joule heat is also counted as positive and $d\dot{Q}_{J}^{h}$ is due to the Joule heat asymmetry at the hot side. Asymmetry of Joule heat distribution and heat conduction will, with falling $\kappa(T)$, as for PbTe and SnSe, favor heat release to the cold side. This will likewise contribute to a higher $\dot{Q}_{in}$ and steeper $\dot{Q}_{in}(I)$, amplifying the same trend as for reversible heat, or will counteract it with rising $\kappa(T)$. Thus, asymmetry of Joule heat distribution will add to the mispoint in $I_{opt,\eta}{}^{CPM}$.

Figure 4a shows that for most materials, $I_{opt,\eta}$ changes for about 1% or less and, consequently, also the deviation of the output power, remain small. However, for PbTe, $\delta I_{opt,\eta}$ reaches 10%. Then the deviation of output power, $\delta P_{\eta max}$, may grow in absolute amount to be as large as $\delta \dot{Q}_{in}$, doubling its effect. Whereas the contribution to $\delta \dot{Q}_{in}$, due to $\delta I_{opt,\eta}$ usually remains insignificant, it becomes relevant for PbTe where it compensates half of $d\dot{Q}_{in}$ related to the distribution of inner heat at an unchanged current, $d\dot{Q}_{\pi\tau}^{h}(I_{opt,\eta}) - d\dot{Q}_{J}^{h}(I_{opt,\eta})$, see Equation (12) and black stars in Figure 4a.

The RD of hot side Joule heat, $\delta \dot{Q}_{J}^{h}$, and Peltier/Thomson heat, $\delta \dot{Q}_{\pi\tau}^{h}$ with $\dot{Q}_{\pi\tau}^{h} = \dot{Q}_{\pi}^{h} - \dot{Q}_{\tau}^{h}$, are shown in Figure 4b. $\delta \dot{Q}_{J}^{h}$ reaches quite significant nominal values (SnSe), mainly due to the low magnitude of $\dot{Q}_{J}^{h}$ itself. For direct comparison to $\delta \dot{Q}_{\pi\tau}^{h}$, the (absolute) deviation $d\dot{Q}_{J}^{h}$ related to $\dot{Q}_{\pi\tau}^{h}$ is plotted and shows that both effects reach the same order of magnitude. Typically, both contributions partly compensate. Furthermore, no general behavior can be observed in their mutual relation over the materials, as in some cases clearly one effect dominates, in others the other.

As seen from Figure 1b, usually, more Joule heat is released to the hot side than to the cold side in a real material, whereas there are symmetric amounts in the CPM case. This contributes to an underestimation of the efficiency in the CPM case, $\delta \eta_{maxJ} > 0$. On the other hand, as explained, the Peltier–Thomson balance tends to an overestimation, $\delta \eta_{max\pi\tau} < 0$, thus, both effects counteract and partially compensate. From Figure 4a, it can be seen that the CPM overestimates the efficiency compared to the real case for all selected materials except $Bi_2Te_3$, which has an exceptionally higher $\kappa_h$ compared to the cold side (Figure A1a in Appendix A.2) together with high Joule release (Figure 4b) and almost compensation of the Peltier–Thomson balance. Thus, the Joule contribution dominates, leading to an underestimation of the efficiency. Additionally, SnSe behaves somewhat differently from the general trend, with a falling $\alpha(T)$ curve (Appendix A.2 Figure A1b) and the over-resistivity at the cold side (Appendix A.2 Figure A1c). Moreover, $\kappa_h$ is much lower than $\kappa_c$. As an effect, Joule heat is preferentially led to the cold side; consequently, hot side Joule heat is greatly overestimated in the CPM (Figure 4b), but as the relative contribution of Joule heat to $Q_{in}$ is small (Figure A2a), the resulting trend towards the overestimation of performance in the CPM remains moderate. On the other hand, as seen from Appendix A.3.2 Figure A1b, Thomson heat is absorbed in the leg as $\alpha(T)$ for SnSe is a falling curve and is mainly bound to the hot side. As seen from Figure 4b, for SnSe, the hot side Peltier–Thomson heat will, unlike for most of the other materials, be overestimated by the CPM. However, the resulting underestimation of efficiency in the CPM will be overcompensated by the counteracting Joule heat distribution.

The first four materials in our list (see Figure 4a) show a minor discrepancy of the CPM with reality. Although Joule heat asymmetry is contributing comparably, from case to case, the dominating source of discrepancy is mostly the uncompensated Peltier heat according to Equation (11)). It is particularly relevant in the cases of n-$Mg_2X$, $Mg_2Si$ and PbTe, which have larger Thomson contributions (Figure A2a), leading to larger discrepancies of the CPM efficiency estimate.

## 2.3. Refining the CPM Efficiency Estimate

Having identified the effects causing a systematic uncertainty in the CPM efficiency estimation, they can be accordingly corrected.

We want to analyze how this can be done practically for the Thomson contribution, $\delta \dot{Q}_{\pi\tau}^h$, by calculating the uncompensated Peltier heat at the hot side. Therefore, we discuss the approach for example materials with dissimilar $\alpha(T)$ characteristics.

The values of $\dot{Q}_{\pi,h}$ and $\dot{Q}_{\pi,h}^{CPM}$ are known from $T_h$, $\alpha_h$ and $\overline{\alpha}$, for a given current, where, as a first approximation, $I_{opt,\eta}^{CPM}$ is used. We have seen that the Thomson heat flowing to the hot side is strictly calculated from the partial T profile $\Delta T_{Thomson}(x)$ by $\dot{Q}_{\tau,h} = -\kappa_h \cdot \frac{d\Delta T_{Thomson}}{dx}\big|_h$. We apply this route to form a reference for an approximate estimation to be developed and, because of this, we omit a numerical calculation of exact T profiles. As derived from Equation (10), we obtain the uncompensated Peltier–Thomson heat from $d\dot{Q}_{\pi\tau}^h = \dot{Q}_{\pi,h} + \dot{Q}_{\tau,h} - \dot{Q}_{\pi,h}^{CPM}$. Neglecting any deviation of current, this can be illustrated in the $\alpha(T)$ diagram based on our interpretation of areas by amounts of reversible heat, see Figure 3. Thus, we aim for a good approximation of the green marked area in Figure 3 by an appropriate and simple approximation. The problem splits into two aspects: finding the temperature $T_{\tau,ex}$ above which the inner Thomson heat is conducted to the hot side and finding a close approximation of the integral. As $\alpha(T)$ may be quite different (see Figure A1b), we meet various situations, represented by different $\Delta T_{Thomson}(x)$ temperature profiles (Figure A2b), among them typical ones with a single maximum according to Thomson heat flowing out to both sides, but also less typical ones with a single minimum (Thomson heat flowing in from both sides) or even two extrema (for $Bi_2Te_3$) where Thomson heat is released to the cold side but absorbed from the hot side. A rule to treat all of the cases likewise is needed. Figure 5a,b and Figure A2c,d accordingly show scenarios where $\alpha(T)$ contains almost linear intervals along with strongly bowed ones, where $\alpha(T)$ is monotonous or contains a maximum, where $\alpha_h$ and $\overline{\alpha}$ are far from each other or close together or where $\alpha(T)$ crosses the $\overline{\alpha}$ horizontal once or twice. The position of the extrema (maxima or minima) of $\Delta T_{Thomson}(x)$ is marked in each diagram by a brown line. Accordingly, the area corresponding to the uncompensated heat might be more complex than is shown in Figure 3, e.g., see Figure 5a. The area to the left of the $\alpha(T)$ curve to the $\alpha$-axis from this point up to the hot side $\alpha_h$ (marked by a red border) represents $\dot{Q}_{\tau,h}$. The fact that the respective area also contains negatively counted parts when $\alpha(T)$ goes through a maximum is also taken into account. Accordingly, the upper slim boat-shaped area in Figure 5a counts as negative; symbolically, it is mirrored in the green area.

However, in such a case, the integration can be simplified, switching from the hot to the cold side, as $\dot{Q}_{\tau,h} = \dot{Q}_\tau - \dot{Q}_{\tau,c}$ and with Equation (10), $\dot{Q}_\tau = \dot{Q}_{\pi,h} - \dot{Q}_{\pi,c} - P_0$. Note that if there are two extrema of $\Delta T_{Thomson}(x)$, then we have two $T_{\tau,ex}$ values where the Thomson heat between both can be neglected as it cancels out completely. Only the intervals outside, $(T_c; T_{\tau,ex})$ or $(T_{\tau,ex}; T_h)$, have to be considered. Among both intervals, the side has to be chosen where $\alpha(T)$ is a monotonous function in the relevant temperature interval, where it is closer to linearity, and possibly where $T_{\tau,ex}$ is closer to $T_h$ or $T_c$.

Applying Equation (11) accordingly to the chosen interval, the integration for $\dot{Q}_\tau$ can be substituted by one for $P_0$, e.g., for the cold side:

$$\dot{Q}_{\tau,c} = \dot{Q}_{\pi,T_{\tau,ex}} - \dot{Q}_{\pi,c} - \int_{T_c}^{T_{\tau,ex}} \alpha dT \tag{13}$$

This facilitates practical execution as $\alpha(T)$ is mostly known as a low-order polynomial, thus integration could be done analytically.

If the Thomson $T$ profile is not known, half of the leg length, $\frac{L}{2}$, can be taken as a first guess of the position for the calculation of $\dot{Q}_{\tau,h}$. The corresponding temperature is marked in the diagrams. This can be a quite good estimate when the Thomson $T$ profile is close to symmetric, as for PbTe (see Figure A2b), but may fail greatly when Thomson heat is strongly asymmetric, as for SnSe. On the contrary, an entropy consideration of Thomson heat in the TE leg (see Appendix A.4.) leads to a rule of thumb for $T_{\tau,ex}$ that is

$$\alpha(T_{\tau,ex}) \approx \bar{\alpha} \tag{14}$$

Indeed, it applies well for all example materials involved here. With this rule, approximation of $\dot{Q}_{\tau,h}$ is facilitated considerably, as just a crossing point of $\alpha(T)$ with its TAv has to be found.

Figure 6 shows the remaining efficiency deviation, $\delta\eta_{max}^{corr}$, corrected by the uncompensated Peltier–Thomson heat calculated from the $\alpha(T)$ graph using the $\frac{L}{2}$ position, using $T_{\tau,ex}$ according to the extremum (maximum) position of $\Delta T_{Thomson}(x)$ but neglecting the current deviation $\delta I_{opt,\eta}$, as well as corrected by the exact deviation $d\dot{Q}_{\pi\tau}^h = \dot{Q}_{\pi,h} - \dot{Q}_{\tau,h} - \dot{Q}_{\pi,h}^{CPM}$. The efficiency estimate by the CPM is greatly improved when the $\Delta T_{Thomson}(x)$ extremum position is used (red dots).

Only occasionally, e.g., when $\alpha(T)$ is close to linear, the $\frac{L}{2}$ position works well for correction but fails for most materials as it does not take into account the asymmetry of heat sources and heat conduction. Similarly, models suggesting half of the Thomson heat on either side for correcting the CPM results [14–16,26,27] will mostly not work sufficiently. The correction employing the $\Delta T_{Thomson}(x)$ peak position is close to the exact numerical correction for most materials as this position considers the asymmetry exactly. The difference between both cases is merely due to the change of the optimum current which is as yet unconsidered by the graphical correction. The remaining discrepancy is due to Joule heat asymmetry.

Whereas we have used exact numerical calculations to demonstrate the principle of the Thomson correction method and to show that the rule $\alpha(T_{\tau,ex}) \approx \bar{\alpha}$ holds well, the suggested practical procedure for the correction of $d\dot{Q}_{\pi\tau}^{h,\,I=const}$ described here, which is based on an analysis of the physical effects behind the deviation of CPM performance estimates, is limited to basic algebraic operations which can be instantaneously calculated by any table calculation software.

## 3. Conclusions

From the study of 2TD and 1TD model materials with one or two selected properties among $\alpha$, $\rho$ and $\kappa$ set as constant, which results in both redistribution of heat between the hot and cold side of the element and the change of spatial averages, we see that in some examples, large deviations in efficiency $\delta\eta_{max}^{model}$ arise as a consequence of considerable modification of the $T$ profile. In comparison to the efficiency deviation between the CPM and real materials $\delta\eta_{max}$ which conserve the spatial property averages and are mostly below 2%, this shows that a change of spatial averages due to an arbitrary modification of the $T$ profile may contribute a strong shift to the efficiency estimate. Thus, conservation of the leg's thermal and electrical resistance is essential for a valid efficiency estimate. However, the shift mainly remains low if only $\rho(T)$ is switched to constant. Nevertheless, it cannot be concluded from this that the temperature dependence of the electrical resistivity plays a minor role in the efficiency estimation by the CPM. The 2TD and 1TD model materials lead to quite good approximations of the partial T profiles $\Delta T_{\text{Joule}}(x)$, $\Delta T_{\text{Thomson}}(x)$ and $\Delta T_{\kappa(T)}(x)$.

It is shown that the deviation of a CPM-based efficiency estimate, $\delta \eta_{\max}$, is not just due to the absence of Thomson heat in the CPM, as the choice of the temperature average of $\alpha(T)$ as a CPM parameter mainly compensates for the absence of Thomson heat. Rather, the discrepancy in efficiency determination in the CPM is shown to be, to a major extent, due to the excess unaccounted heat at the hot side in the CPM $\delta \dot{Q}_{\text{in}}$, which usually leads to overestimation of performance, and, to a minor extent, due to a shift of the optimum current $\delta I_{\text{opt},\eta}$ and, consequently, of the produced electrical power at maximum efficiency, $\delta P_{\eta_{\max}}$. In most cases, the change of the optimum current is small. In materials with rising $\alpha(T)$, less of the released Thomson heat flows back to the hot side than would compensate for the reduced hot side Peltier heat absorption assumed by the CPM. This systematic undercompensation tends towards a higher actual heat intake at the hot side compared to the CPM, thus overestimating efficiency when the CPM is used. Asymmetry of Joule heat usually has an opposite influence but is overcompensated in most cases.

In order to correct for the Peltier–Thomson heat-related deviation $\delta \dot{Q}_{\pi\tau}^h$, a graphical illustration in terms of convective entropy flow based on the $\alpha(T)$ curve is given. It confirms that the rule for the splitting of Thomson heat to the sides $\alpha(T_{\tau,\text{ex}}) \approx \bar{\alpha}$, which results from an entropy consideration, holds well. This enables a valid approximation of $\delta \dot{Q}_{\pi\tau}^h$ with a simple algebraic procedure, omitting the exact numerical calculation of the temperature profile. Although a considerable deformation of the T profile caused by the T dependence of $\kappa(T)$ is observed, it will affect the deviation between the real situation and its CPM approximation simply via a local shift of the thermal and electrical resistivity but will not explicitly contribute to the inflowing heat balance $\dot{Q}_{\text{in}}$.

In summary, the performance of a TE material does not only depend on its averaged material parameters but also on local asymmetry of Thomson and Joule heat, driven by the T dependence of the TE properties. In particular, Thomson heat can show highly asymmetric distribution. Thus, TE device efficiency can be varied beyond the averaged properties, represented by a figure of merit.

**Author Contributions:** Conceptualization, E.M. and P.P.; methodology, E.M. and P.P.; software, P.P.; validation, P.P., E.M. and J.d.B.; formal analysis and investigation, P.P., E.M. and J.d.B.; resources, P.P.; data curation, P.P.; writing—original draft preparation, P.P.; writing—review and editing, P.P., E.M. and J.d.B.; visualization, P.P. and E.M.; supervision, E.M. and J.d.B.; project administration and funding acquisition, E.M. and J.d.B. All authors have read and agreed to the published version of the manuscript.

**Funding:** This research received no external funding.

**Acknowledgments:** We would like to gratefully acknowledge the endorsement from the DLR Executive Board Member for Space Research and Technology and the financial support from the Young Research Group Leader Program. P.P would like to acknowledge the German Academic Exchange Service, DAAD (Fellowship No. 247/2017) for financial support.

**Conflicts of Interest:** The authors declare no conflict of interest.

## Appendix A

*Appendix A.1. Note from Section 2.1*

A very good approximation of the actual T profile and hence the SpAv of $\rho$ and $\kappa$ in accordance with a real material can be calculated in a straightforward way from the T-dependent properties without using an iterative solution [18] for $T(x)$. This may considerably simplify the estimation of appropriate SpAvs as CPM property values. $\Delta T_{\text{Joule}}(x)$ can be obtained analytically from the CPM case, $\Delta T_{\kappa(T)}$ from a integration of the Fourier equation and $\Delta T_{\text{Thomson}}$ and from a 1TD $\alpha(T)$ model by a single integration.

## Appendix A.2. Material Data and Boundary Conditions

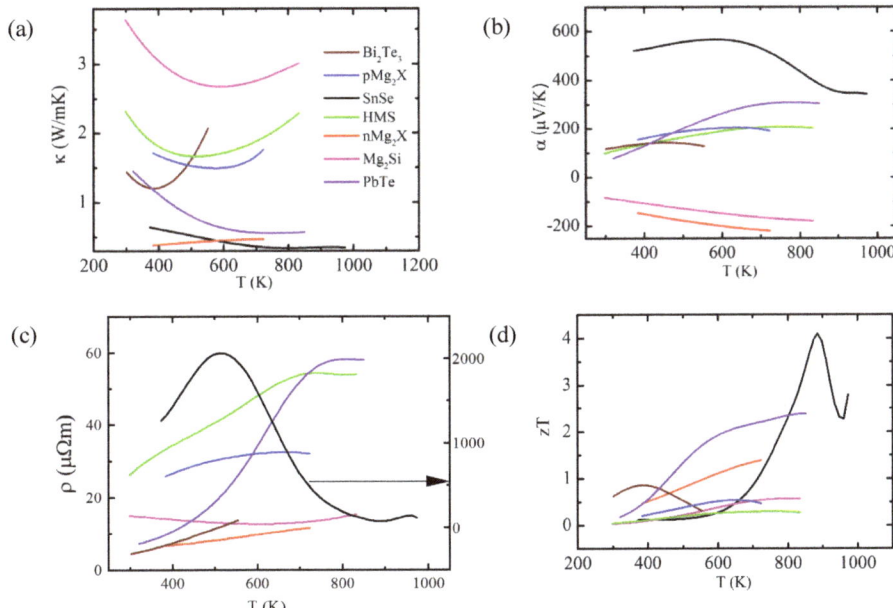

**Figure A1.** Temperature-dependent thermoelectric material properties of representative material classes: (**a**) thermal conductivity, (**b**) Seebeck coefficient, (**c**) electrical resistivity and (**d**) figure of merit. Since SnSe has much higher resistivity, the scale for it is given on the right y-axis. All the raw experimental data taken from the literature [20–24] were fitted with appropriate polynomials (usually 3rd or 4th order). For SnSe, a 9th order polynomial fit was used owing to the complex T dependence and hence shows an unusually high $zT_{max}$. However, this does not affect the physics discussed and hence these fitted data were used throughout the manuscript.

**Table A1.** Temperature range of analysis for all materials of Figure A1.

| Material | Temperature Range of Analysis |
| --- | --- |
| p-Mg$_2$(Si,Sn) | 723 K to 383 K |
| n-Mg$_2$(Si,Sn) | 723 K to 383 K |
| HMS | 833 K to 298 K |
| Mg$_2$Si | 833 K to 298 K |
| p-Bi$_2$Te$_3$ | 553 K to 301 K |
| SnSe | 973 K to 373 K |
| PbTe | 850 K to 320 K |

## Appendix A.3. Additional Information

### Appendix A.3.1. Finding Individual Contributions to the Total T Profile

The partial T profiles are each found by equating $\kappa(T)\frac{\partial^2 T}{\partial x^2}$ in Equation (1) to each of the other corresponding terms, assuming isothermal boundary conditions and fixing all coefficients in the equation according to the total T profile $T(x)$. Thus, solving for the respective partial T profile reduces to a double integration, where the first step provides the total amount of each partial heat contribution to the thermal balance.

As the partial $T$ profiles can have opposite signs in amplitude and partially compensate for many of the common TE materials (however, not always), the $T$ profiles of a real material and the CPM may be quite close, as in the example of n-type $Mg_2X$, Figure 2b.

### Appendix A.3.2. Contributions to $\dot{Q}_{in}$

As both Joule and Thomson heat, after appearing inside the leg, will flow out, physically, as Fourier heat, we have to consider in this discussion the pure Fourier heat $Q_{F,h} = K\Delta T$ (with $K = (\kappa^{-1})^{-1} A/L$), which is merely related to the thermal resistance of the leg and is constant along the leg, separately from the Joule- and Thomson-related contributions. Accordingly, $\dot{Q}_{in}$ is composed of

$$\dot{Q}_{in} = \dot{Q}_{F,h} + \dot{Q}_{\pi,h} - \dot{Q}_{\tau,h} - \dot{Q}_{J,h} \tag{A1}$$

The real Joule- and Thomson-related contributions, $-\dot{Q}_{\tau,h}$ and $-\dot{Q}_{J,h}$, to the inflowing hot side heat are calculated by splitting the overall temperature profile $T(x)$ into additive partial $T$ profiles, each related to one of the individual physical contributions. Partial Thomson $T$ profiles of example materials are plotted in Figure A2b. Evaluating $-\kappa_h \cdot \left(\frac{d}{dx}\Delta T_{Thomson}\right)_h$ and $-\kappa_h \cdot \left(\frac{d}{dx}\Delta T_{Joule}\right)_h$ from the partial $T$ profiles gives $\dot{Q}_{\tau,h}$ and $\dot{Q}_{J,h}$, respectively.

Figure A2a shows the relative contribution of each heat to $\dot{Q}_{in}$: $\frac{\dot{Q}_{F,h}}{\dot{Q}_{in}}, \frac{\dot{Q}_{\pi,h}}{\dot{Q}_{in}}, -\frac{\dot{Q}_{J,h}}{\dot{Q}_{in}}, -\frac{\dot{Q}_{\tau,h}}{\dot{Q}_{in}}$. This comparison reveals that Joule and Thomson heat contribute about 1–5% to $\dot{Q}_{in}$, usually flowing out, with their contributions being roughly of the same order. Figure A2a also shows the fraction of Thomson heat and Joule heat distributed to the hot side ($\frac{\dot{Q}_{\tau,h}}{\dot{Q}_\tau}$ and $\frac{\dot{Q}_{J,h}}{\dot{Q}_J}$).

In order to illustrate example situations of the distribution of Peltier and Thomson heat along the leg, $\alpha(T)$ graphs for p-$Mg_2X$ and $Bi_2Te_3$ are given in Figure A2c,d, respectively. Due to the bowed shape of the $\alpha(T)$ graph and relatively close values of $\alpha_h$ to $\alpha_c$ for p-$Mg_2X$, the difference between $\dot{Q}_{\pi,h}$ and $\dot{Q}_{\pi,h}^{CPM}$ is almost negligible, but $\dot{Q}_{\tau,h}$ amounts to more than twice the amount of $\dot{Q}_{\pi,h} - \dot{Q}_{\pi,h}^{CPM}$. Nevertheless, this did not affect the efficiency deviation $\delta\eta_{max}$ too much, as $\dot{Q}_{\tau,h}$ is quite small in absolute terms. In the case of $Bi_2Te_3$, $\dot{Q}_{\pi,h}^{CPM}$ is even higher than $\dot{Q}_{\pi,h}$ again due to the curved shape of $\alpha(T)$, affecting the position of $\alpha_{TAv}$. However, $\dot{Q}_{\tau,h}$ almost completely compensates for this Peltier heat difference, keeping the influence on the efficiency deviation negligible.

### Appendix A.4. Thomson Heat Distribution and Entropy

With the TEG leg, we discuss the entropy flow in a reversible system of Peltier heat transport and Thomson heat exchange which is running on a non-equilibrium temperature background mainly fixed by the continuous flow of Fourier heat. As released Thomson heat will be transported as Fourier heat but is small in relation to the Fourier heat background (see Figure A2a), which is driven by the temperature difference and the thermal resistance of the TE leg, we will treat the variation of the temperature profile by the conducted Thomson heat as insignificant for the following consideration.

In the steady state, the entropy of the system remains constant; there is a continuous entropy production by the dissipative heat transport from hot to cold and the balancing continuous entropy export by transmitted Fourier heat (plus a negligible fraction arising from outflowing Joule heat). Assuming ideal outer current leads with $\alpha = 0$, there is no other entropy exchange at the hot and cold sides.

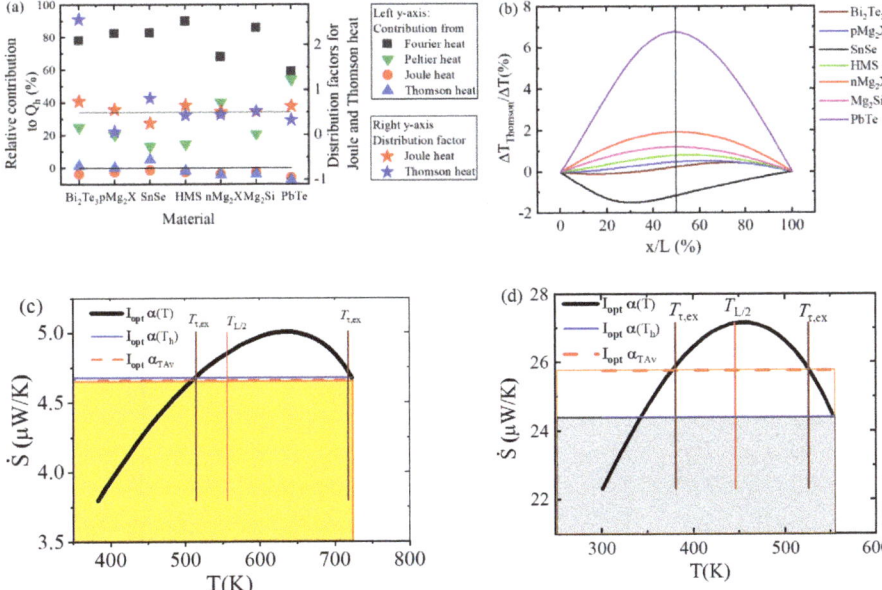

**Figure A2.** (a) Ratio of individual heat contributions to $\dot{Q}_{in}$ (Equation (A1)) calculated from the corresponding partial temperature profiles (for comparison, all quantities are counted as positive when flowing into the element) (left y-axis), and distribution factors (right y-axis) for Thomson and Joule heat. (b) Thomson $T$ profiles for all example materials (c) $\dot{S}(T)$ diagram for p-Mg2X showing the area between $I\alpha(T_h)$ and $I\alpha_{TAv}$ (corresponding to the Peltier heat difference between the CPM and real case), which is very small due to the shape of $\alpha(T)$. The position of the first peak in the Thomson partial $T$ profile is marked as a brown vertical line. (d) $\dot{S}(T)$ diagram for Bi$_2$Te$_3$, where $\alpha_{TAv} > \alpha(T_h)$. Hence, $\dot{Q}_{\pi,h}^{CPM}$ is higher than $\dot{Q}_{\pi,h}$.

In the CPM, we have a constant convective entropy flow $\bar{\alpha}I$ throughout the element, equal to the absorbed and released entropy rate $\bar{\alpha}I$ by absorption and release of Peltier heat at the terminals. In a real material, the absorbed entropy rate $\alpha_h I$ equals the convective entropy flow at the hot side, and, likewise, the amount of $\alpha_c I$ at the cold side. The variation of $\alpha$ along the leg drives local Thomson heat production $d\dot{Q}_\tau = T\frac{d\alpha}{dT}IdT = TId\alpha$, contributing an entropy flow increment $d\dot{S} = Id\alpha$. Thomson heat flows to the hot and cold sides and the related total entropy exchange is $(\alpha_h - \alpha_c)I = \Delta\alpha I$. It distributes by the fraction $x_h$ to the hot and cold sides:

$$\Delta\dot{S}_{\tau,h} = x_h(\alpha_h - \alpha_c)I \text{ and } \Delta\dot{S}_{\tau,c} = (1 - x_h)(\alpha_h - \alpha_c)I. \tag{A2}$$

Driven by the gradient of the partial Thomson temperature profile, all Thomson heat released at one side of a maximum (or minimum) of this profile will be exchanged to this side of the leg. With the temperature $T_{\tau,ex}$ of this position and its Seebeck coefficient $\alpha_{\tau,ex} = \alpha(T_{\tau,ex})$, the shares of the entropy exchange which are bound to each of the sides are

$$\Delta\dot{S}_{\tau,h} = (\alpha_h - \alpha_{\tau,ex})I \text{ and } \Delta\dot{S}_{\tau,c} = (\alpha_{\tau,ex} - \alpha_c)I. \tag{A3}$$

Multiplying both by the respective temperature of the side yields total Thomson heat:

$$\dot{Q}_\tau = T_h(\alpha_h - \alpha_{\tau,ex})I + T_c(\alpha_{\tau,ex} - \alpha_c)I = \{T_h\alpha_h - T_c\alpha_c - \alpha_{\tau,ex}(T_h - T_c)\}I = \Delta\dot{Q}_\pi - I\alpha_{\tau,ex}\Delta T. \tag{A4}$$

Comparing Equation (A4) with the energy balance of reversible heat $\dot{Q}_\tau = \Delta \dot{Q}_\pi - IV_0$, we can conclude that

$$\alpha_{\tau,ex}\Delta T = V_0 = \bar{\alpha}\Delta T, \text{ thus } \alpha_{\tau,ex} = \bar{\alpha} \tag{A5}$$

This gives us a rule for the temperature intervals over which the Thomson heat is flowing to either side of the leg. Consequently, Thomson heat has to be integrated from the crossing point of the curve of the Seebeck coefficient $\alpha(T)$ with its temperature average $\bar{\alpha}$. As a reversible approximation, this result is approximate and not strict as we have neglected here that dissipative processes are involved when Thomson heat is conducted to the leg sides. Below we will analyze these changes and find that these are small, and thus the rule stated here on the position of $\alpha_{\tau,ex}$, although not strict, is a good guide for estimates of the distribution of Thomson heat. Indeed, as observed by comparison to exact numerical calculations, this rule is almost perfectly fulfilled for all the example materials.

Within this reversible approximation, the Thomson heat flowing to the hot side is obtained as $\dot{Q}_{\tau,h} = T_h(\alpha_h - \bar{\alpha})I$. This would be equivalent to a complete compensation of the Peltier heat difference between reality and the CPM, i.e., the vanishing axial redistribution of reversible heat which is consistent with the simplifying assumption that the Thomson heat flowing to the outside is transmitted free of dissipation, i.e., equivalent to reversible heat. Here, the (additional) $T$ gradient related to the flow of Thomson heat is neglected, whereas an underlying $T$ profile related to an independent heat flow (here, the background of Fourier heat transfer) does, in effect, not contribute to its dissipation. We will see below that this happens as Thomson heat flowing to different sides will contribute almost compensating shares to the entropy balance. What is neglected here is that the Thomson heat itself when flowing to the ends of the leg will dissipate, according to the slight shift of the inner $T$ profile it is causing. Above, this $T$ offset was separated and called the partial $T$ profile due to Thomson heat, $\Delta T_{Thomson}(x)$. Additionally, this omission will contribute to a weak deviation from the position rule $\alpha_{\tau,ex} = \bar{\alpha}$.

The dissipative part of the entropy transport to the sides of the leg is related to the $T$ drop or step-up between the location where an increment of Thomson heat $d\dot{Q}_\tau$ is released and the side temperature, $T_h$ or $T_c$. The entropy increment is released over a segment of the leg with the $T$ increment $dT$ is $d\dot{S} = Id\alpha = \frac{d\dot{Q}_\tau}{T}$. With the transfer to the cold side, for example, the transmitted increment of Thomson heat $d\dot{Q}_\tau$ increases its entropy up to $d\dot{S}_c = \frac{d\dot{Q}_\tau}{T_c}$, and the according entropy gain is

$$d\Delta\dot{S}_c = d\dot{S}_c - d\dot{S} = \frac{d\dot{Q}_\tau}{T_c} - \frac{d\dot{Q}_\tau}{T} = \frac{d\dot{Q}_\tau}{TT_c}(T - T_c) = d\dot{S}\frac{T - T_c}{T_c} \tag{A6}$$

Summing over all Thomson heat flowing to that side, we have

$$\Delta\dot{S}_c = \frac{1}{T_c}\int_{I\alpha_c}^{I\bar{\alpha}}(T - T_c)d\dot{S} = \frac{I}{T_c}\int_{\alpha_c}^{\bar{\alpha}}(T - T_c)d\alpha = \frac{I}{T_c}\int_{T_c}^{T_{\bar{\alpha}}}T\frac{d\alpha}{dT}dT - I(\bar{\alpha} - \alpha_c) \tag{A7}$$

Multiplying with the cold side temperature, $\Delta\dot{Q}_{\tau,c} = T_c\Delta\dot{S}_c = \int_{T_c}^{T_{\bar{\alpha}}}T\frac{d\alpha}{dT}dT - I(\bar{\alpha} - \alpha_c)T_c$ gives us the amount of Thomson heat that is just the difference from the Peltier–Thomson heat balance of the CPM, $\Delta\dot{Q}_{\tau,c} = \dot{Q}_{\tau,c} - \left(\dot{Q}_{\pi,c}^{CPM} - \dot{Q}_{\pi,c}\right)$, i.e., the part that we have identified as uncompensated Peltier–Thomson heat in a real material. Note that $\dot{Q}_{\pi,c}^{CPM}$ contains merely completely reversible exchange of Peltier heat. Thus, the incomplete compensation of the Peltier–Thomson heat balance can be understood as an effect of the partly dissipative character of the exchange of the Thomson heat in a real system when conducted to the side. Accordingly, with the same consideration for the hot side, with $d\dot{S}_h = \frac{d\dot{Q}_\tau}{T_h}$, we obtain

$$\Delta \dot{S}_h = d\dot{S}_h - d\dot{S} = \frac{d\dot{Q}_\tau}{T_h} - \frac{d\dot{Q}_\tau}{T} = \frac{I}{T_h} \int_{\bar{\alpha}}^{\alpha_h} (T - T_h) d\alpha = \frac{I}{T_h} \int_{T_{\bar{\alpha}}}^{T_h} T \frac{d\alpha}{dT} dT - I(\alpha_h - \bar{\alpha}) \quad \text{(A8)}$$

i.e., $\Delta \dot{S}_h$ gives a negative contribution to the entropy balance. This sounds contradictory to the second law of thermodynamics but it is not, as the Thomson heat is not really flowing from a lower to a higher temperature but, when released, reduces the $T$ gradient of the underlying background of flowing Fourier heat, thus reducing the Fourier heat flow by the amount of "upstreaming" Thomson heat.

The hot and cold side entropy changes together give

$$\Delta \dot{S} = \Delta \dot{S}_h + \Delta \dot{S}_c = \frac{1}{T_h} \int_{\bar{\alpha}}^{\alpha_h} (T - T_h) d\alpha + \frac{1}{T_c} \int_{\alpha_c}^{\bar{\alpha}} (T - T_c) d\alpha = \frac{1}{T_c} \int_{T_c}^{T_{\bar{\alpha}}} T \frac{d\alpha}{dT} dT + \frac{1}{T_h} \int_{T_{\bar{\alpha}}}^{T_h} T \frac{d\alpha}{dT} dT - I(\alpha_h - \alpha_c). \quad \text{(A9)}$$

With $\frac{1}{T_c} \int_{\alpha_c}^{\bar{\alpha}} T d\alpha \gtrsim \bar{\alpha} - \alpha_c$ and $\frac{1}{T_h} \int_{\bar{\alpha}}^{\alpha_h} T d\alpha \lesssim \alpha_h - \bar{\alpha}$ we get $\frac{1}{T_c} \int_{T_c}^{T_{\bar{\alpha}}} T \frac{d\alpha}{dT} dT + \frac{1}{T_h} \int_{T_{\bar{\alpha}}}^{T_h} T \frac{d\alpha}{dT} dT \approx I(\alpha_h - \alpha_c)$ and thus $\Delta \dot{S} \approx 0$. Hence, assuming $\alpha_{\tau,ex} = \bar{\alpha}$, the entropy balance of the inner Thomson heat transfer as an offset of a much larger background Fourier heat flow is almost zero. This indeed confirms our approach to deduce a rule for the local distribution of Thomson heat based on a reversible approximation, i.e., assuming $\Delta \dot{S} \approx 0$, but also shows that the rule is not completely strict.

**References**

1. Zeier, W.G.; Schmitt, J.; Hautier, G.; Aydemir, U.; Gibbs, Z.M.; Felser, C.; Snyder, G.J.J.N.R.M. Engineering half-Heusler thermoelectric materials using Zintl chemistry. *Nat. Rev. Mater.* **2016**, *1*, 1–10. [CrossRef]
2. Goupil, C.; Seifert, W.; Zabrocki, K.; Müller, E.; Snyder, G.J.J.E. Thermodynamics of thermoelectric phenomena and applications. *Entropy* **2011**, *13*, 1481–1517. [CrossRef]
3. Snyder, G.J.; Toberer, E.S. Complex thermoelectric materials. In *Materials For Sustainable Energy: A Collection of Peer-Reviewed Research and Review Articles from Nature Publishing Group*; World Scientific: Singapore, 2011; pp. 101–110.
4. Rowe, D.M. *Thermoelectrics Handbook: Macro to Nano*; CRC Press: Boca Raton, FA, USA, 2005.
5. Antonova, E.E.; Looman, D.C. Finite elements for thermoelectric device analysis in ANSYS. In Proceedings of the ICT 2005. 24th International Conference on Thermoelectrics, Clemson, SC, USA, 19–23 June 2005; pp. 215–218.
6. Goupil, C. *Continuum Theory and Modeling of Thermoelectric Elements*; John Wiley & Sons: Hoboken, NJ, USA, 2015.
7. Kim, C.N. Development of a numerical method for the performance analysis of thermoelectric generators with thermal and electric contact resistance. *Appl. Therm. Eng.* **2018**, *130*, 408–417. [CrossRef]
8. Hogan, T.; Shih, T. Modeling and characterization of power generation modules based on bulk materials. *Thermoelectr. Handb. Macro Nano* **2006**. [CrossRef]
9. Oliveira, K.S.; Cardoso, R.P.; Hermes, C.J. Two-Dimensional Modeling of Thermoelectric Cells. In Proceedings of the International Refrigeration and Air Conditioning Conference, West Lafayette, IN, USA, 14–17 July 2014.
10. Kim, H.S.; Liu, W.; Ren, Z. The bridge between the materials and devices of thermoelectric power generators. *Energy Environ. Sci.* **2017**, *10*, 69–85. [CrossRef]
11. Kim, H.S.; Liu, W.; Ren, Z.J.J.o.A.P. Efficiency and output power of thermoelectric module by taking into account corrected Joule and Thomson heat. *J. Appl. Phys.* **2015**, *118*, 115103. [CrossRef]
12. Ryu, B.; Chung, J.; Park, S. Thermoelectric efficiency has three Degrees of Freedom. *arXiv* **2018**, arXiv:1810.11148.
13. Sunderland, J.E.; Burak, N.T. The influence of the Thomson effect on the performance of a thermoelectric power generator. *Solid-State Electron.* **1964**, *7*, 465–471. [CrossRef]

14. Min, G.; Rowe, D.M.; Kontostavlakis, K. Thermoelectric figure-of-merit under large temperature differences. *J. Phys. D: Appl. Phys.* **2004**, *37*, 1301. [CrossRef]
15. Chen, J.; Yan, Z.; Wu, L. The influence of Thomson effect on the maximum power output and maximum efficiency of a thermoelectric generator. *J. Appl. Phys.* **1996**, *79*, 8823–8828. [CrossRef]
16. Fraisse, G.; Ramousse, J.; Sgorlon, D.; Goupil, C. Comparison of different modeling approaches for thermoelectric elements. *Energy Convers. Manag.* **2013**, *65*, 351–356. [CrossRef]
17. Sandoz-Rosado, E.J.; Weinstein, S.J.; Stevens, R.J. On the Thomson effect in thermoelectric power devices. *Int. J. Therm. Sci.* **2013**, *66*, 1–7. [CrossRef]
18. Ponnusamy, P.; de Boor, J.; Müller, E. Using the constant properties model for accurate performance estimation of thermoelectric generator elements. *Appl. Energy* **2020**, *262*, 114587. [CrossRef]
19. Zhang, T. Effects of temperature-dependent material properties on temperature variation in a thermoelement. *J. Electron. Mater.* **2015**, *44*, 3612–3620. [CrossRef]
20. Wu, H.; Zhao, L.-D.; Zheng, F.; Wu, D.; Pei, Y.; Tong, X.; Kanatzidis, M.G.; He, J. Broad temperature plateau for thermoelectric figure of merit ZT> 2 in phase-separated PbTe 0.7 S 0.3. *Nat. Commun.* **2014**, *5*, 1–9. [CrossRef]
21. Sankhla, A.; Patil, A.; Kamila, H.; Yasseri, M.; Farahi, N.; Mueller, E.; de Boor, J. Mechanical Alloying of Optimized Mg2 (Si, Sn) Solid Solutions: Understanding Phase Evolution and Tuning Synthesis Parameters for Thermoelectric Applications. *ACS Appl. Energy Mater.* **2018**, *1*, 531–542. [CrossRef]
22. Kamila, H.; Sahu, P.; Sankhla, A.; Yasseri, M.; Pham, H.-N.; Dasgupta, T.; Mueller, E.; de Boor, J. Analyzing transport properties of p-type Mg$_2$Si–Mg$_2$Sn solid solutions: Optimization of thermoelectric performance and insight into the electronic band structure. *J. Mater. Chem. A* **2019**, *7*, 1045–1054. [CrossRef]
23. Kim, H.S.; Kikuchi, K.; Itoh, T.; Iida, T.; Taya, M. Design of segmented thermoelectric generator based on cost-effective and light-weight thermoelectric alloys. *Mater. Sci. Eng. B* **2014**, *185*, 45–52. [CrossRef]
24. Zhao, L.-D.; Lo, S.-H.; Zhang, Y.; Sun, H.; Tan, G.; Uher, C.; Wolverton, C.; Dravid, V.P.; Kanatzidis, M.G. Ultralow thermal conductivity and high thermoelectric figure of merit in SnSe crystals. *Nature* **2014**, *508*, 373. [CrossRef]
25. Seifert, W.; Ueltzen, M.; Müller, E. One-dimensional modelling of thermoelectric cooling. *Phys. Status Solidi A* **2002**, *194*, 277–290. [CrossRef]
26. Lamba, R.; Kaushik, S. Thermodynamic analysis of thermoelectric generator including influence of Thomson effect and leg geometry configuration. *J. Energy Convers. Manag.* **2017**, *144*, 388–398. [CrossRef]
27. Garrido, J.; Casanovas, A.; Manzanares, J.A. Thomson Power in the Model of Constant Transport Coefficients for Thermoelectric Elements. *J. Electron. Mater.* **2019**, *48*, 5821–5826. [CrossRef]

© 2020 by the authors. Licensee MDPI, Basel, Switzerland. This article is an open access article distributed under the terms and conditions of the Creative Commons Attribution (CC BY) license (http://creativecommons.org/licenses/by/4.0/).

*Article*

# Geometry Optimization of Thermoelectric Modules: Deviation of Optimum Power Output and Conversion Efficiency

**Mario Wolf *, Alexey Rybakov, Richard Hinterding and Armin Feldhoff ***

Institute of Physical Chemistry and Electrochemistry, Leibniz University Hannover, Callinstraße 3A, D-30167 Hannover, Germany; alexey.rybakov@pci.uni-hannover.de (A.R.); richard.hinterding@pci.uni-hannover.de (R.H.)
* Correspondence: mario.wolf@pci.uni-hannover.de (M.W.); armin.feldhoff@pci.uni-hannover.de (A.F.)

Received: 28 September 2020; Accepted: 25 October 2020; Published: 29 October 2020

**Abstract:** Besides the material research in the field of thermoelectrics, the way from a material to a functional thermoelectric (TE) module comes alongside additional challenges. Thus, comprehension and optimization of the properties and the design of a TE module are important tasks. In this work, different geometry optimization strategies to reach maximum power output or maximum conversion efficiency are applied and the resulting performances of various modules and respective materials are analyzed. A $Bi_2Te_3$-based module, a half-Heusler-based module, and an oxide-based module are characterized via FEM simulations. By this, a deviation of optimum power output and optimum conversion efficiency in dependence of the diversity of thermoelectric materials is found. Additionally, for all modules, the respective fluxes of entropy and charge as well as the corresponding fluxes of thermal and electrical energy within the thermolegs are shown. The full understanding and enhancement of the performance of a TE module may be further improved.

**Keywords:** thermoelectric materials; energy harvesting; thermoelectric generator; working points; maximum electrical power point

## 1. Introduction

The direct energy conversion from wasted thermal energy into usable electrical energy via thermoelectric (TE) modules has been extensively studied and improved in recent years. Such devices benefit from long-term stability without the need of maintenance and they are quietly operating without moving parts that may get damaged over time [1]. The main parts of research on thermoelectric energy conversion are investigating and improving thermoelectric materials in order to reach high power output and high conversion efficiency on the one hand [2,3] and the scalable and effective manufacturing of devices on the other hand [4,5]. However, up to now, TE modules have not achieved characteristics that justify the investment for a wide commercial usage. Especially, the design of the device, the optimization of the cross-sectional area ratio, and thermal and electrical contact resistivity are crucial factors on the way from a promising material to a functional device with high power output and conversion efficiency [6], even if suitable thermoelectric materials are provided. The aim of the work is to improve the understanding and the optimization of the working principle of TE modules based on finite element method (FEM) simulations of several material combinations with the software ANSYS for various geometry optimization strategies.

*Entropy* **2020**, *22*, 1233; doi:10.3390/e22111233 www.mdpi.com/journal/entropy

## 1.1. Thermoelectric Materials

The thermoelectric energy conversion can be described by the coupling of the flux density of electric charge $j_q$ and the flux density of entropy $j_s$. These fluxes are transmitted by the thermoelectric material tensor, which represents the characteristics of the included thermoelectric materials with a cross-sectional area $A$ and length $l$, when simultaneously placed in a gradient of electrical potential $\nabla \varphi$ and a gradient of temperature $\nabla T$, as shown in Equation (1) [7,8].

$$\begin{pmatrix} j_q \\ j_s \end{pmatrix} = \frac{A}{l} \cdot \begin{pmatrix} \sigma & \sigma \cdot \alpha \\ \sigma \cdot \alpha & \sigma \cdot \alpha^2 + \Lambda_{OC} \end{pmatrix} \cdot \begin{pmatrix} -\nabla \varphi \\ -\nabla T \end{pmatrix} \quad (1)$$

The energy conversion is therefore mainly based on three material parameters: the isothermal electrical conductivity $\sigma$, the Seebeck coefficient $\alpha$ and the entropy conductivity at electrical open-circuit $\Lambda_{OC}$. In principle, all three quantities are tensors themselves, but, for homogeneous materials, they are often treated as scalars [8,9]. The figure of merit $f = zT$ [10,11] shown in Equation (2), which displays the conversion efficiency of a thermoelectric material, is a function of the three material parameters.

$$f = \frac{\sigma \cdot \alpha^2}{\Lambda_{OC}} = \frac{\sigma \cdot \alpha^2}{\lambda_{OC}} \cdot T = zT \quad (2)$$

Consequently, thermoelectric materials are usually desired to have a high power factor $\sigma \alpha^2$ and a simultaneously low open-circuited entropy conductivity $\Lambda_{OC}$. Note that, due to the use of entropy conductivity $\Lambda_{OC}$ instead of the heat conductivity $\lambda_{OC}$, the absolute temperature $T$ does not occur explicitly within the short form of Equation (2), but implicitly within the three material parameter $\sigma(T)$, $\alpha(T)$ and $\Lambda_{OC}(T)$ [11].

Within the thermoelectric materials, the respective flux density of thermal energy $j_{E,th}(x)$ and flux density of electrical energy $j_{E,el}(x)$ at a certain point $x$ across the length of the materials are given as the product of the respective flux density of entropy $j_S(x)$ and flux density of electrical charge $j_q(x)$ and the temperature $T(x)$ and voltage $U(x) = \Delta\varphi(x)$ at this point (Equations (3) and (4)) [8]. Note that this description is analyzed as a function of $x$, along a central line through the respective thermoleg (compare Figure A1), so these values as a function of $x$ are used as scalars.

$$j_{E,th}(x) = j_S(x) \cdot T(x) \quad (3)$$

$$j_{E,el}(x) = j_q(x) \cdot U(x) \quad (4)$$

These descriptions of electrical and thermal phenomena are used as a basis to analyze and improve the understanding of thermoelectric modules within this work. Here, the explicit description of the flux densities of charge and entropy and the resulting flux densities of thermal and electrical energy can be useful in order to further understand and improve the thermoelectric energy conversion.

As thermoelectric materials, various classes of materials have been studied intensively including bismuth telluride [12,13], which is commonly used for thermoelectric modules, other tellurides [14], and selenides [15,16], intermetallic phases, such as as Zintl phases [17–19] and half-Heusler phases [20,21], oxides and oxyselenides [22,23], and conductive polymers [24,25]. Each material class provides a different thermoelectric characteristic, requires special treatments or fabrication and it is suitable in a certain application temperature range [2]. In order to influence and improve the thermoelectric properties, band structure modelling via doping and nanostructuring [26–28], segmentation of thermoelegs [29–31] and the utilization of hybrid materials [32–34] are widely investigated.

The resulting thermoelectric performance of a material is usually described by the $U$-$I_q$-characteristic (voltage-electrical current curve) and the resulting electrical power curve $P_{el}$-$I_q$. Here, two important material working points can be identified: the maximum electrical power point (MEPP) of a respective material (the point where $P_{el} = U \cdot I_q$ reaches its maximum), which is given at half the open-circuited voltage $U_{OC}$, and half the short-circuited current $I_{q,SC}$ and the maximum conversion efficiency point (MCEP) of a respective material, which is a function of the figure of merit $zT$ of the material. The MCEP and MEPP drift apart with increasing figure of merit $zT$ of the respective material, As shown in a previous work [11] (Figure 1). Therefore, optimizing different parameters to influence the materials MEPP and MCEP are important to effectively improve the performance of a resulting TE module. Furthermore, this implies that not only the resulting conversion efficiency based on the figure of merit $zT$, but also the resulting electrical power output, which is a function of the power factor $\sigma \alpha^2$, is a key parameter. In fact, the power factor should have at least the same significance as the figure of merit $zT$, as has been reported before [2,35].

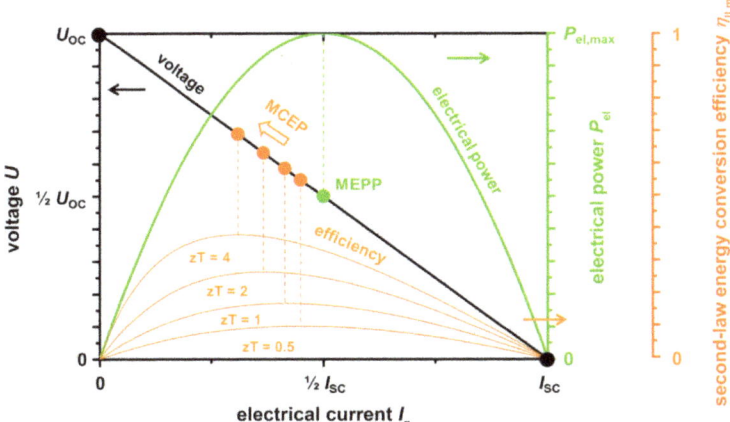

**Figure 1.** Normalized $U$-$I_q$ and $P_{el}$-$I_q$ characteristics of some hypothetic thermoelectric materials with a $zT$ of 0.5, 1, 2, and 4. The second-law energy conversion efficiency $\eta_{II,mat}$ increases with increasing figure of merit $zT$. The maximum conversion efficiency point (MCEP) is a function of the figure of merit $zT$ and, therefore, drifts apart from the maximum electrical power point (MEPP). Working points of short-circuit (SC) with the short-circuit current $I_{q,SC}$ and open-circuit (OC) with the open-circuit voltage $U_{OC}$ are marked. Reworked from [11].

## 1.2. From Material to Device

In this work, the concept of the material working points MEPP and MCEP and the resulting significance of figure of merit $zT$ and power factor $\sigma \alpha^2$ are transferred to a TE module. As described before, the concept and design of a TE module also strongly influence the resulting performance. This is based on several factors:

- The respective thermoelectric materials properties.
- The design of the respective device, the flexibility and the free volume.
- The aimed application temperature range, limiting the options for thermoelectric materials.
- Optimization factors, such as thermal- and electrical-contact resistivity, as well as the cross-sectional area ratio between n- and p-type materials

Especially, the respective geometry of the $p$- and $n$-type materials strongly influence the certain MEPP and MCEPs of the materials and therefore the resulting performance of the TE module [6]. Often, the geometry is optimized to a maximum figure of merit $zT$ and the resulting $A_n/A_p$ ratio is used for simulations for example by Ouyang and Li [30]. For certain materials, this optimization, in fact, leads to overlapping MCEP and MEPPs of the respective materials in a resulting module due to matching values of the thermal conductivity $\lambda_n = \lambda_p$ [36], which, however, is not always the case. Recently, Xing et al. [36] also described that an optimization of TE modules for a high power output and an according materials choice can strongly enhance the resulting properties when compared to an optimization for maximum energy conversion efficiency. This corresponds to the assertion of the significance of the power factor. Therefore, in this work, an analysis of different material combinations in a TE module is provided, based on the analogous description of $j_{E,th}(x)$ and $j_{E,el}(x)$ shown above for three different optimization strategies: for maximum $zT$, for matching $I_{q,SC}$ (and, therefore, overlapping material working points), and for maximum electrical power output. For this purpose, FEM simulations of various modules are provided both based on materials with similar ($Bi_2Te_3$-based TE module and half-Heusler-based TE module), as well as with very different thermoelectric properties (oxide-based TE module) of the $n$- and $p$-type materials.

## 2. Methods and Simulation

### 2.1. Materials and Modules for FEM Simulations

Table 1 shows the used materials. For all thermoelectric materials, literature data have been used. The exact input values are shown in Tables A1–A3 in Appendix A. As a connector, a metal conductor made of copper with 0.5 mm height, an electrical conductivity of $4.85 \times 10^8$ S m$^{-1}$ and a thermal conductivity of 400 W m$^{-1}$ K$^{-1}$ was used. Figure 2 shows the resulting TE modules used for FEM simulations.

**Table 1.** Material combinations for the simulated modules with according literature for the thermoelectric properties. The exact input values are shown in Tables A1–A3 in Appendix A. For all modules, a stable temperature difference of 50 K has been assumed. For the calculation of the $A_n/A_p$ ratios, a linear behavior has been assumed and the calculation was done with the medium values of the respective temperature range.

| Module | $p$-Type | $n$-Type | $T_{hot}/K$ | $T_{cold}/K$ |
|---|---|---|---|---|
| Module 1 | $Bi_{0.5}Sb_{1.5}Te_3$ [37] | $Bi_2Te_{3-x}Sb_x$ [38] | 348 | 298 |
| Module 2 | $FeNb_{0.88}Hf_{0.12}Sb$ [39] | $Hf_{0.6}Zr_{0.4}NiSn_{0.995}Sb_{0.005}$ [40] | 1000 | 950 |
| Module 3 | $Ca_3Co_4O_9$ [41] | $In_{1.95}Sn_{0.05}O_3$ [41] | 1075 | 1025 |

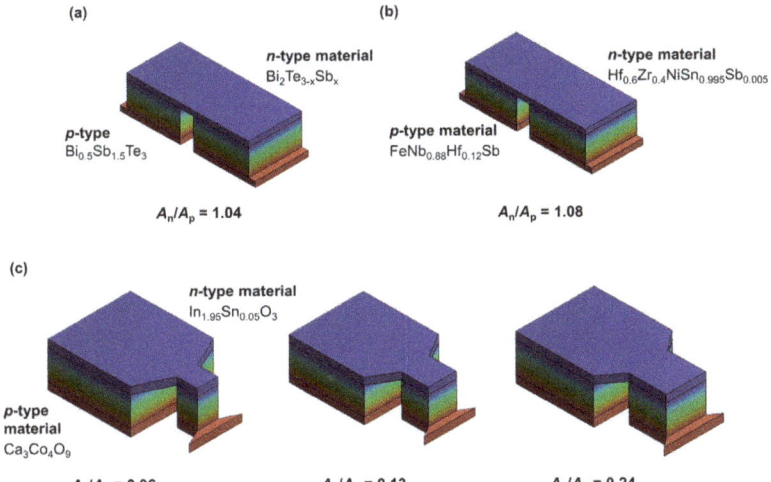

**Figure 2.** Resulting modules characterized via finite elemente simulations (FEM)-simulations. (a) Bi$_2$Te$_3$-based TE module 1, (b) half-Heusler-based TE module 2, and (c) oxide-based TE module 3 with three different $A_n/A_p$ ratios. The colors refer to the respective temperatures (red: hot side, blue: cold side). Note that the effective area $A_n + A_p$ is constant for all modules and $A_n/A_p$ ratios. As connector, the characteristics of copper has been used in the simulation.

## 2.2. Optimization of Geometry

The $A_n/A_p$ ratios for the simulated modules have been calculated for three different optimizations: First, according to a $zT$ optimization for maximum energy conversion efficiency that has been derived and used before (Equation (5)) [30]. Here, $\rho_n$ and $\rho_p$ are the specific electrical resistivity and $\lambda_n$ and $\lambda_p$ the heat conductivity of the $n$- and $p$-type materials, respectively:

$$[\frac{A_n}{A_p}]_{zT} = \sqrt{\frac{\rho_n}{\rho_p} \cdot \frac{\lambda_p}{\lambda_n}} \tag{5}$$

Second, the $[\frac{A_n}{A_p}]_{\text{matching } I_{q,\text{sc}}}$ ratio for overlapping material working points was calculated according to Equation (6) (compare Equations (A1)–(A7) in Appendix B). Here, $\alpha_n$ and $\alpha_p$ are the Seebeck coefficient of the $n$-type and $p$-type materials, respectively:

$$[\frac{A_n}{A_p}]_{\text{matching } I_{q,\text{sc}}} = \frac{\alpha_p}{|\alpha_n|} \cdot \frac{\rho_n}{\rho_p} \tag{6}$$

Third, an optimization for maximum power output was conducted according to Xing et al. [36] via Equation (7) (compare Equations (A8)–(A14) in Appendix C):

$$[\frac{A_n}{A_p}]_{\text{power}} = \sqrt{\frac{\rho_n}{\rho_p}} \tag{7}$$

Additionally, the areas of the $n$- and $p$-type materials have been chosen for the same effective area $A_n + A_p$ for all modules. The maximum first-law energy conversion efficiency $\eta_{\text{I,TEG,max}}$ for all optimized geometries have been calculated from the thermoelectric properties of the materials [9,11,30] (compare

Equations (A15)–(A19) in Appendix D). The length of all thermolegs was chosen to be $l$ = 2 mm, as otherwise there would have been too many varying parameters and a fixed and matching length for $n$- and $p$-type is reasonable for a functional TE module.

### 2.3. Simulation Parameters

The software ANSYS Mechanical (Version 2020 R1), which is based on the finite element method, is used in order to simulate the TE modules. Here, a steady-state thermal-electrical conduction analysis that allows for a simultaneous solution of thermal and electrical fields was chosen. After setting the material parameters for the $n$- and $p$-type thermolegs, the following boundary conditions for the simulation were set: the temperature of the cold junction, the ambient temperature that is equal to the temperature of the cold junction, the side at zero potential, and the side that determines the value of the electric current; all of the remaining faces were set for free convection in air with the heat transfer coefficient with a typical value of 20 W m$^{-2}$ K$^{-1}$ [42].

The simulation process was divided into two stages. First, a $U$-$I_q$ curve was taken in order to evaluate the general characteristics of the TE module. By changing the value of the electrical current that can flow through the TE module, the effect of the external load on the voltage is simulated. Using the $U$-$I_q$ curve, the electrical power $P_{el}$ was calculated and a $P_{el}$-$I_q$ curve was constructed to determine the MEPP. Then, to study the specific characteristics of the TE module at the MEPP, the following four distributions were simulated: temperature, flux density of thermal energy, electrical voltage, and flux density of charge. From each distribution, the values alongside the center of the thermoleg have been calculated. For these positions inside the leg, the local entropy flux density was calculated from the local temperature and local flux density of thermal energy according to Equation (3). The values of the electrical voltage and the local flux density of electrical charge were used in order to calculate the flux density of the electrical energy according to Equation (4). As a result, a description of all parameters as a function of the position $x$, along a central line through the respective thermoleg, is received. The corresponding images of the distribution of the temperature $T(x)$, the voltage $U(x)$, flux density of thermal energy $j_{E,th}(x)$, and flux density of electrical charge $j_q(x)$ within the thermolegs are shown in Figures A2–A6 in the Appendix E.

### 2.4. Notes on Limitations

For all material parameters, a linear behavior within the applied temperature range has been assumed and the average value has been used for the calculation of the $A_n/A_p$ ratio. Over a relatively small temperature difference of 50 K, the assumption of linear behavior of the thermoelectric parameters can be made, but, for exact simulations, the respective behavior has to be analyzed in detail for each specific case. Because the maximum temperature difference in the simulation was only 50 K and the maximum application temperature was about 1000 K, the dominant mechanism of heat transfer is convection, so the influence of thermal radiation was not considered. Note that, for temperatures above 1000 K and if ceramic substrates are used on top and at the bottom, the thermal radiation becomes increasingly important and has to be considered if an application at higher temperatures is aimed. For all of the simulated modules, an active cooling with a stable temperature difference of 50 K was assumed. Although a matching length $l$ for both thermolegs is reasonable, this may also be optimized, since the length strongly influences the $U$–$I_q$-curve as well as the temperature difference, if no active cooling with a stable temperature difference is applied. Additionally, as mentioned before, the electric and thermal contact resistivity between each individual thermoleg and the connector is an important parameter, which has to be investigated and optimized for each individual case. To allow for comparison, ideal contacts are assumed in this work. The results in this work are specifically shown for a thermoelectric module in generator mode, which, however, may also apply for the entropy pump mode in thermoelectric coolers. For the thermoelectric materials, the respective material working points are also correlated to the material properties [11], but, for the thermoelectric modules in entropy pump mode, this is yet to be proven.

## 3. Results and Discussion

As material combinations, a $Bi_2Te_3$-based TE module (module 1), a half-Heusler-based TE module (module 2) and an oxide-based TE module (module 3) were chosen. The respective optimized geometries $A_n/A_p$ for maximum $zT$, matching $I_{q,SC}$ and for maximum electrical power are shown in Table 2. For module 1 and 2, all of the optimizations led to very similar $A_n/A_p$ ratios. Therefore, only the $zT$-optimized modules have been simulated. For module 3, the resulting $A_n/A_p$ ratios vary widely, so simulations of this module were done for all the calculated optimized geometries.

Table 2. Resulting optimized geometries according to the $zT$ optimization, matching $I_{q,SC}$ and power optimization. For the values in brackets, no simulations were carried out, due to insignificant deviation from the $zT$ optimization.

| Module | $[\frac{A_n}{A_p}]_{zT}$ | $[\frac{A_n}{A_p}]_{matching\ I_{q,SC}}$ | $[\frac{A_n}{A_p}]_{power}$ |
|---|---|---|---|
| Module 1 | 1.0345 | (1.0745) | (1.0459) |
| Module 2 | 1.0831 | (1.0969) | (1.0308) |
| Module 3 | 0.0596 | 0.1306 | 0.2433 |

### 3.1. Similar Material Properties

For the materials that were chosen for module 1 and 2, the optimizations of the $A_n/A_p$ ratios for maximum $zT$, matching material working points and for maximum power output all result in ratios near 1, with only a slight variation. This is a result of the fairly similar thermoelectric properties of the respective n- and p-types. Therefore, a fixed $A_n/A_p$ ratio of 1.04 and 1.08 are used for the simulations of module 1 and module 2, respectively. Note that, although the calculated optimum $A_n/A_p$ ratios for module 1 and 2 all are close together, they are not the same, meaning that an optimization for maximum power output may still result in a slightly higher power output of the respective module compared to a $zT$ optimization. However, the effect is much stronger for the oxide-based module 3, which is the reason why this module is analyzed in depth for all three optimized geometries.

#### 3.1.1. $Bi_2Te_3$-Based TE Module

For module 1, $Bi_2Te_{3-x}Sb_x$ [38] and $Bi_{0.5}Sb_{1.5}Te_3$ [37] were chosen as n- and p-type materials, respectively. As $A_n/A_p$ ratio, the $zT$-optimized ratio of 1.04 was used in the simulation. Figure 3 shows the simulated $U$-$I_q$ characteristics and the electrical power output of the $Bi_2Te_3$-based TE module and the respective thermoelectric parameters across the length of the respective legs. The working points of the p- and n-type material with a $zT$-optimized $A_n/A_p$ ratio show a good overlap. This results in a high electrical power output of the TE module with a maximum power density $w_{el,max,TEG}$ of approximately 124.5 mW cm$^{-2}$ at the applied temperature difference of 50 K. The individual fluxes that are within in the p-type and n-type thermolegs are shown in Figure 3c–h. The temperature is set to be 348 K at the hot side and 298 K at the cold side. The entropy flux density $j_S(x)$ and therefore also the thermal energy flux density $j_{E,th}(x)$ are very similar in the respective legs, due to the similar thermal conductivity of the chosen materials. At the applied temperature difference of 50 K, a voltage $U(x)$ of 11 mV is achieved in one thermocouple. Analogous to the entropy flux density, the electrical flux density $j_q(x)$ is also similar in the p-type and n-type thermolegs. In one thermocouple, this results in an electrical energy flux density $j_{E,el}(x)$ of 2.4 × 10$^{-3}$ W m$^{-2}$. Note that the dashed lines presented in Figure 3c–h represent the metallic connector between the p-type and n-type materials, so both materials are not in direct contact. Thus the different fluxes do not necessarily have the same value at the dashed line.

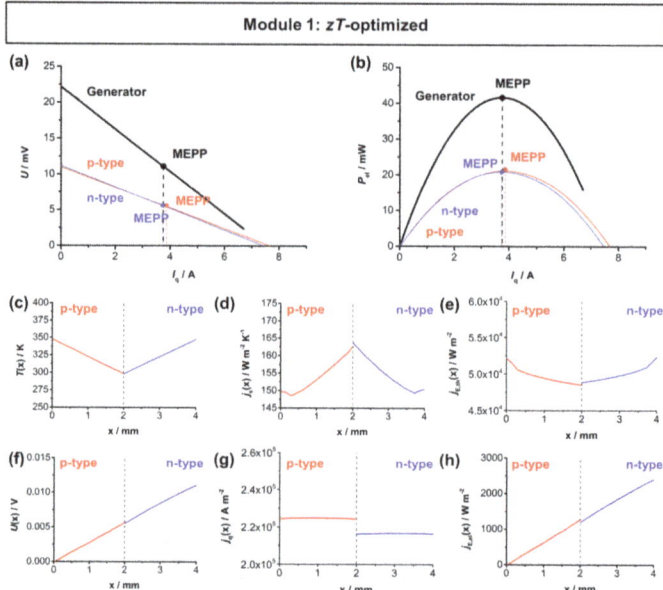

**Figure 3.** FEM simulations of module 1 (p-type $Bi_{0.5}Sb_{1.5}Te_3$ and n-type $Bi_2Te_{3-x}Sb_x$) with a hot side temperature of 348 K and cold side temperature of 298 K. (**a**) $U$-$I_q$ characteristics and (**b**) electrical power output $P_{el}$-$I_q$ of the module. The respective MEPPs of the materials overlap and result in a high power output of the TE module. Thermoelectric characteristics of the respective materials as a function of the length of the respective legs: (**c**) temperature $T(x)$, (**d**) entropy flux density $j_S(x)$, (**e**) thermal energy flux density $j_{E,th}(x)$, (**f**) voltage $U(x)$, (**g**) electrical flux density $j_q(x)$, and (**h**) electrical energy flux density $j_{E,el}(x)$ trend throughout one thermocouple. Note that the dashed line in (**c**–**h**) represents the metallic connector between the p-type and n-type materials. The simulated distributions are shown in Figure A2 in Appendix E.

3.1.2. Half-Heusler-Based TE Module

Figure 4 shows the simulated $U$-$I_q$ characteristics and the electrical power output of the half-Heusler-based TE module and the respective thermoelectric parameters across the length of the respective legs. For n- and p-type materials, $Hf_{0.6}Zr_{0.4}NiSn_{0.995}Sb_{0.005}$ [40] and $FeNb_{0.88}Hf_{0.12}Sb$ [39] were chosen.

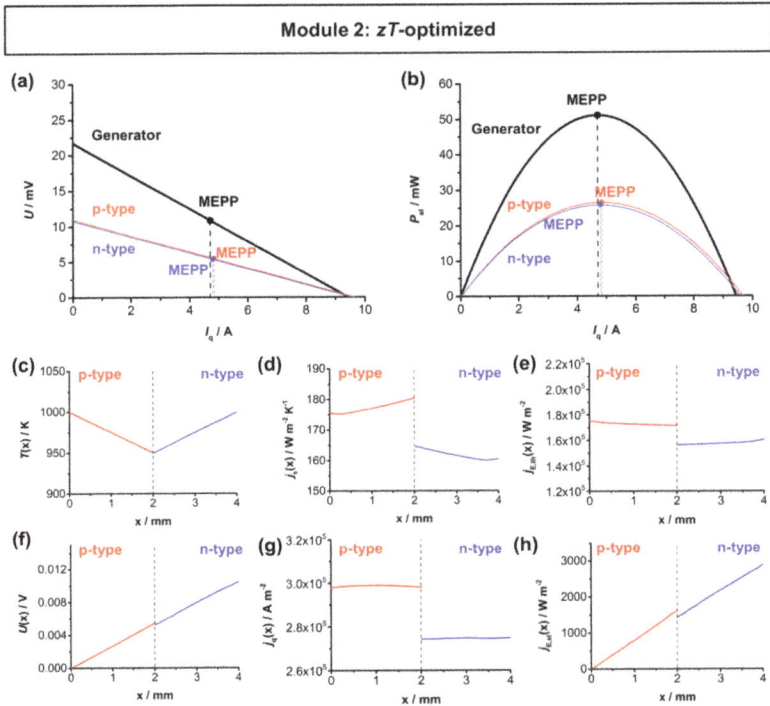

**Figure 4.** FEM simulations of module 2 (p-type FeNb$_{0.88}$Hf$_{0.12}$Sb and n-type Hf$_{0.6}$Zr$_{0.4}$NiSn$_{0.995}$Sb$_{0.005}$) with a hot side temperature of 1000 K and cold side temperature of 950 K. (**a**) $U$-$I_q$ characteristics and (**b**) electrical power output $P_{el}$-$I_q$ of the module. The respective MEPPs of the materials overlap and result in a high power output of the module. Thermoelectric characteristics of the respective materials as a function of the length of the respective legs: (**c**) temperature $T(x)$, (**d**) entropy flux density $j_S(x)$, (**e**) thermal energy flux density $j_{E,th}(x)$, (**f**) voltage $U(x)$, (**g**) electrical flux density $j_q(x)$, and (**h**) electrical energy flux density $j_{E,el}(x)$ trend throughout one thermocouple. Note that the dashed line in (**c**–**h**) represent the metallic connector between the p-type and n-type materials. The simulated distributions are shown in Figure A3 in Appendix E.

Analogous to the Bi$_2$Te$_3$-based module, the materials exhibit similar thermoelectric properties and the resulting $A_n/A_p$ ratio is still near 1. For the simulations, the $zT$-optimized $A_n/A_p$ ratio of 1.08 was used. The material working points also show a good overlap as a result of the $zT$ optimization. Therefore, the module's MEPP and MCEP are also close together. The TE module reaches a high electrical power output of approximately 51.1 mW. With an effective area of 0.334 cm$^2$, this corresponds to a similarly high maximum power density $w_{el,max,TEG}$ of 153.14 mW cm$^{-2}$, which is slightly higher compared to the Bi$_2$Te$_3$-based module 1. The individual fluxes within in the p-type and n-type thermolegs are shown in Figure 4c–h. The temperature is set to be 1000 K at the hot side and 950 K at the cold side. At the applied 50 K temperature difference, a voltage $U(x)$ of 10.54 mV can be reached, which is slightly lower compared to the Bi$_2$Te$_3$-based module, as a result of the slightly lower Seebeck coefficient of the n-type material. The entropy flux density $j_S(x)$ of the p-type is slightly higher when compared to the n-type thermoleg, due to the higher thermal conductivity of the p-type material. Analogously, the electrical flux density $j_q(x)$ is also slightly higher in the $p$-type material, due to the higher electrical conductivity of the $p$-type material. In one thermocouple, a thermal energy flux density $j_{E,th}(x)$ of $16 \times 10^3$ W m$^{-2}$ and an electrical energy flux density $j_{E,el}(x)$ of $2.9 \times 10^3$ W m$^{-2}$ are reached, both being higher when compared to the Bi$_2$Te$_3$-based module, due to the higher values of electrical and thermal conductivity of the respective materials.

Table 3 summarizes the simulated characteristics of the Bi$_2$Te$_3$-based and half-Heusler-based TE modules. The respective material working points are close together, which results in a high electrical power output and conversion efficiency of both modules. However, the Bi$_2$Te$_3$-based module 1 reaches a higher conversion efficiency of 2.5%, while the half-Heusler based module 2 reaches a higher power output of up to 153 mW cm$^{-2}$. This is the expected behavior, due to the higher power factor, but simultaneously higher thermal conductivity of the half-Heusler materials. This also displays the aforementioned importance of the power factor (for power output), which, for certain applications, may be equally important as the figure of merit $zT$ (for efficiency).

**Table 3.** Resulting maximum electrical power output $P_{el,max,TEG}$, electrical power density $\omega_{el,max,TEG}$ and maximum first-law energy conversion efficiency $\eta_{I,TEG,max}$ of module 1 (Bi$_2$Te$_3$) and module 2 (half-Heusler materials) for $zT$-optimized geometry.

| Module | Module MEPP/A | $P_{el,max,TEG}$/mW | $\omega_{el,max,TEG}$/mW cm$^{-2}$ | $\eta_{I,TEG,max}$ |
|---|---|---|---|---|
| Module 1 | 3.75 | 41.60 | 124.50 | 2.50 |
| Module 2 | 4.72 | 51.10 | 153.14 | 0.97 |

*3.2. Dissimilar Material Properties*

For $n$- and $p$-type materials of module 3, In$_{1.995}$Sn$_{0.05}$O$_3$ and Ca$_3$Co$_4$O$_9$ [41] were chosen. For these materials, the optimizations of the $A_n/A_p$ ratios for maximum $zT$, matching $I_{q,SC}$ and for maximum power output result in dissimilar ratios of 0.06, 0.13, and 0.24, respectively. Therefore, modules with all calculated $A_n/A_p$ ratios were simulated.

Oxide-Based TE Module

Figure 5 shows the simulated $U$-$I_q$ characteristics and the electrical power output of the Ca$_3$Co$_4$O$_9$-In$_{1.95}$Sn$_{0.05}$O$_2$ TE module and the thermoelectric parameters across the length of the respective legs. Here, the $zT$ optimization of the $A_n/A_p$ ratio does not result in an overlap of the respective material working points. The short-circuited electrical current $I_{q,SC}$ of the $p$-type Ca$_3$Co$_4$O$_9$ is approximately twice the short-circuited current $I_{q,SC}$ of the $n$-type In$_{1.95}$Sn$_{0.05}$O$_2$. Therefore, the resulting MEPP of the TE module is located between the respective material working points, and the power output of the module is only slightly higher when compared to the power output of the $p$-type Ca$_3$Co$_4$O$_9$ leg. With an effective area of 0.3332 cm$^2$ the simulated TE module reaches a maximum electrical power density $\omega_{el,max,TEG}$ of approximately 4.5 mW cm$^{-2}$. The individual fluxes within in the $p$-type and $n$-type thermolegs are shown in Figure 4c–h. The temperature difference was again set to 50 K, with a hot side temperature of 1075 K and a cold side temperature of 1025 K. The strong difference of the Seebeck coefficient of $n$- and $p$-type materials is displayed in the distribution of the voltage $U(x)$. In the $p$-type material, a voltage of 6.6 mV is reached, while, in the $n$-type material, the voltage only increases by 1 mV to 7.6 mV. The strong difference of thermoelectric properties of $p$- and $n$-type materials is also displayed in the flux density of charge and flux density of entropy, both being higher in the $n$-type In$_{1.995}$Sn$_{0.05}$O$_3$ due to the higher electrical and thermal conductivity. Therefore, the same behavior is noticeable in the flux densities of thermal energy and electrical energy.

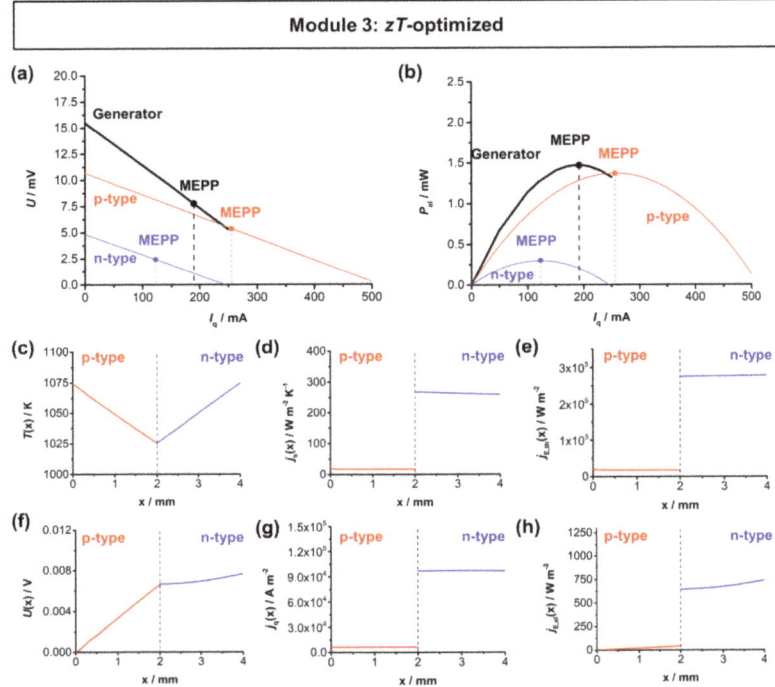

**Figure 5.** FEM simulations of the $zT$-optimized module 3 ($p$-type $Ca_3Co_4O_9$ and $n$-type $In_{1.95}Sn_{0.05}O_3$) with a hot side temperature of 1050 K and cold side temperature of 1000 K. (**a**) $U$-$I_q$ characteristics and (**b**) electrical power output $P_{el}$-$I_q$ of the module. Thermoelectric characteristics of the respective materials as a function of the length of the respective legs: (**c**) temperature $T(x)$, (**d**) entropy flux density $j_S(x)$, (**e**) thermal energy flux density $j_{E,th}(x)$, (**f**) voltage $U(x)$, (**g**) electrical flux density $j_q(x)$, and (**h**) electrical energy flux density $j_{E,el}(x)$ trend throughout one thermocouple. Note, that the dashed line in (**c**–**h**) represent the metallic connector between the $p$-type and $n$-type materials. The simulated distributions are shown in Figure A4 in Appendix E.

Figure 6 shows the simulated $U$-$I_q$ characteristics and the electrical power output of the $Ca_3Co_4O_9$–$In_{1.95}Sn_{0.05}O_2$ TE module and the respective thermoelectric parameters across the length of the respective legs for an optimized $A_n/A_p$ ratio for matching $I_{q,SC}$. As a result of this optimization, the module MEPP is also similar to the both materials' working points and the power output of the module is already significantly higher than of the respective materials. With an effective area of 0.333 cm$^2$ a maximum electrical power density $w_{el,max,TEG}$ of approximately 5.64 mW cm$^{-2}$ can be reached. The individual fluxes within in the $p$-type and $n$-type thermolegs for the module optimized for matching $I_{q,SC}$ are shown in Figure 6c–h. When compared to the $zT$ optimization, the larger area of the $n$-type $In_{1.995}Sn_{0.05}O_3$ results in a bigger impact of the material, displayed in a higher value of voltage reached in the $n$-type material. Additionally, both the entropy flux density (slightly) as well as the electrical flux density (significantly) of the $n$-type material are lower, due to the larger area, which results in the same trend for the flux densities of thermal energy and electrical energy.

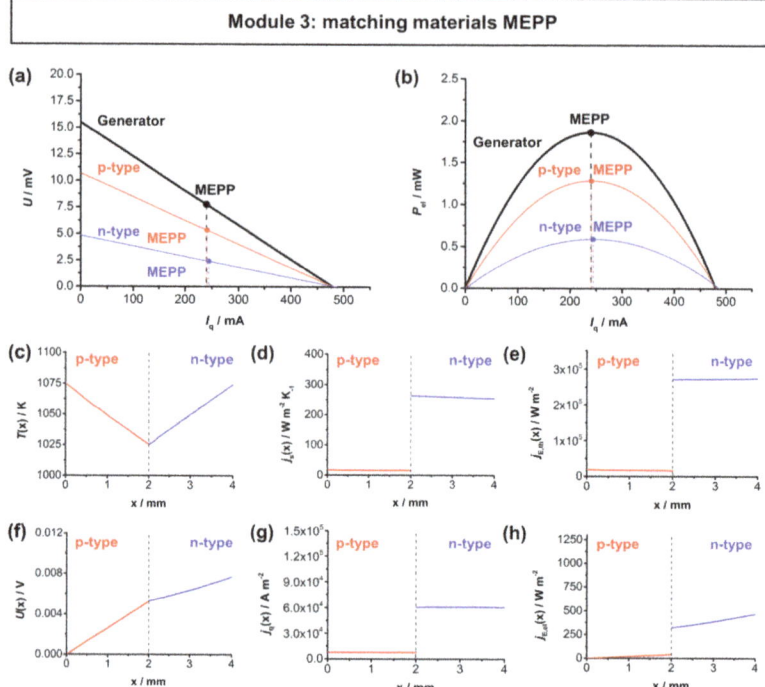

**Figure 6.** FEM simulations of the module 3 (p-type $Ca_3Co_4O_9$ and n-type $In_{1.95}Sn_{0.05}O_3$) with matching $I_{q,SC}$ with a hot side temperature of 1050 K and cold side temperature of 1000 K. (**a**) $U$–$I_q$ characteristics and (**b**) electrical power output $P_{el}$–$I_q$ of the module. Thermoelectric characteristics of the respective materials as a function of the length of the respective legs: (**c**) temperature $T(x)$, (**d**) entropy flux density $j_S(x)$, (**e**) thermal energy flux density $j_{E,th}(x)$, (**f**) voltage $U(x)$, (**g**) electrical flux density $j_q(x)$, and (**h**) electrical energy flux density $j_{E,el}(x)$ trend throughout one thermocouple. Note, that the dashed line in (**c–h**) represent the metallic connector between the p-type and n-type materials. The simulated distributions are shown in Figure A5 in Appendix E.

Finally, in Figure 7, the power optimization of the $A_n/A_p$ ratio according to Equation (7) is shown. Again, the material working points do not overlap, but as a result of the increasing cross-sectional area of the n-type $In_{1.95}Sn_{0.05}O_2$, the electrical power output of the the n-type material is significantly higher compared to the other two optimization strategies. In fact, both materials reach a similar electrical power output $P_{el,max}$ of about 1 mW, resulting in a maximum electrical power output $P_{el,max,TEG}$ of about 2 mW for the module. This corresponds to a maximum electrical power density $w_{el,max,TEG}$ of 5.89 mW cm$^{-2}$. The individual fluxes within in the p-type and n-type thermolegs for the power-optimized module are shown in Figure 7c–h. Here, the trend from the module optimized for matching $I_{q,SC}$ continues. The larger area of the n-type material results in a higher voltage $U(x)$ and as well as decreasing flux densities of entropy $j_S(x)$ (slightly lower) and charge $j_q(x)$ (significantly lower).

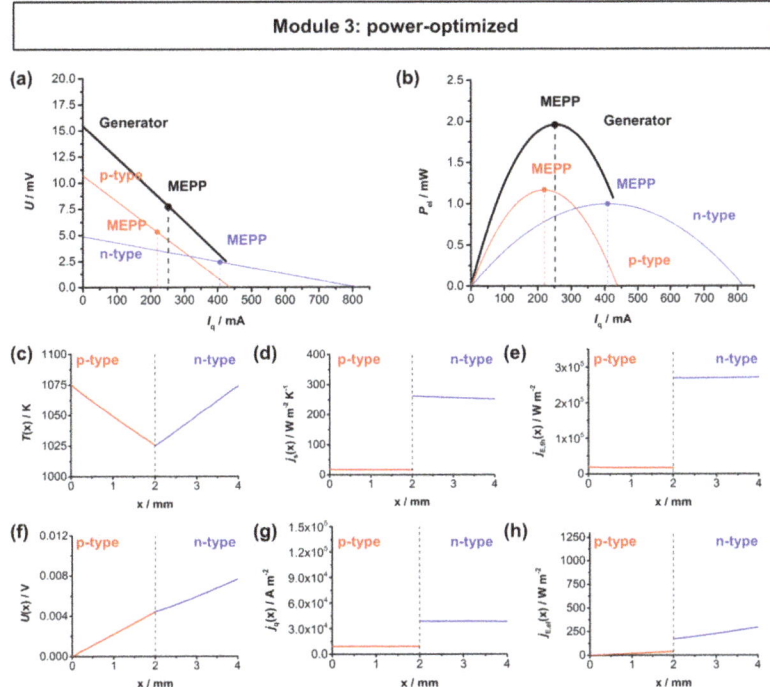

**Figure 7.** FEM simulations of the power-optimized module 3 (p-type $Ca_3Co_4O_9$ and n-type $In_{1.95}Sn_{0.05}O_3$) with a hot side temperature of 1050 K and cold side temperature of 1000 K. (**a**) $U$-$I_q$ characteristics and (**b**) electrical power output $P_{el}$-$I_q$ of the module. Thermoelectric characteristics of the respective materials as a function of the length of the respective legs: (**c**) temperature $T(x)$, (**d**) entropy flux density $j_S(x)$, (**e**) thermal energy flux density $j_{E,th}(x)$, (**f**) voltage $U(x)$, (**g**) electrical flux density $j_q(x)$, and (**h**) electrical energy flux density $j_{E,el}(x)$ trend throughout one thermocouple. Note, that the dashed line in (**c**–**h**) represent the metallic connector between the p-type and n-type materials. The simulated distributions are shown in Figure A6 in Appendix E.

Table 4 summarizes the simulated characteristics of the $zT$-optimized $Ca_3Co_4O_9$-$In_{1.95}Sn_{0.05}O_2$ TE module, the optimized module for matching $I_{q,SC}$, as well as for the power-optimized geometry. The module with power-optimized $A_n/A_p$ ratio reaches a maximum power density of 5.89 mW cm$^{-2}$, which is slightly higher compared to the module with overlapping material working points and about 30% higher when compared to the module with $zT$-optimized geometry. Additionally, the maximum first-law energy conversion efficiency $\eta_{I,TEG,max}$ for all three optimized geometries have been calculated. As expected, the $zT$-optimized module reaches the highest $\eta_{I,TEG,max}$ with 0.13%, while the module optimized for matching $I_{q,SC}$ and the power-optimized module show slightly lower efficiencies of 0.11% and 0.09%, respectively. This shows the contrary trend of a higher efficiency (for the $zT$-optimized module) and of higher power density (for the power-optimized module).

**Table 4.** Resulting maximum electrical power output $P_{el,max,TEG}$, electrical power density $w_{el,max,TEG}$, and maximum first-law energy conversion efficiency $\eta_{I,TEG,max}$ of module 3 with optimized geometry for maximum $zT$, matching $I_{q,SC}$ and maximum power output. The resulting power density increases due to the overlapping material working points.

| Module | Module MEPP/mA | $P_{el,max,TEG}$/mW | $w_{el,max,TEG}$/mW cm$^{-2}$ | $\eta_{I,TEG,max}$ |
|---|---|---|---|---|
| $zT$-optimized | 189.90 | 1.50 | 4.51 | 0.13% |
| same $I_{q,SC}$ | 239.89 | 1.86 | 5.64 | 0.11% |
| power-optimized | 252.00 | 1.96 | 5.89 | 0.09% |

As a result, module 3 is build based on the same materials with identical thermoelectric properties, but it is either optimized for maximum $zT$, matching $I_{q,SC}$ or maximum power output. Figure 8 summarizes the results of all three optimization strategies. The $zT$ optimization leads to a module with the highest conversion efficiency, but the lowest electrical power output. Contrary, the power optimization leads to a module with the the highest electrical power output, but the lowest conversion efficiency. The module with optimized geometry for matching $I_{q,SC}$ is in between, but closer to the maximum electrical power output. This also corresponds to the results of Xing et al. [36], who observed a similar increase in the maximum electrical power output with a respective geometry optimization when compared to an optimization for maximum $zT$. Note that this correlation between the deviation of optimum power output and optimum conversion efficiency is here shown on the example of module 3, but also applies for the other TE modules. As shown in Table 2, the optimum $A_n/A_p$ ratio for the $Bi_2Te_3$-based module 1 and the half-Heusler-based module 2 also varies slightly for the different optimization strategies. Therefore, also for quite similar thermoelectric materials, a slight deviation between optimum power output and energy conversion efficiency can be expected.

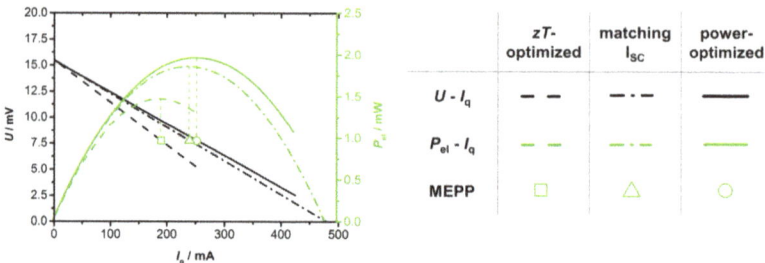

**Figure 8.** Comparison of all three optimization strategies for module 3 (dash: $zT$-optimized, dash-dot: matching $I_{q,SC}$, line: power-optimized). The power-optimized module shows a significantly higher power output when compared to $zT$-optimized module. The module with overlapping material working points is in between, but closer to the maximum power output.

## 4. Conclusions

Three different optimization strategies for the $A_n/A_p$ ratio were applied, whereas, for certain modules, they all resulted in different geometries. For module 3, based on strongly dissimilar thermoelectric properties of the $p$-type $Ca_3Co_4O_9$ and the $n$-type $In_{1.95}Sn_{0.05}O_3$, the geometry optimizations show strongly dissimilar $A_n/A_p$ ratios. Here, a strong deviation between high conversion efficiency (with $zT$-optimized geometry) and high power output (with power-optimized geometry) was found. The power optimization resulted in a 30% higher power output compared to the $zT$-optimized counterpart. For modules with more similar thermoelectric properties of the *n*- and *p*-type, which, in this work, are the $Bi_2Te_3$-based module 1 and the half-Heusler-based module 2, the respective optimum geometries only differ slightly, but also show this deviation in the geometry optimization. This emphasizes that, for TE module concepts, various optimization strategies may be applied, either to target high conversion efficiency or high power output. This phenomena correlates to the diversity of the thermoelectric materials that were used for the TE module. Additionally, this also underlines the similar importance of the power factor of thermoelectric materials, to target a high power output, when compared to the figure of merit $zT$.

**Author Contributions:** M.W., R.H. and A.F. worked on the conceptualization. A.R. developed the methods and carried out the simulations. M.W. wrote the original draft. All authors critically revised and edited the manuscript draft. A.F. is responsible for the acquisition of funding. All authors have read and agreed to the published version of the manuscript.

**Funding:** This work was funded by the Deutsche Forschungsgemeinschaft (DFG, German Research Foundation)—project number 325156807. The publication of this article was funded by the Open Access fund of Leibniz University Hannover.

**Conflicts of Interest:** The authors declare no conflict of interest.

## Abbreviations

The following abbreviations are used in this manuscript:

| | |
|---|---|
| TE module | thermoelectric module |
| TEG | thermoelectric generator |
| MCEP | maximum conversion efficiency point |
| MEPP | maximum electrical power point |
| OC | (electrical) open-circuit |
| SC | (electrical) short-circuit |

## Symbols

The following symbols are used in this manuscript:

*Geometry*

| | |
|---|---|
| $A$ | cross-sectional area of thermoelectric material |
| $A_n$ | cross-sectional area of *n*-type material |
| $A_p$ | cross-sectional area of *p*-type material |
| $l$ | length of thermoelectric material |
| $l_n$ | length of *n*-type material |
| $l_p$ | length of *p*-type material |
| $\frac{A_n}{A_p}$ | ratio of the cross-sectional areas of the *n*-type and *p*-type materials |
| $[\frac{A_n}{A_p}]_{zT}$ | ratio of the cross-sectional areas of the *n*-type and *p*-type materials for maximum $zT$ |
| $[\frac{A_n}{A_p}]_{\text{matching } I_{q,SC}}$ | ratio of the cross-sectional areas of the *n*-type and *p*-type materials for matching $I_{q,SC}$ |
| $[\frac{A_n}{A_p}]_{\text{power}}$ | ratio of the cross-sectional areas of the *n*-type and *p*-type materials for maximum power |

*Material properties*

| | |
|---|---|
| $\alpha$ | Seebeck coefficient |
| $\alpha_n$ | Seebeck coefficient of *n*-type material |
| $\alpha_p$ | Seebeck coefficient of *p*-type material |
| $\lambda_n$ | heat conductivity of *n*-type material |
| $\lambda_p$ | heat conductivity of *p*-type material |
| $\lambda_{OC}$ | heat conductivity under electrically open-circuited (OC) conditions |
| $\Lambda_{OC}$ | entropy conductivity under electrically open-circuited (OC) conditions |
| $\rho$ | specific electrical resistivity |
| $\rho_n$ | specific electrical resistivity of *n*-type material |
| $\rho_p$ | specific electrical resistivity of *p*-type material |
| $R_n$ | resistance of *n*-type material |
| $R_p$ | resistance of *p*-type material |
| $\sigma$ | isothermal electrical conductivity |
| $f$ | figure of merit (as introduced by Zener [43] |
| $zT$ | figure of merit (as introduced by Ioffe [10]) |

*Thermodynamic potentials*

| | |
|---|---|
| $\varphi$ | electric potential |
| $T$ | absolute temperature |
| $T_{cold}$ | temperature of the thermoelectric material at its cold side |
| $T_{hot}$ | temperature of the thermoelectric material at its hot side |
| $\nabla T$ | gradient of the temperature |
| $\Delta T$ | difference of temperature (along the thermoelectric material) |
| $U$ | voltage |
| $U_{OC}$ | voltage at electrically open-circuited (OC) conditions |

*Fluxes*

| | |
|---|---|
| $i$ | normalized electrical current |
| $I_q$ | electrical current |
| $I_{q,SC}$ | electrical current at electrically short-circuited (SC) conditions |
| $j_q$ | electrical flux density |
| $j_s$ | entropy flux density |
| $j_{E,el}$ | electrical energy flux density |
| $j_{E,th}$ | thermal energy flux density |
| $q$ | electric charge |
| $S$ | entropy |

## Performance

| | |
|---|---|
| $P_{el,max}$ | maximum electrical power output of the thermoelectric material (at MEPP) |
| $P_{el,max,TEG}$ | maximum electrical power output of the module (at MEPP) |
| $w_{el,max,TEG}$ | maximum electrical power density of the module (at MEPP) |
| $I_{q,MEPP}$ | current $I_q$ at the MEPP |
| $I_{q,MEPP,n}$ | current $I_q$ at the MEPP of the $n$-type material |
| $I_{q,MEPP,p}$ | current $I_q$ at the MEPP of the $p$-type material |
| $U_{MEPP,TEG}$ | voltage $U$ at the MEPP of the TE module |
| $I_{q,MEPP,TEG}$ | current $I_q$ at the MEPP of the TE module |
| $R_{TEG}$ | internal resistance of the TE module |
| $\eta_{II,mat}$ | second-law energy conversion efficiency of a thermoelectric material |
| $\eta_{I,TEG,max}$ | maximum first-law energy conversion efficiency of the TE module |
| $\eta_{Carnot}$ | Carnot efficiency of the TE module |
| $\eta_{II,TEG,max}$ | maximum second-law energy conversion efficiency of the TE module |

## Appendix A. Input Data for FEM-Simulation

**Table A1.** Thermoelectric parameters of module 1 ($p$-type $Bi_{0.5}Sb_{1.5}Te_3$ [37], $n$-type $Bi_2Te_{3-x}Sb_x$ [38]) used for FEM simulations. For all material parameters a linear behavior within the applied temperature range has been assumed and the average value has been used for the calculation of the $A_n/A_p$ ratio.

| | $p$-Type | | $n$-Type | |
|---|---|---|---|---|
| T/K | 348 | 298 | 348 | 298 |
| $\sigma$/S cm$^{-1}$ | 760 | 990 | 711 | 875 |
| $\alpha$/$\mu$V K$^{-1}$ | 227 | 213 | $-228$ | $-220$ |
| $\lambda_{OC}$/W m$^{-1}$ K$^{-1}$ | 1.31 | 1.39 | 1.40 | 1.35 |
| $\Lambda_{OC}$/W m$^{-1}$ K$^{-2}$ | $3.76 \times 10^{-3}$ | $4.66 \times 10^{-3}$ | $4.02 \times 10^{-3}$ | $4.53 \times 10^{-3}$ |

**Table A2.** Thermoelectric parameters of module 2 ($p$-type $FeNb_{0.88}Hf_{0.12}Sb$ [39], $n$-type $Hf_{0.6}Zr_{0.4}NiSn_{0.995}Sb_{0.005}$ [40]) used for FEM simulations. For all material parameters a linear behavior within the applied temperature range has been assumed and the average value has been used for the calculation of the $A_n/A_p$ ratio.

| | $p$-Type | | $n$-Type | |
|---|---|---|---|---|
| T/K | 1000 | 950 | 1000 | 950 |
| $\sigma$/S cm$^{-1}$ | 1053 | 1158 | 960 | 1000 |
| $\alpha$/$\mu$V K$^{-1}$ | 223 | 217 | $-212$ | $-219$ |
| $\lambda_{OC}$/W m$^{-1}$ K$^{-1}$ | 4.33 | 4.44 | 4.16 | 3.90 |
| $\Lambda_{OC}$/W m$^{-1}$ K$^{-2}$ | $4.33 \times 10^{-3}$ | $4.76 \times 10^{-3}$ | $4.16 \times 10^{-3}$ | $4.11 \times 10^{-3}$ |

**Table A3.** Thermoelectric parameters of module 3 ($p$-type $Ca_3Co_4O_9$ [41], $n$-type $In_{1.95}Sn_{0.05}O_3$ [41]) used for FEM simulations. For all material parameters a linear behavior within the applied temperature range has been assumed and the average value has been used for the calculation of the $A_n/A_p$ ratio.

| | $p$-Type | | $n$-Type | |
|---|---|---|---|---|
| T/K | 1075 | 1025 | 1075 | 1025 |
| $\sigma$/S cm$^{-1}$ | 29.63 | 31.48 | 609.26 | 448.15 |
| $\alpha$/$\mu$V K$^{-1}$ | 202.83 | 225.74 | $-100.94$ | $-92.45$ |
| $\lambda_{OC}$/W m$^{-1}$ K$^{-1}$ | 0.63 | 0.66 | 10.70 | 10.94 |
| $\Lambda_{OC}$/W m$^{-1}$ K$^{-2}$ | $0.59 \times 10^{-3}$ | $0.64 \times 10^{-3}$ | $9.95 \times 10^{-3}$ | $10.67 \times 10^{-3}$ |

## Appendix B. $A_n/A_p$ Optimization for Matching Short-Circuit Current

Here, the optimized $A_n/A_p$ ratio for matching $I_{q,SC}$ is derived. The idea of this optimization is as follows: the working points of the respective thermoelectric materials overlap, if the flux of charge in both materials is the same. Then, the working points of the materials overlap, so

$$I_{q,MEPP,n} = I_{q,MEPP,p} \tag{A1}$$

By including

$$I_{q,MEPP,n} = \frac{|\alpha_n|(\Delta T)^2}{2R_n} \tag{A2}$$

and

$$I_{q,MEPP,p} = \frac{\alpha_p (\Delta T)^2}{2R_p} \tag{A3}$$

with the electrical resistance of the materials

$$R_n = \rho_n \cdot \frac{l_n}{A_n} \tag{A4}$$

and

$$R_p = \rho_p \cdot \frac{l_p}{A_p} \tag{A5}$$

the following relation is received:

$$\frac{|\alpha_n| \cdot (\Delta T)^2}{2\rho_n \frac{l_n}{A_n}} = \frac{\alpha_p \cdot (\Delta T)^2}{2\rho_p \frac{l_p}{A_p}} \tag{A6}$$

After rearrangement and with the assumed same length of the thermolegs $l_n = l_p$ the result for a $A_n/A_p$ ratio for matching $I_{q,SC}$ is:

$$\frac{A_n}{A_p} = \frac{\alpha_p}{|\alpha_n|} \cdot \frac{\rho_n}{\rho_p} \tag{A7}$$

## Appendix C. $A_n/A_p$ Optimization for Maximum Power

The maximum power output of a TE module is a function of the electrical current $I_q$ of the module at the MEPP $I_{q,MEPP,TEG}$ and the voltage $U$ of the module at the MEPP $U_{MEPP,TEG}$, which are calculated according to Equations (A8) and (A9):

$$I_{q,MEPP,TEG} = \frac{(\alpha_p - \alpha_n) \cdot \Delta T}{2R} \tag{A8}$$

$$U_{MEPP,TEG} = \frac{(\alpha_p - \alpha_n) \cdot \Delta T}{2} \tag{A9}$$

From this, the maximum electrical power output of a module at the MEPP can be derived as

$$P_{el,max,TEG} = \frac{(\alpha_p - \alpha_n)^2 \cdot (\Delta T)^2}{4R_{TEG}} \tag{A10}$$

with the internal electrical resistance of the module $R_{TEG}$

$$R_{TEG} = \rho_p \frac{l_p}{A_p} + \rho_n \frac{l_n}{A_n} \tag{A11}$$

Considering, that the effecive area $A$ is a sum of the cross-sectional areas $A_n$ and $A_p$, Equation (A11) can be differentiated and has to be equal 0 for its maximum. So

$$-(\alpha_p - \alpha_n)^2 (\Delta T)^2 \cdot \frac{\frac{\rho_p l_p}{(A+A_p)^2} - \frac{\rho_n l_n}{A_p^2}}{(\frac{\rho_p l_p}{A_p} + \frac{\rho_n l_n}{A-A_p})^2} = 0 \tag{A12}$$

This Equation (A12) is zero, if the numerator of the fraction is zero, so

$$\frac{\rho_p l_p}{(A+A_p)^2} - \frac{\rho_n l_n}{A_p^2} = 0 \tag{A13}$$

After rearrangement, the optimum $A_n/A_p$ ratio for maximum power output is received as

$$\frac{A_n}{A_p} = \sqrt{\frac{\rho_n}{\rho_p}} \tag{A14}$$

The final Equation (A14) derived corresponds to the reported ratio for maximum power output of Xing et al. [36].

### Appendix D. Efficiency of the Module

The maximum first-law efficiency $\eta_{I,TEG,max}$ of a module is the product of the Carnot efficiency $\eta_{Carnot}$ and the second-law efficiency $\eta_{II,TEG,max}$ [9,11] and can be determined as

$$\eta_{I,TEG,max} = \eta_{Carnot} \cdot \eta_{II,TEG,max} = \frac{T_{hot} - T_{cold}}{T_{hot}} \cdot \frac{\sqrt{1+Z\bar{T}} - 1}{\sqrt{1+Z\bar{T}} + 1} \tag{A15}$$

Here, $\bar{T}$ is the average temperature and $Z$ is a function of the materials thermoelectric parameters:

$$Z = \frac{\alpha^2}{R \cdot K} \tag{A16}$$

with

$$\alpha = (\alpha_p - \alpha_n) \tag{A17}$$

$$R = \frac{1}{\sigma_p} \cdot \frac{l_p}{A_p} + \frac{1}{\sigma_n} \cdot \frac{l_n}{A_n} \tag{A18}$$

and

$$K = \lambda_p \cdot \frac{A_p}{l_p} + \lambda_n \cdot \frac{A_n}{l_n} \tag{A19}$$

So, the maximum first-law efficiency $\eta_{I,TEG,max}$ of a module can be determined as a function of the materials thermoelectric parameter and the respective cross-sectional areas $A_n$ and $A_p$ [30].

## Appendix E. Simulated Module Fluxes

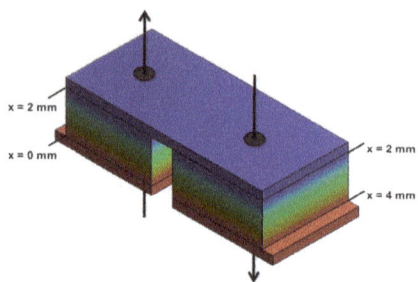

**Figure A1.** Analyzed path $x$ along a central line through the respective thermoleg on the example of module 1.

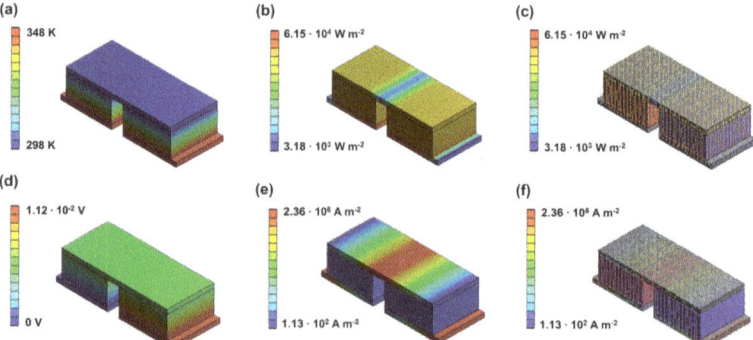

**Figure A2.** Distribution of (a) temperature $T(x)$, (b,c) flux density of thermal energy $j_{E,\text{th}}(x)$, (d) voltage $U(x)$ and (e,f) flux density of electrical charge $j_q(x)$ in module 1.

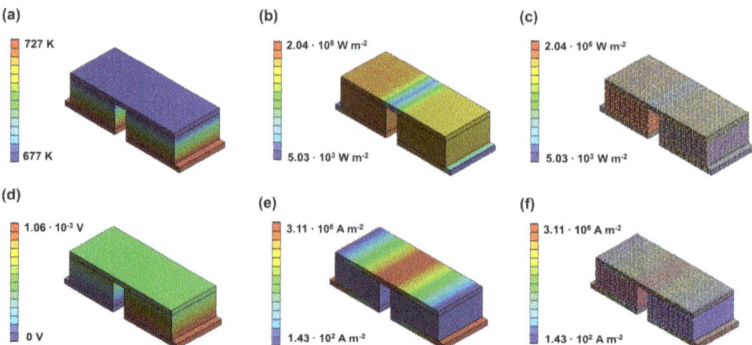

**Figure A3.** Distribution of (a) temperature $T(x)$, (b,c) flux density of thermal energy $j_{E,\text{th}}(x)$, (d) voltage $U(x)$ and (e,f) flux density of electrical charge $j_q(x)$ in module 2.

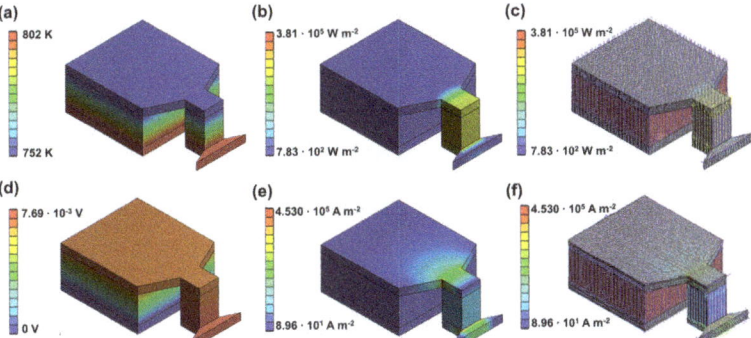

**Figure A4.** Distribution of (**a**) temperature $T(x)$, (**b**,**c**) flux density of thermal energy $j_{E,th}(x)$, (**d**) voltage $U(x)$ and (**e**,**f**) flux density of electrical charge $j_q(x)$ in module 3 with $zT$-optimized geometry.

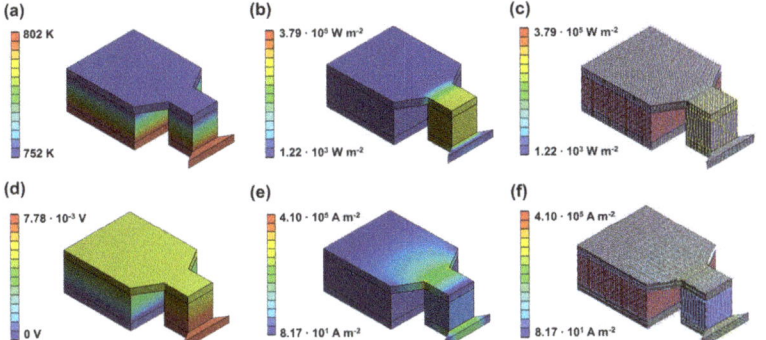

**Figure A5.** Distribution of (**a**) temperature $T(x)$, (**b**,**c**) flux density of thermal energy $j_{E,th}(x)$, (**d**) voltage $U(x)$ and (**e**,**f**) flux density of electrical charge $j_q(x)$ in module 3 with geometry for matching $I_{q,SC}$.

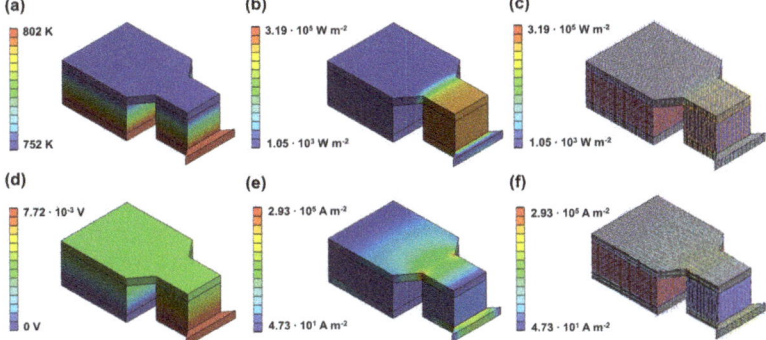

**Figure A6.** Distribution of (**a**) temperature $T(x)$, (**b**,**c**) flux density of thermal energy $j_{E,th}(x)$, (**d**) voltage $U(x)$ and (**e**,**f**) flux density of electrical charge $j_q(x)$ in module 3 with power-optimized geometry.

## References

1. Kishore, R.A.; Marin, A.; Wu, C.; Kumar, A.; Priya, S. *Energy Harvesting—Materials, Physics, and System Design with Practical Examples*; DEStech Publications: Boston, MA, USA; Berlin, Germany, 2019.
2. Wolf, M.; Hinterding, R.; Feldhoff, A. High Power Factor vs. High zT—A Review of Thermoelectric Materials for High-Temperature Application. *Entropy* **2019**, *21*, 1058. [CrossRef]
3. Gayner, C.; Kar, K.K. Recent Advances in Thermoelectric Materials. *Prog. Mater. Sci.* **2016**, *83*, 330–382. [CrossRef]
4. He, R.; Schierning, G.; Nielsch, K. Thermoelectric Devices: A Review of Devices, Architectures, and Contact Optimization. *Adv. Mater. Technol.* **2018**, *3*. [CrossRef]
5. Liu, X.; Wang, Z. Printable Thermoelectric Materials and Applications. *Front. Mater.* **2019**, *6*, 1–5. [CrossRef]
6. He, W.; Zhang, G.; Zhang, X.; Ji, J.; Li, G.; Zhao, X. Recent Development and Application of Thermoelectric Generator and Cooler. *Appl. Energy* **2015**, *143*, 1–25. [CrossRef]
7. Fuchs, H.U. A Direct Entropic Approach to Uniform and Spatially Continuous Dynamical Models of Thermoelectric Devices. *Energy Harvest. Syst.* **2014**, *1*, 1–18. [CrossRef]
8. Feldhoff, A. Thermoelectric Material Tensor Derived from the Onsager-de Groot-Callen Model. *Energy Harvest. Syst.* **2015**, *2*, 5–13. [CrossRef]
9. Fuchs, H. *The Dynamics of Heat—A Unified Approach to Thermodynamics and Heat Transfer*, 2nd ed.; Springer: New York, NY, USA, 2010. [CrossRef]
10. Ioffe, A.F. *Semiconductor Thermoelements, and Thermoelectric Cooling*, 1st ed.; Info-search Ltd.: London, UK, 1957. [CrossRef]
11. Feldhoff, A. Power Conversion and its Efficiency in Thermoelectric Materials. *Entropy* **2020**, *22*, 803. [CrossRef]
12. Mamur, H.; Bhuiyan, M.R.; Korkmaz, F.; Nil, M. A Review on Bismuth Telluride ($Bi_2Te_3$) Nanostructure for Thermoelectric Applications. *Renew. Sustain. Energy Rev.* **2018**, *82*, 4159–4169. [CrossRef]
13. Guo, W.; Ma, J.; Zheng, W. $Bi_2Te_3$ Nanoflowers Assembled of Defective Nanosheets with Enhanced Thermoelectric Performance. *J. Alloy. Compd.* **2016**, *659*, 170–177. [CrossRef]
14. Zhang, J.; Wu, D.; He, D.; Feng, D.; Yin, M.; Qin, X.; He, J. Extraordinary Thermoelectric Performance Realized in n-Type PbTe through Multiphase Nanostructure Engineering. *Adv. Mater.* **2017**, *29*, 1–7. [CrossRef] [PubMed]
15. Zhao, L.D.; Lo, S.H.; Zhang, Y.; Sun, H.; Tan, G.; Uher, C.; Wolverton, C.; Dravid, V.P.; Kanatzidis, M.G. Ultralow Thermal Conductivity and High Thermoelectric Figure of Merit in SnSe Crystals. *Nature* **2014**, *508*, 373–377. [CrossRef] [PubMed]
16. Peng, K.; Lu, X.; Zhan, H.; Hui, S.; Tang, X.; Wang, G.; Dai, J.; Uher, C.; Wang, G.; Zhou, X. Broad Temperature Plateau for High zTs in Heavily Doped p-Type SnSe Single Crystals. *Energy Environ. Sci.* **2016**, *9*, 454–460. [CrossRef]
17. Shuai, J.; Mao, J.; Song, S.; Zhang, Q.; Chen, G.; Ren, Z. Recent Progress and Future Challenges on Thermoelectric Zintl Materials. *Mater. Today Phys.* **2017**, *1*, 74–95. [CrossRef]
18. Sun, J.; Singh, D.J. Thermoelectric Properties of $AMg_2X_2$, $AZn_2Sb_2$ (A = Ca, Sr, Ba; X = Sb, Bi), and $Ba_2ZnX_2$ (X = Sb, Bi) Zintl Compounds. *J. Mater. Chem. A* **2017**, *5*, 8499–8509. [CrossRef]
19. Chen, X.; Wu, H.; Cui, J.; Xiao, Y.; Zhang, Y.; He, J.; Chen, Y.; Cao, J.; Cai, W.; Pennycook, S.J.; et al. Extraordinary Thermoelectric Performance in n-Type Manganese Doped $Mg_3Sb_2$ Zintl: High Band Degeneracy, Tuned Carrier Scattering Mechanism and Hierarchical Microstructure. *Nano Energy* **2018**, *52*, 246–255. [CrossRef]
20. Zhu, H.; He, R.; Mao, J.; Zhu, Q.; Li, C.; Sun, J.; Ren, W.; Wang, Y.; Liu, Z.; Tang, Z.; et al. Discovery of ZrCoBi Based Half Heuslers with High Thermoelectric Conversion Efficiency. *Nat. Commun.* **2018**, *9*, 1–9. [CrossRef]
21. Poon, S.J. Half-Heusler Compounds: Promising Materials For Mid-To-High Temperature Thermoelectric Conversion. *J. Phys. D Appl. Phys.* **2019**, *52*, 493001. [CrossRef]
22. Yin, Y.; Tudu, B.; Tiwari, A. Recent Advances in Oxide Thermoelectric Materials and Modules. *Vacuum* **2017**, *146*, 356–374. [CrossRef]
23. Zhang, X.; Chang, C.; Zhou, Y.; Zhao, L.D. BiCuSeO Thermoelectrics: An Update on Recent Progress and Perspective. *Materials* **2017**, *10*, 198. [CrossRef]

24. Cowen, L.M.; Atoyo, J.; Carnie, M.J.; Baran, D.; Schroeder, B.C. Review—Organic Materials for Thermoelectric Energy Generation. *ECS J. Solid State Sci. Technol.* **2017**, *6*, N3080–N3088. [CrossRef]
25. Boudouris, B.W.; Yee, S. Structure, Properties and Applications of Thermoelectric Polymers. *J. Appl. Polym. Sci.* **2017**, *134*. [CrossRef]
26. Ashalley, E.; Chen, H.; Tong, X.; Li, H.; Wang, Z.M. Bismuth Telluride Nanostructures: Preparation, Thermoelectric Properties and Topological Insulating Effect. *Front. Mater. Sci.* **2015**, *9*, 103–125. [CrossRef]
27. Gharsallah, M.; Serrano-Sánchez, F.; Bermúdez, J.; Nemes, N.M.; Martínez, J.L.; Elhalouani, F.; Alonso, J.A. Nanostructured $Bi_2Te_3$ Prepared by a Straightforward Arc-Melting Method. *Nanoscale Res. Lett.* **2016**, *11*, 4–10. [CrossRef]
28. Kim, K.; Kim, G.; Lee, H.; Lee, K.H.; Lee, W. Band Engineering and Tuning Thermoelectric Transport Properties of p-type $Bi_{0.52}Sb_{1.48}Te_3$ by Pb Doping for Low-Temperature Power Generation. *Scr. Mater.* **2018**, *145*, 41–44. [CrossRef]
29. Ming, T.; Wu, Y.; Peng, C.; Tao, Y. Thermal Analysis on a Segmented Thermoelectric Generator. *Energy* **2015**, *80*, 388–399. [CrossRef]
30. Ouyang, Z.; Li, D. Modelling of Segmented High-Performance Thermoelectric Generators with Effects of Thermal Radiation, Electrical and Thermal Contact Resistances. *Sci. Rep.* **2016**, *6*, 1–12. [CrossRef]
31. Korotkov, A.S.; Loboda, V.V.; Makarov, S.B.; Feldhoff, A. Modeling Thermoelectric Generators Using the ANSYS Software Platform: Methodology, Practical Applications, and Prospects. *Russ. Microelectron.* **2017**, *46*, 131–138. [CrossRef]
32. Oshima, K.; Inoue, J.; Sadakata, S.; Shiraishi, Y.; Toshima, N. Hybrid-Type Organic Thermoelectric Materials Containing Nanoparticles as a Carrier Transport Promoter. *J. Electron. Mater.* **2017**, *46*, 3207–3214. [CrossRef]
33. Culebras, M.; Igual-Muñoz, A.M.; Rodríguez-Fernández, C.; Gómez-Gómez, M.I.; Gómez, C.; Cantarero, A. Manufacturing Te/PEDOT Films for Thermoelectric Applications. *ACS Appl. Mater. Interfaces* **2017**, *9*, 20826–20832. [CrossRef]
34. Wolf, M.; Menekse, K.; Mundstock, A.; Hinterding, R.; Nietschke, F.; Oeckler, O.; Feldhoff, A. Low Thermal Conductivity in Thermoelectric Oxide-Based Multiphase Composites. *J. Electron. Mater.* **2019**, *48*, 7551–7561. [CrossRef]
35. Narducci, D. Do we Really Need High Thermoelectric Figures of Merit? A Critical Appraisal to the Power Conversion Efficiency of Thermoelectric Materials. *Appl. Phys. Lett.* **2011**, *99*. [CrossRef]
36. Xing, Z.; Liu, R.; Liao, J.; Wang, C.; Zhang, Q.; Song, Q.; Xia, X.; Zhu, T.; Bai, S.; Chen, L. A-Device-To-Material Strategy Guiding the "Double-High" Thermoelectric Module. *Joule* **2020**. [CrossRef]
37. Poudel, B.; Hao, Q.; Ma, Y.; Lan, Y.; Minnich, A.; Yu, B.; Yan, X.; Wang, D.; Muto, A.; Vashaee, D.; et al. High-Thermoelectric Performance of Nanostructured Bismuth Antimony Telluride Bulk Allys. *Science* **2008**, *320*, 634–638. [CrossRef]
38. Kim, H.S.; Kikuchi, K.; Itoh, T.; Iida, T.; Taya, M. Design of Segmented Thermoelectric Generator Based on Cost-Effective and Light-Weight Thermoelectric Alloys. *Mater. Sci. Eng. B Solid State Mater. Adv. Technol.* **2014**, *185*, 45–52. [CrossRef]
39. Fu, C.; Bai, S.; Liu, Y.; Tang, Y.; Chen, L.; Zhao, X.; Zhu, T. Realizing High Figure of Merit in Heavy-Band p-Type half-Heusler Thermoelectric Materials. *Nat. Commun.* **2015**, *6*, 1–7. [CrossRef] [PubMed]
40. Chen, L.; Gao, S.; Zeng, X.; Mehdizadeh Dehkordi, A.; Tritt, T.M.; Poon, S.J. Uncovering High Thermoelectric Figure of Merit in (Hf,Zr)NiSn half-Heusler Alloys. *Appl. Phys. Lett.* **2015**, *107*, 041902. [CrossRef]
41. Bittner, M.; Geppert, B.; Kanas, N.; Singh, S.P.; Wiik, K.; Feldhoff, A. Oxide-Based Thermoelectric Generator for High-Temperature Application Using p-Type $Ca_3Co_4O_9$ and n-Type $In_{1.95}Sn_{0.05}O_3$ Legs. *Energy Harvest. Syst.* **2016**, *3*, 213–222. [CrossRef]
42. Bergman, T.L.; Lavine, A.S.; Incropera, F.P.; DeWitt, D.P. *Fundamentals of Heat and Mass Transfer*, 7th ed.; John Wiley & Sons: Hoboken, NJ, USA, 2011; p. 8.
43. Zener, C. Putting Electrons to Work. *Trans. ASM* **1961**, *53*, 1052–1068.

© 2020 by the authors. Licensee MDPI, Basel, Switzerland. This article is an open access article distributed under the terms and conditions of the Creative Commons Attribution (CC BY) license (http://creativecommons.org/licenses/by/4.0/).

*Article*

# Non-Equilibrium Quantum Brain Dynamics: Super-Radiance and Equilibration in 2 + 1 Dimensions

**Akihiro Nishiyama** [1,*], **Shigenori Tanaka** [1,*] **and Jack A. Tuszynski** [2,3,4,*]

1. Graduate School of System Informatics, Kobe University, 1-1 Rokkodai, Nada-ku, Kobe 657-8501, Japan
2. Department of Oncology, University of Alberta, Cross Cancer Institute, Edmonton, AB T6G 1Z2, Canada
3. Department of Physics, University of Alberta, Edmonton, AB T6G 2J1, Canada
4. DIMEAS, Corso Duca degli Abruzzi, 24, Politecnico di Torino, 10129 Turin, TO, Italy
* Correspondence: anishiyama@people.kobe-u.ac.jp (A.N.); tanaka2@kobe-u.ac.jp (S.T.); jackt@ualberta.ca (J.A.T.)

Received: 15 October 2019; Accepted: 26 October 2019; Published: 30 October 2019

**Abstract:** We derive time evolution equations, namely the Schrödinger-like equations and the Klein–Gordon equations for coherent fields and the Kadanoff–Baym (KB) equations for quantum fluctuations, in quantum electrodynamics (QED) with electric dipoles in 2 + 1 dimensions. Next we introduce a kinetic entropy current based on the KB equations in the first order of the gradient expansion. We show the H-theorem for the leading-order self-energy in the coupling expansion (the Hartree–Fock approximation). We show conserved energy in the spatially homogeneous systems in the time evolution. We derive aspects of the super-radiance and the equilibration in our single Lagrangian. Our analysis can be applied to quantum brain dynamics, that is QED, with water electric dipoles. The total energy consumption to maintain super-radiant states in microtubules seems to be within the energy consumption to maintain the ordered systems in a brain.

**Keywords:** non-equilibrium quantum field theory; quantum brain dynamics; Kadanoff–Baym equation; entropy; super-radiance

---

## 1. Introduction

Numerous attempts to understand memory in a brain have been made over one hundred years starting at the end of 19th century. Nevertheless, the concrete mechanism of memory still remains an open question in conventional neuroscience [1–3]. Conventional neuroscience is based on classical mechanics with neurons connected by synapses. However, we still cannot answer how limited connections between neurons describe mass excitations in a brain in classical neuron doctrine.

Quantum field theory (QFT) of the brain or quantum brain dynamics (QBD), is one of the hypotheses expected to describe the mechanism of memory in the brain [4–6]. Experimentally, several properties of memory, namely the diversity, the long-term but imperfect stability and nonlocality (Memory is diffused and non-localized in several domains in a brain. It does not disappear due to the destruction in a particular local domain. The term 'nonlocality' does not indicate nonlocality in entanglement in quantum mechanics.), are suggested in [7–9]. The QBD can describe these properties by adopting infinitely physically or unitarily inequivalent vacua in QFT, distinguished from quantum mechanics which cannot describe unitarily inequivalence. Unitarily inequivalence represents the emergence of the diversity of phases and allows the possibility of spontaneous symmetry breaking (SSB) [10–13]. The vacua or the ground states appearing in SSB describe the stability of the states. Furthermore, the QFT can describe both microscopic degrees of freedom and macroscopic matter [10]. To describe stored information, we can adopt the macroscopic ordered states in QFT with SSB involving long-range correlation via Nambu–Goldstone (NG) quanta. In 1967, Ricciardi and Umezawa proposed a quantum field theoretical approach to describe memory in a brain [14]. They adopted the SSB with long-range correlations mediated by NG quanta in QFT. Stuart et al. developed QBD by assuming a brain as a mixed system of classical neurons and quantum degrees of freedom, namely corticons and exchange bosons [15,16]. The vacua appearing in SSB, the macroscopic order, are interpreted as the memory storage in QBD. The finite number of excitations of NG modes represents the memory retrieval. Around the same time, Fröhlich proposed the application of a theory of electric dipoles to the study of biological systems [17–22]. He suggested a theory of the emergence of a giant dipole in open systems with breakdown of rotational symmetry of dipoles where dipoles are aligned in the same direction (the ordered states with coherent wave propagation of dipole oscillation in the Fröhrich condensate). In 1976, Davydov and Kislukha studied a theory of solitary wave propagation in protein chains, called the Davydov soliton [23]. It is found that the theory by Fröhlich and that by Davydov represent static and dynamical properties in the nonlinear Schödinger equation with an equivalent quantum Hamiltonian, respectively [24]. In the 1980s, Del Giudice et al. applied a theory of water electric dipoles to biological systems [25–28]. In particular, the derivation of laser-like behavior is a suggestive study. In the 1990s, Jibu and Yasue gave a concrete picture of corticons and exchange bosons, namely water electric dipole fields and photon fields [4,29–32]. The QBD is nothing but quantum electrodynamics (QED) with water electric dipole fields. When electric dipoles are aligned in the same directions coherently, the polaritons, NG bosons in SSB of rotational symmetry, emerge. The dynamical order in the vacua in SSB is maintained by long-range correlation of the massless NG bosons. In QED, the NG bosons are absorbed by photons and then photons acquire mass due to the Higgs mechanism and can stay in coherent domains. The massive photons are called evanescent photons. The size of a coherent domain is in the order of 50 µm. Furthermore, two quantum mechanisms of information transfer and integration among coherent domains are suggested. The first one is to use the super-radiance and the self-induced transparency via microtubules connecting two coherent domains [31]. Super-radiance is the phenomenon indicating coherent photon emission with correlation among not only photons but also atoms (or dipoles) [33–37]. The atoms (or dipoles) cooperatively decay in a short time interval due to correlation; coherent photons with intensity proportional to the square of the number of atoms (or dipoles) are emitted. The pulse wave photons in super-radiance propagate through microtubules without decay. Then the self-induced transparency appears, since microtubules are perfectly transparent in the propagation. The second one is to use the quantum tunneling effect among coherent domains surrounded by incoherent domains [32]. The effect is essentially equivalent to the Josephson effect between two superconducting domains separated by a normal domain. Del Giudice et al. studied this effect in biological systems [28]. In 1995, Vitiello has shown that a huge memory capacity can be realized by regarding a brain as an open dissipative system and doubling the degrees of freedom with mathematical techniques in thermo-field dynamics [38]. In dissipative model of a brain, each memory state evolves in classical deterministic trajectory like a chaos [39]. The overlap among distinct memory states is zero at any time in the infinite volume limit. However, finite volume effects allow states to overlap one another, which might represent association of memories [6]. In 2003, exclusion zone (EZ) water was discovered experimentally [40]. The properties of EZ water correspond to those of coherent water [41].

However, we have never seen the dynamical memory formations based on QBD at the physiological temperature in the presence of thermal effects written by quantum fluctuations. Hence, there are still criticisms related with the decoherence phenomena (We should use the mass of polaritons in estimating the critical temperature of ordered states, not that of water molecules themselves.)in memory formations in QBD [42]. So, we need to derive time evolution equations of coherent fields and quantum fluctuations and show numerical simulations of memory formation processes in non-equilibrium situations to check whether or not memory in QBD is robust against thermal effects. Futhermore, in 2012 Craddock et al. suggested the mechanism of memory coding in microtubules with phosphorylation by Ca$^{2+}$ calmodulin kinase II [43]. It will be an interesting topic to investigate how water electric dipoles and evanescent photons are affected by phosphorylated microtubules.

The aim of this paper is to derive time evolution equations, namely the Schrödinger-like equations for coherent dipole fields, the Klein–Gordon equations for coherent photon fields, the Kadanoff–Baym equations for quantum fluctuations [44–46], with the two-particle-irreducible effective action technique with Keldysh formalism [47–51]. We derive both the equilibration for quantum fluctuations and the super-radiance for background coherent fields from the single Lagrangian in quantum electrodynamics (QED) with electric dipole fields. We arrive at the Maxwell–Bloch equations for the super-radiance by starting with QED with electric dipole fields in 2 + 1 dimensions. When we consider electric fields in super-radiance, we only need two spatial dimensions, one axis for the amplitude and another axis for the propagation. Hence we have discussed the case in 2 + 1 dimensions in this paper. By using our equations for super-radiance in this paper, we can describe information transfer via microtubules. Then, microtubule-associated proteins can make an important contribution to information transfer with interconnections among microtubules. We also derive the Higgs mechanism and the tachyonic instability for coherent fields in the Klein–Gordon equation for coherent electric fields. In two energy level approximation for electric dipole fields, namely with the ground state and the first excited states, the Higgs mechanism appears in normal population in which the probability amplitude in the ground state is larger than that in the first excited states. The penetrating length in the Meissner effect due to the Higgs mechanism is 6.3 µm derived by using coefficients in 2 + 1 dimensions and the number density of liquid water molecules in 3 + 1 dimensions. On the other hand, the tachyonic instability appears in inverted population in which the probability amplitudes in the first excited states are larger than that in the ground state. Then the electric field increases exponentially while the system is in inverted population. The increase stops at times when normal population is realized. Our analysis also contains the dynamics of quantum fluctuations in non-equilibrium cases. We also derive the Kadanoff–Baym equations for quantum fluctuations with the leading-order self-energy in the coupling expansion. The Kadanoff–Baym equations describe the entropy producing dynamics during equilibration as shown in the proof of the H-theorem. Entropy production stops when the Bose–Einstein distribution is realized. By combining time evolution equations (the Klein–Gordon equations for coherent electric fields and the Schrödinger-like equations for coherent electric dipole fields) and the Kadanoff–Baym equations for quantum fluctuations, we can describe the dynamical behavior of dipoles with thermal effects written by quantum fluctuations. Our analysis will be applied to memory formation processes in QBD. In particular, by extending our method to the case in open systems (networks), we can also trace dynamical memory recalling processes with excitations of particles in coherent domains via quantum tunneling processes, which are described by the Kadanoff–Baym equations. We can perform the simulations of the dynamical recalling processes in QBD with our equations to understand our thinking processes.

This paper is organized as follows. In Section 2, we introduce the two-particle-irreducible effective action in the closed-time path contour to describe non-equilibrium phenomena and derive time evolution equations. In Section 3, we introduce a kinetic entropy current in the first order of the gradient expansion, and show the H-theorem in the leading-order approximation of the coupling expansion. In Section 4, we show the time evolution equations, the conserved total energy and the potential energy in spatially homogeneous systems in an isolated system. In Section 5, we derive the super-radiance by analyzing the time evolution equations for coherent fields. In Section 6, we discuss our results. In Section 7, we provide the concluding remarks. In the Appendix A, we show how quantum fluctuations appear as additional terms in the Klein–Gordon equations. In this paper, the labels $i, j = 1$ and 2 represent $x$ and $y$ directions in space, the labels $a, b, c, d = 1, 2$ represent two contours in the closed-time path, the labels $\alpha = -1, 1$ represent the angular momentum of electric dipoles. The speed of light, the Planck constant divided by $2\pi$ and the Boltzmann constant are set to be 1 in this paper. We adopt the metric tensor $\eta^{\mu\nu} = \text{diag}(1, -1, -1)$ with $\mu, \nu = 0, 1, 2$.

## 2. The Two-Particle-Irreducible Effective Action and Time Evolution Equations

We begin with the following Lagrangian density to describe quantum electrodynamics (QED) with electric dipoles in $2+1$ dimensions in the background field method [52–55],

$$\mathcal{L}[\Psi^*(x,\theta),\Psi(x,\theta),A(x),a(x)] = -\frac{1}{4}F^{\mu\nu}[A+a]F_{\mu\nu}[A+a] - \frac{(\partial^\mu a_\mu)^2}{2\alpha_1}$$
$$+ \int_0^{2\pi} d\theta \left[ \Psi^* i \frac{\partial}{\partial x^0} \Psi + \frac{1}{2m} \Psi^* \nabla_i^2 \Psi \right.$$
$$\left. + \frac{1}{2I} \Psi^* \frac{\partial^2}{\partial \theta^2} \Psi - 2ed_e \Psi^* u^i \Psi F^{0i}[A+a] \right], \quad (1)$$

where $A$ is the background coherent photon fields, $a$ is the quantum fluctuations of photon fields, $F^{\mu\nu}[A] = \partial^\mu A^\nu - \partial^\nu A^\mu$ is the field strength, the $\alpha_1$ is a gauge fixing parameter, the $m$ is the mass of a dipole, the $I$ is the moment of inertia, $u^i = (\cos\theta, \sin\theta)$ is the direction of dipoles and $2ed_e$ is the absolute value of dipole vector. The variable $\theta$ represents the degrees of freedom of rotation of dipoles in $2+1$ dimensions. The dipole–photon interaction term $-2ed_e\Psi^* u^i \Psi F^{0i}[A+a]$ has the similar form to that in [27]. We shall expand the electric dipole fields $\Psi$ and $\Psi^*$ by the angular momentum and consider only the ground state and the first excited states in energy-levels. Then we can write them as,

$$\Psi(x,\theta) = \frac{1}{\sqrt{2\pi}}\left(\psi_0(x) + \psi_1(x)e^{i\theta} + \psi_{-1}(x)e^{-i\theta}\right),$$
$$\Psi^*(x,\theta) = \frac{1}{\sqrt{2\pi}}\left(\psi_0^*(x) + \psi_1^*(x)e^{-i\theta} + \psi_{-1}^*(x)e^{i\theta}\right), \quad (2)$$

in $2+1$ dimensions. (In $3+1$ dimensions, we might expand $\Psi$ and $\Psi^*$ by spherical harmonics.) We can rewrite the terms in the above Lagrangian as,

$$\int d\theta \Psi^*(x,\theta) i \frac{\partial}{\partial x^0} \Psi(x,\theta) = \psi_0^* i \frac{\partial}{\partial x^0}\psi_0 + \psi_1^* i \frac{\partial}{\partial x^0}\psi_1 + \psi_{-1}^* i \frac{\partial}{\partial x^0}\psi_{-1}, \quad (3)$$

$$\int d\theta \frac{1}{2m}\Psi^*\nabla_i^2\Psi = \frac{1}{2m}\left[\psi_0^*\nabla_i^2\psi_0 + \psi_1^*\nabla_i^2\psi_1 + \psi_{-1}^*\nabla_i^2\psi_{-1}\right], \quad (4)$$

$$\int d\theta \frac{1}{2I}\Psi^*\frac{\partial^2}{\partial\theta^2}\Psi = \frac{-1}{2I}\left[\psi_1^*\psi_1 + \psi_{-1}^*\psi_{-1}\right]. \quad (5)$$

We also write the dipole–photon interaction term with electric fields $F^{0i} = -E_i$ by,

$$\int d\theta 2ed_e \Psi^* u^i \Psi E_i = ed_e \int d\theta \left[(E_1 - iE_2)\Psi^* e^{i\theta}\Psi + (E_1 + iE_2)\Psi^* e^{-i\theta}\Psi\right]$$
$$= ed_e \left[(E_1 - iE_2)(\psi_0^*\psi_{-1} + \psi_1^*\psi_0) + (E_1 + iE_2)(\psi_0^*\psi_1 + \psi_{-1}^*\psi_0)\right], \quad (6)$$

with the direction of dipoles $u^i = (\cos\theta, \sin\theta)$.

Next, we show two-particle-irreducible (2PI) effective action [47–49] for electric dipole fields and photon fields. Starting with the above Lagrangian density, we write the generating functional with the gauge fixing condition for quantum fluctuation,

$$\text{gauge fixing}: a^0 = 0, \quad (7)$$

and perform the Legendre transformations. Then we arrive at,

$$
\begin{aligned}
\Gamma_{2PI}[A, \bar{a}^i \bar{\psi}, \bar{\psi}^*] &= \int_C d^{d+1}x \left[ -\frac{1}{4} F^{\mu\nu}[A+\bar{a}] F_{\mu\nu}[A+\bar{a}] + i\bar{\psi}_0^* \frac{\partial}{\partial x_0} \bar{\psi}_0 + \sum_{\alpha=-1,1} i\bar{\psi}_\alpha^* \frac{\partial}{\partial x_0} \bar{\psi}_\alpha \right. \\
&+ \frac{1}{2m} \left( \bar{\psi}_0^* \nabla_i^2 \bar{\psi}_0 + \sum_{\alpha=-1,1} \bar{\psi}_\alpha^* \nabla_i^2 \bar{\psi}_\alpha \right) - \frac{1}{2I} \sum_{\alpha=-1,1} \bar{\psi}_\alpha^* \bar{\psi}_\alpha \\
&+ e d_e \sum_{\alpha=-1,1} \left[ (E_1 + i\alpha E_2)(\bar{\psi}_0^* \bar{\psi}_\alpha + \bar{\psi}_{-\alpha}^* \bar{\psi}_0) \right] \\
&\left. + i \operatorname{Tr} \ln \Delta^{-1} + i \operatorname{Tr} \Delta_0^{-1} \Delta + \frac{i}{2} \operatorname{Tr} \ln D^{-1} + \frac{i}{2} \operatorname{Tr} D_0^{-1} D + \frac{\Gamma_2[\Delta, D]}{2} \right],
\end{aligned}
\tag{8}
$$

where the $C$ represents the Keldysh contour [50,51] shown in Figure 1, the spatial dimension $d = 2$, the bar represents the expectation value $\langle \cdot \rangle$ with the density matrix. The $3 \times 3$ matrix $i\Delta_0^{-1}(x,y)$ is defined as follows,

$$
\begin{aligned}
i\Delta_0^{-1}(x,y) &\equiv \left. \frac{\delta^2 \int_x \mathcal{L}}{\delta \psi^*(y) \delta \psi(x)} \right|_{a=0} \\
&= \begin{bmatrix} i\frac{\partial}{\partial x^0} + \frac{\nabla_i^2}{2m} - \frac{1}{2I} & ed_e(E_1 + iE_2) & 0 \\ ed_e(E_1 - iE_2) & i\frac{\partial}{\partial x^0} + \frac{\nabla_i^2}{2m} & ed_e(E_1 + iE_2) \\ 0 & ed_e(E_1 - iE_2) & i\frac{\partial}{\partial x^0} + \frac{\nabla_i^2}{2m} - \frac{1}{2I} \end{bmatrix} \delta_C^{d+1}(x-y),
\end{aligned}
\tag{9}
$$

for $-1, 0$ and $1$, and the $iD_{0,ij}^{-1}(x,y)$ is written by,

$$
\begin{aligned}
iD_{0,ij}^{-1}(x,y) &\equiv \frac{\delta^2 \int_x \mathcal{L}}{\delta a^i(x) \delta a^j(y)} \\
&= -\delta_{ij} \partial_x^2 \delta_C^{d+1}(x-y),
\end{aligned}
\tag{10}
$$

where $i$ and $j$ run over spatial components $1, \cdots, d = 2$ in $2+1$ dimensions. The $3 \times 3$ matrix $\Delta(x,y)$ is,

$$
\Delta(x,y) = \begin{bmatrix} \Delta_{-1-1}(x,y) & \Delta_{-10}(x,y) & \Delta_{-11}(x,y) \\ \Delta_{0-1}(x,y) & \Delta_{00}(x,y) & \Delta_{01}(x,y) \\ \Delta_{1-1}(x,y) & \Delta_{10}(x,y) & \Delta_{11}(x,y) \end{bmatrix},
\tag{11}
$$

where $\Delta_{-10}(x,y) = \langle T_C \delta \psi_{-1}(x) \delta \psi_0^*(y) \rangle$ with time-ordered product $T_C$ in the closed-time path contour. The Green's function of dipole fields $\Delta_{-10}(x,y)$ is also written by the $2 \times 2$ matrix $\Delta_{-10}^{ab}(x,y)$ with $a, b = 1, 2$ in the contour. The Green's function for photon fields $D_{ij}(x,y)$ represents,

$$
D_{ij}(x,y) = \langle T_C a_i(x) a_j(y) \rangle.
\tag{12}
$$

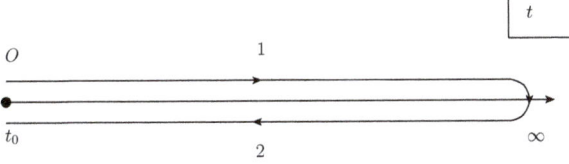

**Figure 1.** Closed-time path contour $C$. The label "1" represents the path from $t_0$ to $\infty$ and the label "2" represents the path from $\infty$ to $t_0$.

Finally we write time evolution equations for coherent fields and quantum fluctuations. The 2PI effective action satisfies the following equations,

$$\left.\frac{\delta \Gamma_{2PI}}{\delta \Delta}\right|_{\bar{a}=0} = 0, \tag{13}$$

$$\left.\frac{\delta \Gamma_{2PI}}{\delta D}\right|_{\bar{a}=0} = 0, \tag{14}$$

$$\left.\frac{\delta \Gamma_{2PI}}{\delta \bar{a}^i}\right|_{\bar{a}=0} = \left.\frac{\delta \Gamma_{2PI}}{\delta A^i}\right|_{\bar{a}=0} = 0, \tag{15}$$

$$\left.\frac{\delta \Gamma_{2PI}}{\delta \bar{\psi}^{(*)}_{-1,0,1}}\right|_{\bar{a}=0} = 0, \tag{16}$$

due to the Legendre transformation of the generating functional. Equation (13) is written by,

$$i\Delta_0^{-1} - i\Delta^{-1} - i\Sigma = 0, \tag{17}$$

with $i\Sigma \equiv -\frac{1}{2}\frac{\delta \Gamma_2}{\delta \Delta}$. The matrix of self-energy $\Sigma$ can be written by diagonal elements,

$$\Sigma = \mathrm{diag}(\Sigma_{-1-1}, \Sigma_{00}, \Sigma_{11}), \tag{18}$$

since we can neglect the off-diagonal elements which are higher order of the coupling expansion. Equation (17) represents the Kadanoff–Baym equations for electric dipole fields in the two-energy-level approximation in $2+1$ dimensions. Similarly, the Kadanoff–Baym equation for photon fields in Equation (14) is written by,

$$iD_0^{-1} - iD^{-1} - i\Pi = 0, \tag{19}$$

with $i\Pi \equiv -\frac{\delta \Gamma_2}{\delta D}$. Equation (15) is given by,

$$\partial^\nu F_{\nu i} = J_i, \tag{20}$$

with,

$$J_1(x) = -ed_e \frac{\partial}{\partial x^0} \sum_{\alpha=-1,1} \left( \Delta_{0\alpha}(x,x) + \Delta_{\alpha 0}(x,x) + \bar{\psi}_0(x)\bar{\psi}_\alpha^*(x) + \bar{\psi}_\alpha(x)\bar{\psi}_0^*(x) \right), \tag{21}$$

$$J_2(x) = -ed_e \frac{\partial}{\partial x^0} \sum_{\alpha=-1,1} \left( -i\alpha(\Delta_{0\alpha}(x,x) - \Delta_{\alpha 0}(x,x) + \bar{\psi}_0(x)\bar{\psi}_\alpha^*(x) - \bar{\psi}_\alpha(x)\bar{\psi}_0^*(x)) \right). \tag{22}$$

Equation (20) represents the Klein–Gordon equations for spatial dimensions $i=1$ and 2. Equation (16) is written by,

$$\left(i\frac{\partial}{\partial x^0} + \frac{\nabla_i^2}{2m}\right)\bar{\psi}_0 + \sum_{\alpha=-1,1} ed_e(E_1 + i\alpha E_2)\bar{\psi}_\alpha = 0, \tag{23}$$

$$\left(i\frac{\partial}{\partial x^0} + \frac{\nabla_i^2}{2m} - \frac{1}{2I}\right)\bar{\psi}_\alpha + ed_e(E_1 - i\alpha E_2)\bar{\psi}_0 = 0, \tag{24}$$

and their complex conjugates. They are Schrödinger-like equations for coherent dipole fields. Equations (23) and (24) and their complex conjugates give the following probability conservation,

$$\frac{\partial}{\partial x^0}\left(\bar{\psi}_0^*\bar{\psi}_0 + \sum_{\alpha=-1,1}\bar{\psi}_\alpha^*\bar{\psi}_\alpha\right) + \frac{1}{2mi}\nabla_i\left(\bar{\psi}_0^*\nabla_i\bar{\psi}_0 - \bar{\psi}_0\nabla_i\bar{\psi}_0^* + \sum_{\alpha=-1,1}(\bar{\psi}_\alpha^*\nabla_i\bar{\psi}_\alpha - \bar{\psi}_\alpha\nabla_i\bar{\psi}_\alpha^*)\right) = 0. \quad (25)$$

We shall define $J_0(x)$ as,

$$J_0(x) = -ed_e\frac{\partial}{\partial x^1}\sum_{\alpha=-1,1}\left(\Delta_{0\alpha}(x,x) + \Delta_{\alpha 0}(x,x) + \bar{\psi}_0(x)\bar{\psi}_\alpha^*(x) + \bar{\psi}_\alpha(x)\bar{\psi}_0^*(x)\right)$$
$$-ed_e\frac{\partial}{\partial x^2}\left(-i\alpha(\Delta_{0\alpha}(x,x) - \Delta_{\alpha 0}(x,x) + \bar{\psi}_0(x)\bar{\psi}_\alpha^*(x) - \bar{\psi}_\alpha(x)\bar{\psi}_0^*(x))\right). \quad (26)$$

Then since we can use $\partial_0 J_0 - \nabla_i J_i = 0$ with $i = 1, 2$,

$$\partial_0 J_0 = \nabla_i J_i = -\partial^i\partial^\nu F_{\nu i} = \partial^\mu\partial^\nu F_{\nu\mu} - \partial^i\partial^\nu F_{\nu i} = \partial^0\partial^\nu F_{\nu 0},$$
$$\text{or, } \partial^\nu F_{\nu 0} = J_0, \quad (27)$$

where the time dependent term in the time integral might be interpreted as an initial charge, but it is set to be zero. This equation represents the Poisson equation for scalar potential $A^0$ given by $\nabla^2 A^0 = \nabla \cdot \mu$ with the vector of dipole moments $-\mu$ on the right-hand side in Equation (26). (Since the Fourier transformed $\tilde{A}^0(\mathbf{q})$ is written by $\tilde{A}^0(\mathbf{q}) \propto (q^i\tilde{\mu}_i)/\mathbf{q}^2$ with $\mu_i = \tilde{\mu}_i\delta(\mathbf{r})$, the electric field $E_j = -\nabla_j A^0(\mathbf{r})$ is proportional to $\int_\mathbf{q} e^{i\mathbf{q}\cdot\mathbf{r}}\frac{q^jq^i\tilde{\mu}_i}{\mathbf{q}^2}$. If we can also apply the analysis in this section to the case in 3 + 1 dimensions, we find $E_j \propto \partial_j\partial_i\frac{\tilde{\mu}_i}{r}$. Then we obtain dipole–dipole interaction potential $-\tilde{\mu}_j E_j \sim \left[\frac{\tilde{\mu}_j\tilde{\mu}_j}{r^3} - \frac{3(r_i\tilde{\mu}_i)(r_j\tilde{\mu}_j)}{r^5}\right]$ in 3 + 1 dimensions.)

## 3. Kinetic Entropy Current in the Kadanoff–Baym Equations and the H-Theorem

In this section, we derive a kinetic entropy current from the Kadanoff–Baym equations with first order approximation of the gradient expansion and show the H-theorem for the leading-order approximations in the coupling expansion based on [56–58]. The analysis in this section is similar to that in open systems (the central region connected to the left and the right region) [59]. Since $(-1, 1)$ and $(1, -1)$ components in $i\Delta_0^{-1}(x, y)$ in Equation (9) are zero, the same procedures to rewrite the Kadanoff–Baym equations as those in open systems [59–63] can be adopted. We set $t_0 \to -\infty$.

First, we shall write the Kadanoff–Baym equations in Equation (17) for each components. By multiplying the matrix $\Delta$ from the right in Equation (17) and taking the $(0, 0)$ component, we can write it as,

$$i\left(\Delta_{0,00}^{-1} - \Sigma_{00}\right)\Delta_{00} + \sum_{\alpha=-1,1}ed_e(E_1 + i\alpha E_2)\Delta_{\alpha 0} = i\delta_C(x - y), \quad (28)$$

where the $(0, 0)$ component of the matrix $\Delta_0^{-1}$ represents $i\Delta_{0,00}^{-1}(x, y) = \left(i\frac{\partial}{\partial x^0} + \frac{\nabla_i^2}{2m}\right)\delta_C(x - y)$. By taking $(\alpha, 0)$ component, we can write it as,

$$i(\Delta_{0,\alpha\alpha}^{-1} - \Sigma_{\alpha\alpha})\Delta_{\alpha 0} + ed_e(E_1 - i\alpha E_2)\Delta_{00} = 0. \quad (29)$$

It is convenient to introduce the Green's functions $\Delta_{g,\alpha\alpha}$ as,

$$i\Delta_{g,\alpha\alpha}^{-1} = i\Delta_{0,\alpha\alpha}^{-1} - i\Sigma_{\alpha\alpha}. \quad (30)$$

Then by using Equations (29) and (30), we can write $\Delta_{\alpha 0}$ as,

$$\Delta_{\alpha 0}(x,y) = -\frac{ed_e}{i}\int_C dw \Delta_{g,\alpha\alpha}(x,w)(E_1(w) - i\alpha E_2(w))\Delta_{00}(w,y). \tag{31}$$

Equation (31) means the propagation from $y$ to $x$ with zero angular momentum, change of angular momentum at $w$ and the propagation from $w$ to $x$ with angular momentum $\alpha = \pm 1$. By using Equation (31), we can rewrite Equation (28) as,

$$i\int_C dw(\Delta_{0,00}^{-1}(x,w) - \Sigma_{00}(x,w))\Delta_{00}(w,y)$$
$$+i\sum_{\alpha=-1,1}(ed_e)^2 \int_C dw(E_1(x) + i\alpha E_2(x))\Delta_{g,\alpha\alpha}(x,w)(E_1(w) - i\alpha E_2(w))\Delta_{00}(w,y) = i\delta_C(x-y). \tag{32}$$

The second term on the left-hand side in Equation (32) represents the propagation from $y$ to $w$ with zero angular momentum, the change of the angular momentum to $\alpha = \pm 1$ at $w$ due to the coherent electric fields, the propagation from $w$ to $x$ and the change of the angular momentum from $\alpha = \pm 1$ to zero due to the coherent electric fields. In a similar way to $\phi^4$ theory in open systems [59], we can derive,

$$i\int_C dw \Delta_{00}(x,w)(\Delta_{0,00}^{-1}(w,y) - \Sigma_{00}(w,y))$$
$$+i\sum_{\alpha=-1,1}(ed_e)^2 \int_C dw \Delta_{00}(x,w)(E_1(w) + i\alpha E_2(w))\Delta_{g,\alpha\alpha}(w,y)(E_1(y) - i\alpha E_2(y)) = i\delta_C(x-y), \tag{33}$$

where we have used,

$$\Delta_{0\alpha}(x,y) = -\frac{1}{i}\int_C dw \Delta_{00}(x,w)(ed_e)(E_1(w) + i\alpha E_2(w))\Delta_{g,\alpha\alpha}(w,y). \tag{34}$$

The $(\alpha, \alpha)$ components of the Kadanoff–Baym equations are written by,

$$i\int_C dw \left(\Delta_{0,\alpha\alpha}^{-1}(x,w) - \Sigma_{\alpha\alpha}(x,w)\right)\Delta_{\alpha\alpha}(w,y)$$
$$+i(ed_e)^2 \int_C dw(E_1(x) - i\alpha E_2(x))\Delta_{00}(x,w)(E_1(w) + i\alpha E_2(w))\Delta_{g,\alpha\alpha}(w,y) = i\delta_C(x-y), \tag{35}$$

and,

$$i\int_C dw \Delta_{\alpha\alpha}(x,w)\left(\Delta_{0,\alpha\alpha}^{-1}(w,y) - \Sigma_{\alpha\alpha}(w,y)\right)$$
$$+i(ed_e)^2 \int_C dw \Delta_{g,\alpha\alpha}(x,w)(E_1(w) - i\alpha E_2(w))\Delta_{00}(w,x)(E_1(x) + i\alpha E_2(x)) = i\delta_C(x-y), \tag{36}$$

where we have used Equations (31) and (34).

Next, we shall perform the Fourier transformation ($\int d(x-y)e^{ip\cdot(x-y)}$) with the relative coordinate $x - y$ of the $(0,0)$ and $(\alpha, \alpha)$ components of the Kadanoff–Baym equations. We use the $2 \times 2$ matrix notation in the closed-time path with $a,b,c,d = 1,2$. Equations (32) and (33) are transformed as,

$$i\left(\Delta_{0,00}^{-1}(p) - \Sigma_{00}(X,p)\sigma_z + \sum_\alpha U_{\alpha\alpha}(X,p)\sigma_z\right)^{ac} \circ \Delta_{00}^{cb}(X,p) = i\sigma_z^{ab}, \tag{37}$$

$$i\Delta_{00}^{ac}(X,p) \circ \left(\Delta_{0,00}^{-1}(p) - \sigma_z\Sigma_{00}(X,p) + \sigma_z \sum_\alpha U_{\alpha\alpha}(X,p)\right)^{cb} = i\sigma_z^{ab}, \tag{38}$$

where $X = \frac{x+y}{2}$, $\sigma_z = \text{diag}(1,-1)$,

$$i\Delta_{0,00}^{-1}(p) = p^0 - \frac{\mathbf{p}^2}{2m}, \tag{39}$$

and the $U_{\alpha\alpha}(X,p)$ is the Fourier transformation,

$$\begin{aligned}
U_{\alpha\alpha}(X,p) &= (ed_e)^2 \int d(x-y) e^{ip\cdot(x-y)} (E_1(x) + i\alpha E_2(x)) \Delta_{g,\alpha\alpha}(x,y)(E_1(y) - i\alpha E_2(y)) \\
&= (ed_e)^2 \mathbf{E}(X)^2 \Delta_{g,\alpha\alpha}(X, p + \alpha\partial\zeta) + \left(\frac{\partial^2}{\partial X^2}\right),
\end{aligned} \tag{40}$$

with the definition of $\zeta$ and $|\mathbf{E}|$,

$$E_1(x) + i\alpha E_2(x) = |\mathbf{E}(x)| e^{i\alpha\zeta(x)}, \tag{41}$$

and,

$$(U_{\alpha\alpha}(X,p)\sigma_z)^{ac} = U_{\alpha\alpha}^{ad}(X,p)\sigma_z^{dc}, \tag{42}$$

The $\circ$ is expanded by the derivative of $X$ [64–67] as,

$$H(X,p) \circ I(X,p) = H(X,p)I(X,p) + \frac{i}{2}\{H,I\} + \left(\frac{\partial^2}{\partial X^2}\right), \tag{43}$$

with the definition of the Poisson bracket,

$$\{H,I\} \equiv \frac{\partial H}{\partial p^\mu}\frac{\partial I}{\partial X_\mu} - \frac{\partial H}{\partial X^\mu}\frac{\partial I}{\partial p_\mu}. \tag{44}$$

We find that the $U_{\alpha\alpha}$ represents the change of momenta of dipoles as shown in Figure 2a.

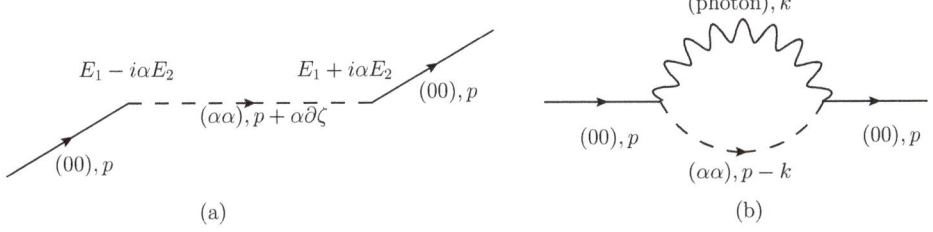

(a)           (b)

**Figure 2.** Diagrams of (a) $U_{\alpha\alpha}(X,p)$ and (b) self-energy $\Sigma_{00}(X,p)$.

In a similar way to [59], in the 0th and the first order in the gradient expansion in Equations (37) and (38), we can derive the following retarded Green's function,

$$\Delta_{00,R}(X,p) = \frac{-1}{p^0 - \frac{\mathbf{p}^2}{2m} - \Sigma_{00,R} + \sum_{\alpha=-1,1} U_{\alpha\alpha,R}}, \tag{45}$$

with the retarded parts (the subscript 'R') $\Delta_{00,R} = i(\Delta_{00}^{11} - \Delta_{00}^{12})$, $\Sigma_{00,R} = i(\Sigma_{00}^{11} - \Sigma_{00}^{12})$ and $U_{\alpha\alpha,R} = i(U_{\alpha\alpha}^{11} - U_{\alpha\alpha}^{12})$. By taking the imaginary part of the retarded Green's function $\Delta_{00,R}(X,p)$, we can derive the spectral function $\rho_{00} = i(\Delta_{00}^{21} - \Delta_{00}^{12}) = 2i\text{Im}\Delta_{00,R}(X,p)$ which represents the information of dispersion relations. Similarly, the $(\alpha,\alpha)$ components of the Kadanoff–Baym equations are written as,

$$i\left(\Delta_{0,\alpha\alpha}^{-1}(p) - \Sigma_{\alpha\alpha}(X,p)\sigma_z\right) \circ \Delta_{\alpha\alpha}(X,p) + iV_{\alpha\alpha}(X,p)\sigma_z \circ \Delta_{g,\alpha\alpha}(X,p) = i\sigma_z, \tag{46}$$

and,

$$i\Delta_{\alpha\alpha}(X,p) \circ \left(\Delta_{0,\alpha\alpha}^{-1}(p) - \sigma_z \Sigma_{\alpha\alpha}(X,p)\right) + i\Delta_{g,\alpha\alpha}(X,p) \circ \sigma_z V_{\alpha\alpha}(X,p) = i\sigma_z, \qquad (47)$$

where,

$$i\Delta_{0,\alpha\alpha}^{-1}(p) = p^0 - \frac{\mathbf{p}^2}{2m} - \frac{1}{2I}, \qquad (48)$$

and,

$$\begin{aligned}
V_{\alpha\alpha}(X,p) &= (ed_e)^2 \int d(x-y) e^{ip\cdot(x-y)} (E_1(x) - i\alpha E_2(x)) \Delta_{00}(x,y)(E_1(y) + i\alpha E_2(y)) \\
&= (ed_e)^2 \mathbf{E}(X)^2 \Delta_{00}(X, p - \alpha \partial \zeta) + \left(\frac{\partial^2}{\partial X^2}\right).
\end{aligned} \qquad (49)$$

We can also write for $\Delta_{g,\alpha\alpha}^{cb}(X,p)$ as,

$$i \left(\Delta_{0,\alpha\alpha}^{-1}(p) - \Sigma_{\alpha\alpha}(X,p)\sigma_z\right)^{ac} \circ \Delta_{g,\alpha\alpha}^{cb}(X,p) = i\sigma_z^{ab}, \qquad (50)$$

$$\Delta_{g,\alpha\alpha}^{ac}(X,p) \circ i \left(\Delta_{0,\alpha\alpha}^{-1}(p) - \sigma_z \Sigma_{\alpha\alpha}(X,p)\right)^{cb} = i\sigma_z^{ab}. \qquad (51)$$

In the 0th and the first order in the gradient expansion in Equations (46) and (47), we can derive,

$$\Delta_{\alpha\alpha,R} = \Delta_{g,\alpha\alpha,R} + \Delta_{g,\alpha\alpha,R} V_{\alpha\alpha,R} \Delta_{g,\alpha\alpha,R} \qquad (52)$$

with $\Delta_{\alpha\alpha,R} = i(\Delta_{\alpha\alpha}^{11} - \Delta_{\alpha\alpha}^{12})$ and $V_{\alpha\alpha,R} = i(V_{\alpha\alpha}^{11} - V_{\alpha\alpha}^{12})$. Here we have used the solution in the 0th and the first order in the gradient expansion in Equations (50) and (51) given by,

$$\Delta_{g,\alpha\alpha,R} = \frac{-1}{p^0 - \frac{\mathbf{p}^2}{2m} - \frac{1}{2I} - \Sigma_{\alpha\alpha,R}}, \qquad (53)$$

with $\Sigma_{\alpha\alpha,R} = i(\Sigma_{\alpha\alpha}^{11} - \Sigma_{\alpha\alpha}^{12})$. The derivation is the same as [59]. The imaginary part of the retarded Green's function $\Delta_{\alpha\alpha,R}(X,p)$ multiplied by $2i$ represents the spectral function $\rho_{\alpha\alpha} = i(\Delta_{\alpha\alpha}^{21} - \Delta_{\alpha\alpha}^{12}) = 2i\text{Im}\Delta_{\alpha\alpha,R}(X,p)$ which represents the information of dispersion relations. In addition, the Kadanoff–Baym equations for photons in Equation (19) are written by,

$$i \left(D_{0,ij}^{-1}(k) - \Pi_{ij}(X,k)\sigma_z\right)^{ac} \circ D_{jl}^{cb}(X,k) = i\delta_{il}\sigma_z^{ab}, \qquad (54)$$

$$iD_{ij}^{ac}(X,k) \circ \left(D_{0,jl}^{-1}(k) - \sigma_z \Pi_{jl}(X,k)\right)^{cb} = i\delta_{il}\sigma_z^{ab}, \qquad (55)$$

with,

$$iD_{0,ij}^{-1}(k) = k^2 \delta_{ij}. \qquad (56)$$

Next we shall derive the self-energy in the leading-order (LO) of the coupling expansion in Equation (6). The $(a,b) = (1,2)$ and $(2,1)$ component of $i\frac{\Gamma_2}{2}$ are given by,

$$\begin{aligned}
i\frac{\Gamma_{2,LO}}{2} = -\frac{1}{2}(ed_e)^2 \int du dw \sum_{\alpha=-1,1} \Big( &\Delta_{\alpha\alpha}^{21}(w,u)\Delta_{00}^{12}(u,w)(1,-\alpha i)_j \partial_u^0 \partial_w^0 \left(D_{jl}^{12}(u,w) + D_{lj}^{21}(w,u)\right)(1,\alpha i)_l^{\dagger} \\
+&\Delta_{\alpha\alpha}^{12}(w,u)\Delta_{00}^{21}(u,w)(1,-\alpha i)_j \partial_u^0 \partial_w^0 \left(D_{jl}^{21}(u,w) + D_{lj}^{12}(w,u)\right)(1,\alpha i)_l^{\dagger} \Big),
\end{aligned} \qquad (57)$$

where $t$ represents the transposition. It is convenient to rewrite,

$$D_{ij}^{ab}(k) = \left(\delta_{ij} - \frac{k_i k_j}{k^2}\right) D_T^{ab}(k) + \frac{k_i k_j}{k^2} D_L^{ab}(k), \tag{58}$$

$$\Pi_{ij}^{ab}(k) = \left(\delta_{ij} - \frac{k_i k_j}{k^2}\right) \Pi_T^{ab}(k) + \frac{k_i k_j}{k^2} \Pi_L^{ab}(k), \tag{59}$$

where $T$ and $L$ represent the transverse and the longitudinal part, respectively. The LO self-energy $i\Pi_{ji}^{21}(y,x) = -\frac{\delta \Gamma_{2,LO}}{\delta D_{ij}^{12}(x,y)}$ is,

$$i\Pi_{jl}^{21}(y,x) = -i(ed_e)^2 \sum_{\alpha=-1,1} \left( \partial_x^0 \partial_y^0 \left( \Delta_{\alpha\alpha}^{21}(y,x) \Delta_{00}^{12}(x,y) \right) (1, -\alpha i)_l (1, \alpha i)_j^t \right.$$
$$\left. + \partial_x^0 \partial_y^0 \left( \Delta_{00}^{21}(y,x) \Delta_{\alpha\alpha}^{12}(x,y) \right) (1, -\alpha i)_j (1, \alpha i)_l^t \right). \tag{60}$$

By Fourier-transforming with the relative coordinate $x - y$ and multiplying $\delta_{ij} - \frac{k_i k_j}{k^2}$ or $\frac{k_i k_j}{k^2}$, we arrive at,

$$\Pi_T^{21}(X,k) = -(ed_e)^2 \left(k^0\right)^2 \int_p \sum_{\alpha=-1,1} \left( \Delta_{\alpha\alpha}^{21}(X,k+p) \Delta_{00}^{12}(X,p) + \Delta_{00}^{21}(X,k+p) \Delta_{\alpha\alpha}^{12}(X,p) \right)$$
$$+ \left(\frac{\partial^2}{\partial X^2}\right), \tag{61}$$

$$\Pi_L^{21}(X,k) = \Pi_T^{21}(X,k), \tag{62}$$

with $\int_p = \int \frac{d^{d+1}p}{(2\pi)^{d+1}}$. The second equation is due to the spatial dimension $d = 2$. Similarly, we arrive at,

$$\Pi_T^{12}(X,k) = -(ed_e)^2 \left(k^0\right)^2 \int_p \sum_{\alpha=-1,1} \left( \Delta_{\alpha\alpha}^{12}(X,k+p) \Delta_{00}^{21}(X,p) + \Delta_{00}^{12}(X,k+p) \Delta_{\alpha\alpha}^{21}(X,p) \right)$$
$$+ \left(\frac{\partial^2}{\partial X^2}\right), \tag{63}$$

$$\Pi_L^{12}(X,k) = \Pi_T^{12}(X,k). \tag{64}$$

The Fourier transformation of the LO self-energy $i\Sigma_{00}^{12}(x,y) = -\frac{1}{2} \frac{\delta \Gamma_{2,LO}}{\delta \Delta_{00}^{21}(y,x)}$ is,

$$\Sigma_{00}^{12}(X,p) = -(ed_e)^2 \int_k \sum_{\alpha=-1,1} \left(k^0\right)^2 \Delta_{\alpha\alpha}^{12}(X,p-k) \left[D_T^{12}(X,k) + D_L^{12}(X,k)\right] + \left(\frac{\partial^2}{\partial X^2}\right). \tag{65}$$

Similarly,

$$\Sigma_{00}^{21}(X,p) = -(ed_e)^2 \int_k \sum_{\alpha=-1,1} \left(k^0\right)^2 \Delta_{\alpha\alpha}^{21}(X,p-k) \left[D_T^{21}(X,k) + D_L^{21}(X,k)\right] + \left(\frac{\partial^2}{\partial X^2}\right). \tag{66}$$

This self-energy is shown in Figure 2b. Similarly we can derive,

$$\Sigma_{\alpha\alpha}^{12}(X,p) = -(ed_e)^2 \int_k \left(k^0\right)^2 \Delta_{00}^{12}(X,p-k) \left[D_T^{12}(X,k) + D_L^{12}(X,k)\right] + \left(\frac{\partial^2}{\partial X^2}\right), \tag{67}$$

and,

$$\Sigma^{21}_{\alpha\alpha}(X,p) = -(ed_e)^2 \int_k \left(k^0\right)^2 \Delta^{21}_{00}(X, p-k)\left[D^{21}_T(X,k) + D^{21}_L(X,k)\right] + \left(\frac{\partial^2}{\partial X^2}\right). \tag{68}$$

Finally we derive a kinetic entropy current in the first order approximation in the gradient expansion and show the H-theorem in the LO approximation in the coupling expansion. By taking a difference of Equations (32) and (33), we arrive at,

$$i\left\{p^0 - \frac{\mathbf{p}^2}{2m}, \Delta^{ab}_{00}\right\} = i\left[\left(\Sigma_{00} - \sum_\alpha U_{\alpha\alpha}\right)\sigma_z \circ \Delta_{00}\right]^{ab} - i\left[\Delta_{00} \circ \sigma_z \left(\Sigma_{00} - \sum_\alpha U_{\alpha\alpha}\right)\right]^{ab}. \tag{69}$$

We use the Kadanoff–Baym Ansatz $\Delta^{12}_{00} = \frac{\rho_{00}}{i} f_{00}$, $\Delta^{21}_{00} = \frac{\rho_{00}}{i}(f_{00}+1)$, $\Sigma^{12}_{00} = \frac{\Sigma_{00,\rho}}{i}\gamma_{00}$, $\Sigma^{21}_{00} = \frac{\Sigma_{00,\rho}}{i}(\gamma_{00}+1)$, $U^{12}_{\alpha\alpha} = \frac{U_{\alpha\alpha,\rho}}{i}\gamma_{U,\alpha\alpha}$ and $U^{21}_{\alpha\alpha} = \frac{U_{\alpha\alpha,\rho}}{i}(\gamma_{U,\alpha\alpha}+1)$ with $\rho_{00} = i(\Delta^{21}_{00} - \Delta^{12}_{00}) = 2i\mathrm{Im}\Delta_{00,R}$, $\Sigma_{00,\rho} = i(\Sigma^{21}_{00} - \Sigma^{12}_{00}) = 2i\mathrm{Im}\Sigma_{00,R}$ and $U_{\alpha\alpha,\rho} = i(U^{21}_{\alpha\alpha} - U^{12}_{\alpha\alpha}) = 2i\mathrm{Im}U_{\alpha\alpha,R}$ where we just rewrite the (1,2) and the (2,1) components with the spectral parts $\rho_{00}$, $\Sigma_{00,\rho}$ and $U_{\alpha\alpha,\rho}$ and distribution functions $f_{00}$, $\gamma_{00}$ and $\gamma_{U,\alpha\alpha}$. The distribution functions $f_{00}$, $\gamma_{00}$ and $\gamma_{U,\alpha\alpha}$ approach the Bose–Einstein distributions near equilibrium states. In the first order approximation in the gradient expansion in Equation (69) for $(a,b) = (1,2)$ and $(2,1)$, we can derive,

$$f_{00} = \gamma_{00} + O\left(\frac{\partial}{\partial X}\right), \quad \text{and} \quad f_{00} = \gamma_{U,\alpha\alpha} + O\left(\frac{\partial}{\partial X}\right). \tag{70}$$

(Rewrite $(a,b) = (1,2)$ and $(2,1)$ components in Equation (69), then we can show the collision terms $\Delta^{21}_{00}\Sigma^{12}_{00} - \Delta^{12}_{00}\Sigma^{21}_{00} \propto f_{00} - \gamma_{00} = O\left(\frac{\partial}{\partial X}\right)$ and $f_{00} - \gamma_{U,\alpha\alpha} = O\left(\frac{\partial}{\partial X}\right)$.) We shall multiply $\ln \frac{i\Delta^{12}_{00}}{\rho_{00}}$ in $(a,b) = (1,2)$ component in Equation (69) and $\ln \frac{i\Delta^{21}_{00}}{\rho_{00}}$ in $(2,1)$ component in Equation (69), take the difference of them and integrate with $\int_p$. By the use of Equation (70), we arrive at,

$$\partial_\mu s^\mu_{\text{matter},00} = -\int_p \left(\Sigma^{21}_{00}(X,p)\Delta^{12}_{00}(X,p) - \Sigma^{12}_{00}(X,p)\Delta^{21}_{00}(X,p)\right)\ln\frac{\Delta^{12}_{00}(X,p)}{\Delta^{21}_{00}(X,p)}$$
$$+ \sum_\alpha \int_p \left(U^{21}_{\alpha\alpha}(X,p)\Delta^{12}_{00}(X,p) - U^{12}_{\alpha\alpha}(X,p)\Delta^{21}_{00}(X,p)\right)\ln\frac{\Delta^{12}_{00}(X,p)}{\Delta^{21}_{00}(X,p)}, \tag{71}$$

with the definition of entropy current $s^\mu_{\text{matter},00}$ for $(0,0)$ component,

$$s^\mu_{\text{matter},00} \equiv \int_p\left[\left(\delta^\mu_0 + \frac{\delta^\mu_i p^i}{m} - \frac{\partial \mathrm{Re}(\Sigma_{00,R} - \sum_\alpha U_{\alpha\alpha,R})}{\partial p_\mu}\right)\frac{\rho_{00}}{i}\right.$$
$$\left.+ \frac{\partial \mathrm{Re}\Delta_{00,R}}{\partial p_\mu}\frac{\Sigma_{00,\rho} - \sum_\alpha U_{\alpha\alpha,\rho}}{i}\right]\sigma[f_{00}], \tag{72}$$

$$\sigma[f_{00}] \equiv (1+f_{00})\ln(1+f_{00}) - f_{00}\ln f_{00}. \tag{73}$$

We can derive the Boltzmann entropy $\int_p [(1+n)\ln(1+n) - n\ln n]$ with the number density $n(X,\mathbf{p})$ in the quasi-particle limit $\mathrm{Im}U_{\alpha\alpha,R} = \mathrm{Im}\Sigma_{00,R} \to 0$ in the same way as in [58]. Similarly, we can derive a kinetic entropy current for $(\alpha\alpha)$ components. >From Equations (46) and (47), we can derive

$$i\left\{p^0 - \frac{\mathbf{p}^2}{2m} - \frac{1}{2I}, \Delta^{ab}_{\alpha\alpha}\right\} = i\left[\Sigma_{\alpha\alpha}\sigma_z \circ \Delta_{\alpha\alpha} - \Delta_{\alpha\alpha}\circ\sigma_z\Sigma_{\alpha\alpha}\right]^{ab}$$
$$-i\left[V_{\alpha\alpha}\sigma_z \circ \Delta_{g,\alpha\alpha} - \Delta_{g,\alpha\alpha}\circ\sigma_z V_{\alpha\alpha}\right]^{ab}. \tag{74}$$

We use the Kadanoff–Baym Ansatz $\Delta_{\alpha\alpha}^{12} = \frac{\rho_{\alpha\alpha}}{i} f_{\alpha\alpha}$, $\Delta_{\alpha\alpha}^{21} = \frac{\rho_{\alpha\alpha}}{i}(f_{\alpha\alpha}+1)$, $\Delta_{g,\alpha\alpha}^{12} = \frac{\Delta_{g,\alpha\alpha,\rho}}{i}\gamma_{g,\alpha\alpha}$, $\Delta_{g,\alpha\alpha}^{21} = \frac{\Delta_{g,\alpha\alpha,\rho}}{i}(\gamma_{g,\alpha\alpha}+1)$, $\Sigma_{\alpha\alpha}^{12} = \frac{\Sigma_{\alpha\alpha,\rho}}{i}\gamma_{\alpha\alpha}$, $\Sigma_{\alpha\alpha}^{21} = \frac{\Sigma_{\alpha\alpha,\rho}}{i}(\gamma_{\alpha\alpha}+1)$, $V_{\alpha\alpha}^{12} = \frac{V_{\alpha\alpha,\rho}}{i}\gamma_{V,\alpha\alpha}$ and $V_{\alpha\alpha}^{21} = \frac{V_{\alpha\alpha,\rho}}{i}(\gamma_{V,\alpha\alpha}+1)$ with $\rho_{\alpha\alpha} = i(\Delta_{\alpha\alpha}^{21} - \Delta_{\alpha\alpha}^{12}) = 2i\mathrm{Im}\Delta_{\alpha\alpha,R}$, $\Sigma_{\alpha\alpha,\rho} = i(\Sigma_{\alpha\alpha}^{21} - \Sigma_{\alpha\alpha}^{12}) = 2i\mathrm{Im}\Sigma_{\alpha\alpha,R}$ and $V_{\alpha\alpha,\rho} = i(V_{\alpha\alpha}^{21} - V_{\alpha\alpha}^{12}) = 2i\mathrm{Im}V_{\alpha\alpha,R}$. In Equation (74), we can show,

$$f_{\alpha\alpha} \sim \gamma_{\alpha\alpha}, \quad \gamma_{g,\alpha\alpha} \sim \gamma_{V,\alpha\alpha}, \tag{75}$$

for distribution functions $f_{\alpha\alpha}$, $\gamma_{\alpha\alpha}$ and $\gamma_{V,\alpha\alpha}$ by writing the $(a,b)=(1,2)$ and $(2,1)$ components in the Kadanoff–Baym equations (74). We can also show,

$$\gamma_{\alpha\alpha} \sim \gamma_{g,\alpha\alpha}, \tag{76}$$

from Equations (50) and (51). We shall multiply $\ln \frac{i\Delta_{\alpha\alpha}^{12}}{\rho_{\alpha\alpha}}$ in $(a,b)=(1,2)$ component in Equation (74) and $\ln \frac{i\Delta_{\alpha\alpha}^{21}}{\rho_{\alpha\alpha}}$ in $(2,1)$ component in Equation (74), take the difference of them and integrate with $\int_p$. By using Equations (75) and (76), we arrive at,

$$\begin{aligned}\partial_\mu s^\mu_{\mathrm{matter},\alpha\alpha} &= -\int_p \left(\Sigma_{\alpha\alpha}^{21}(X,p)\Delta_{\alpha\alpha}^{12}(X,p) - \Sigma_{\alpha\alpha}^{12}(X,p)\Delta_{\alpha\alpha}^{21}(X,p)\right)\ln\frac{\Delta_{\alpha\alpha}^{12}(X,p)}{\Delta_{\alpha\alpha}^{21}(X,p)} \\ &+ \int_p \left(V_{\alpha\alpha}^{21}(X,p)\Delta_{g,\alpha\alpha}^{12}(X,p) - V_{\alpha\alpha}^{12}(X,p)\Delta_{g,\alpha\alpha}^{21}(X,p)\right)\ln\frac{\Delta_{\alpha\alpha}^{12}(X,p)}{\Delta_{\alpha\alpha}^{21}(X,p)},\end{aligned} \tag{77}$$

with the definitions of entropy current $s^\mu_{\mathrm{matter},\alpha\alpha}$ for $(\alpha\alpha)$ components,

$$\begin{aligned}s^\mu_{\mathrm{matter},\alpha\alpha} &\equiv \int_p \left[\left(\delta_0^\mu + \frac{\delta_i^\mu p^i}{m} - \frac{\partial\mathrm{Re}\Sigma_{\alpha\alpha,R}}{\partial p_\mu}\right)\frac{\rho_{\alpha\alpha}}{i} + \frac{\partial\mathrm{Re}\Delta_{\alpha\alpha,R}}{\partial p_\mu}\frac{\Sigma_{\alpha\alpha,\rho}}{i} \right. \\ &\left. + \frac{\partial\mathrm{Re}V_{\alpha\alpha,R}}{\partial p_\mu}\frac{\Delta_{g,\alpha\alpha,\rho}}{i} - \frac{\partial\mathrm{Re}\Delta_{g,\alpha\alpha,R}}{\partial p_\mu}\frac{V_{\alpha\alpha,\rho}}{i}\right]\sigma[f_{\alpha\alpha}].\end{aligned} \tag{78}$$

In this derivation, we have used the same way as that in open systems in [59]. We can also derive the following equations for the Kadanoff–Baym equations for photons with the Kadanoff–Baym Ansatz $D_T^{21} = \frac{\rho_T}{i}(1+f_T)$, $D_T^{12} = \frac{\rho_T}{i}f_T$, $D_L^{21} = \frac{\rho_L}{i}(1+f_L)$ and $D_L^{12} = \frac{\rho_L}{i}f_L$ with distribution functions $f_T$ and $f_L$ and spectral functions $\rho_T$ and $\rho_L$,

$$\begin{aligned}\partial_\mu s^\mu_{\mathrm{photon}} &= -\frac{1}{2}\int_k \left[\Pi_T^{21}(X,k)D_T^{12}(X,k) - \Pi_T^{12}(X,k)D_T^{21}(X,k)\right]\ln\frac{D_T^{12}(X,k)}{D_T^{21}(X,k)} \\ &- \frac{1}{2}\int_k \left[\Pi_L^{21}(X,k)D_L^{12}(X,k) - \Pi_L^{12}(X,k)D_L^{21}(X,k)\right]\ln\frac{D_L^{12}(X,k)}{D_L^{21}(X,k)},\end{aligned} \tag{79}$$

with the entropy current for photons,

$$\begin{aligned}s^\mu_{\mathrm{photon}} &\equiv \int_k \left[\left(k^\mu - \frac{1}{2}\frac{\partial\mathrm{Re}\Pi_{T,R}}{\partial k_\mu}\right)\frac{D_{T,\rho}}{i} + \frac{1}{2}\frac{\partial\mathrm{Re}D_{T,R}}{\partial k_\mu}\frac{\Pi_{T,\rho}}{i}\right]\sigma[f_T] \\ &+ \int_k \left[\left(k^\mu - \frac{1}{2}\frac{\partial\mathrm{Re}\Pi_{L,R}}{\partial k_\mu}\right)\frac{D_{L,\rho}}{i} + \frac{1}{2}\frac{\partial\mathrm{Re}D_{L,R}}{\partial k_\mu}\frac{\Pi_{L,\rho}}{i}\right]\sigma[f_L].\end{aligned} \tag{80}$$

As a result, the total entropy current $s^\mu = s^\mu_{\text{matter},00} + \sum_\alpha s^\mu_{\text{matter},\alpha\alpha} + s^\mu_{\text{photon}}$ satisfies,

$$\begin{aligned}
\partial_\mu s^\mu &= (ed_e)^2 \int_{p,k} \left(k^0\right)^2 \sum_\alpha \left[\Delta^{21}_{\alpha\alpha}(p-k)\Delta^{12}_{00}(p)D^{21}_T(k) - \Delta^{12}_{\alpha\alpha}(p-k)\Delta^{21}_{00}(p)D^{12}_T(k)\right] \\
&\quad \times \ln \frac{\Delta^{21}_{\alpha\alpha}(p-k)\Delta^{12}_{00}(p)D^{21}_T(k)}{\Delta^{12}_{\alpha\alpha}(p-k)\Delta^{21}_{00}(p)D^{12}_T(k)} \\
&\quad + (ed_e)^2 \int_{p,k} \left(k^0\right)^2 \sum_\alpha \left[\Delta^{21}_{\alpha\alpha}(p-k)\Delta^{12}_{00}(p)D^{21}_L(k) - \Delta^{12}_{\alpha\alpha}(p-k)\Delta^{21}_{00}(p)D^{12}_L(k)\right] \\
&\quad \times \ln \frac{\Delta^{21}_{\alpha\alpha}(p-k)\Delta^{12}_{00}(p)D^{21}_L(k)}{\Delta^{12}_{\alpha\alpha}(p-k)\Delta^{21}_{00}(p)D^{12}_L(k)} \\
&\quad + (ed_e)^2 (\mathbf{E}(X))^2 \sum_\alpha \int_p \left(\Delta^{21}_{g,\alpha\alpha}(p+\alpha\partial\zeta)\Delta^{12}_{00}(p) - \Delta^{12}_{g,\alpha\alpha}(p+\alpha\partial\zeta)\Delta^{21}_{00}\right) \\
&\quad \times \ln \frac{\Delta^{21}_{g,\alpha\alpha}(p+\alpha\partial\zeta)\Delta^{12}_{00}(p)}{\Delta^{12}_{g,\alpha\alpha}(p+\alpha\partial\zeta)\Delta^{21}_{00}(p)} \geq 0,
\end{aligned} \qquad (81)$$

where we have used the inequality $(x-y)\ln\frac{x}{y} \geq 0$ for real variables $x$ and $y$ with $x > 0$ and $y > 0$. The equality is satisfied in $f_{00} = f_{\alpha\alpha} = f_T = f_L = \frac{1}{e^{p^0/T}-1}$. Here we have used $\frac{\Delta^{21}_{\alpha\alpha}}{\Delta^{12}_{\alpha\alpha}} \sim \frac{\Delta^{21}_{g,\alpha\alpha}}{\Delta^{12}_{g,\alpha\alpha}}$ with $\gamma_{g,\alpha\alpha} \sim f_{\alpha\alpha}$ in first order in the gradient expansion. We have shown the H-theorem in the LO approximation in the coupling expansion and in the first order approximation in the gradient expansion. There is no violation in the second law in thermodynamics in the dynamics.

## 4. Time Evolution Equations in Spatially Homogeneous Systems and Conserved Energy

In this section, we write time evolution equations in spatially homogeneous systems and show a concrete form of the conserved energy density.

It is convenient to introduce the statistical functions $F_{00} = \frac{\Delta^{21}_{00}+\Delta^{12}_{00}}{2}$, $F_{\alpha\alpha} = \frac{\Delta^{21}_{\alpha\alpha}+\Delta^{12}_{\alpha\alpha}}{2}$, $F_T = \frac{D^{21}_T+D^{12}_T}{2}$, $F_L = \frac{D^{21}_L+D^{12}_L}{2}$, which represent the information of how many particles are occupied in $(p^0, \mathbf{p})$ (particle distributions) and statistical parts, $U_{\alpha\alpha,F} = \frac{U^{21}_{\alpha\alpha}+U^{12}_{\alpha\alpha}}{2}$, $V_{\alpha\alpha,F} = \frac{V^{21}_{\alpha\alpha}+V^{12}_{\alpha\alpha}}{2}$, $\Delta_{g,\alpha\alpha,F} = \frac{\Delta^{21}_{g,\alpha\alpha}+\Delta^{12}_{g,\alpha\alpha}}{2}$, $\Sigma_{00,F} = \frac{\Sigma^{21}_{00}+\Sigma^{12}_{00}}{2}$, $\Sigma_{\alpha\alpha,F} = \frac{\Sigma^{21}_{\alpha\alpha}+\Sigma^{12}_{\alpha\alpha}}{2}$, $\Pi_{T,F} = \frac{\Pi^{21}_T+\Pi^{12}_T}{2}$ and $\Pi_{L,F} = \frac{\Pi^{21}_L+\Pi^{12}_L}{2}$. The variables of these functions are $(X^0, p^0, \mathbf{p})$ with the center-of-mass coordinate $X^0 = \frac{x^0+y^0}{2}$ and $p$ given by the Fourier transformation with the relative coordinate $x - y$ in variables $(x, y)$ in Green's functions and self-energy in Section 2. The statistical functions and parts are real at any time when we start with real statistical functions at initial time. The spectral functions are given by taking the difference of $(2, 1)$ and $(1, 2)$ components multiplied by $i$, namely $\rho_{00} = i(\Delta^{21}_{00} - \Delta^{12}_{00})$. They represent the information of which states can be occupied by particles in $(p^0, \mathbf{p})$ (dispersion relations). The spectral parts in self-energy are given by taking the difference of $(2, 1)$ and $(1, 2)$ components multiplied by $i$ (and written by the subscript $\rho$), namely $\Delta_{g,\alpha\alpha,\rho} = i(\Delta^{21}_{g,\alpha\alpha} - \Delta^{12}_{g,\alpha\alpha})$, $\Sigma_{00,\rho} = i(\Sigma^{21}_{00} - \Sigma^{12}_{00})$ and so on. The spectral functions and parts are pure imaginary at any time when we start with pure imaginary spectral functions at initial time. We can use the real statistical parts labeled by the subscripts $F$ and the pure imaginary spectral parts labeled by the subscript $\rho$ in self-energy in the time evolution. We use the subscript 'R', 'F' and '$\rho$' to represent the retarded, statistical and spectral parts in self-energy, respectively.

The Kadanoff–Baym equation for the statistical and spectral functions are given by,

$$\left\{ p^0 - \frac{\mathbf{p}^2}{2m} - \text{Re}\Sigma_{00,R} + \sum_{\alpha=-1,1} \text{Re}U_{\alpha\alpha,R}, F_{00} \right\} + \left\{ \text{Re}\Delta_{00,R}, \Sigma_{00,F} - \sum_\alpha U_{\alpha\alpha,F} \right\}$$
$$= \frac{1}{i}(F_{00}\Sigma_{00,\rho} - \rho_{00}\Sigma_{00,F}) - \frac{1}{i}\sum_\alpha (F_{00}U_{\alpha\alpha,\rho} - \rho_{00}U_{\alpha\alpha,F}), \qquad (82)$$

$$\left\{p^0 - \frac{\mathbf{p}^2}{2m} - \text{Re}\Sigma_{00,R} + \sum_{\alpha=-1,1} \text{Re}U_{\alpha\alpha,R}, \rho_{00}\right\} + \left\{\text{Re}\Delta_{00,R}, \Sigma_{00,\rho} - \sum_{\alpha} U_{\alpha\alpha,\rho}\right\} = 0, \quad (83)$$

$$\left\{p^0 - \frac{\mathbf{p}^2}{2m} - \frac{1}{2I} - \text{Re}\Sigma_{\alpha\alpha,R}, F_{\alpha\alpha}\right\} + \left\{\text{Re}\Delta_{\alpha\alpha,R}, \Sigma_{\alpha\alpha,F}\right\} + \left\{\text{Re}V_{\alpha\alpha,R}, \Delta_{g,\alpha\alpha,F}\right\} - \left\{\text{Re}\Delta_{g,\alpha\alpha,R}, V_{\alpha\alpha,F}\right\}$$
$$= \frac{1}{i}\left(F_{\alpha\alpha}\Sigma_{\alpha\alpha,\rho} - \rho_{\alpha\alpha}\Sigma_{\alpha\alpha,F}\right) - \frac{1}{i}\left(\Delta_{g,\alpha\alpha,F}V_{\alpha\alpha,\rho} - \Delta_{g,\alpha\alpha,\rho}V_{\alpha\alpha,F}\right), \quad (84)$$

$$\left\{p^0 - \frac{\mathbf{p}^2}{2m} - \frac{1}{2I} - \text{Re}\Sigma_{\alpha\alpha,R}, \rho_{\alpha\alpha}\right\} + \left\{\text{Re}\Delta_{\alpha\alpha,R}, \Sigma_{\alpha\alpha,\rho}\right\}$$
$$+ \left\{\text{Re}V_{\alpha\alpha,R}, \Delta_{g,\alpha\alpha,\rho}\right\} - \left\{\text{Re}\Delta_{g,\alpha\alpha,R}, V_{\alpha\alpha,\rho}\right\} = 0, \quad (85)$$

$$\left\{p^0 - \frac{\mathbf{p}^2}{2m} - \frac{1}{2I} - \text{Re}\Sigma_{\alpha\alpha,R}, \Delta_{g,\alpha\alpha,F}\right\} + \left\{\text{Re}\Delta_{g,\alpha\alpha,R}, \Sigma_{\alpha\alpha,F}\right\}$$
$$= \frac{1}{i}\left(\Delta_{g,\alpha\alpha,F}\Sigma_{\alpha\alpha,\rho} - \Delta_{g,\alpha\alpha,\rho}\Sigma_{\alpha\alpha,F}\right), \quad (86)$$

$$\left\{p^0 - \frac{\mathbf{p}^2}{2m} - \frac{1}{2I} - \text{Re}\Sigma_{\alpha\alpha,R}, \Delta_{g,\alpha\alpha,\rho}\right\} + \left\{\text{Re}\Delta_{g,\alpha\alpha,R}, \Sigma_{\alpha\alpha,\rho}\right\} = 0, \quad (87)$$

$$\left\{p^2 - \text{Re}\Pi_{R,T}, F_T\right\} + \left\{\text{Re}D_{R,T}, \Pi_{F,T}\right\} = \frac{1}{i}\left(F_T\Pi_{\rho,T} - \rho_T\Pi_{F,T}\right), \quad (88)$$

$$\left\{p^2 - \text{Re}\Pi_{R,T}, \rho_T\right\} + \left\{\text{Re}D_{R,T}, \Pi_{\rho,T}\right\} = 0, \quad (89)$$

and longitudinal parts given by changing the label $T$ to $L$ in Equations (88) and (89). We can use Equation (69) in the previous section to derive Equations (82) and (83), for example.

We can write,

$$U_{\alpha\alpha,F}(X,p) = (ed_e)^2 \mathbf{E}(X)^2 \Delta_{g,\alpha\alpha,F}(p + \alpha\partial\zeta), \quad U_{\alpha\alpha,\rho}(X,p) = (ed_e)^2 \mathbf{E}(X)^2 \Delta_{g,\alpha\alpha,\rho}(p + \alpha\partial\zeta), \quad (90)$$
$$V_{\alpha\alpha,F}(X,p) = (ed_e)^2 \mathbf{E}(X)^2 F_{00}(p - \alpha\partial\zeta), \quad V_{\alpha\alpha,\rho}(X,p) = (ed_e)^2 \mathbf{E}(X)^2 \rho_{00}(p - \alpha\partial\zeta). \quad (91)$$

In case we start with initial condition $E_2(X^0 = 0) = 0$, $\partial_0 E_2(X^0 = 0) = 0$ and symmetric Green's functions for $\alpha \to -\alpha$ in spatially homogeneous systems, we can use $\partial\zeta = 0$ in the above equations at any time. We can write the self-energy as,

$$\Sigma_{00,F}(p) = -(ed_e)^2 \sum_{\alpha=-1,1} \int_k \left(k^0\right)^2 \left[F_{\alpha\alpha}(p-k)(F_T(k) + F_L(k)) + \frac{1}{4}\frac{\rho_{\alpha\alpha}(p-k)}{i}\frac{\rho_T(k) + \rho_L(k)}{i}\right], \quad (92)$$

$$\Sigma_{00,\rho}(p) = -(ed_e)^2 \sum_{\alpha=-1,1} \int_k \left(k^0\right)^2 \left[F_{\alpha\alpha}(p-k)(\rho_T(k) + \rho_L(k)) + \rho_{\alpha\alpha}(p-k)(F_T(k) + F_L(k))\right], \quad (93)$$

$$\Sigma_{\alpha\alpha,F}(p) = -(ed_e)^2 \int_k \left(k^0\right)^2 \left[F_{00}(p-k)(F_T(k) + F_L(k)) + \frac{1}{4}\frac{\rho_{00}(p-k)}{i}\frac{\rho_T(k) + \rho_L(k)}{i}\right], \quad (94)$$

$$\Sigma_{\alpha\alpha,\rho}(p) = -(ed_e)^2 \int_k \left(k^0\right)^2 \left[F_{00}(p-k)(\rho_T(k) + \rho_L(k)) + \rho_{00}(p-k)(F_T(k) + F_L(k))\right], \quad (95)$$

$$\Pi_{T,F}(k) = \Pi_{L,F}(k) = -(ed_e)^2 \left(k^0\right)^2 \sum_{\alpha=-1,1} \int_p \left[ F_{\alpha\alpha}(k+p)F_{00}(p) - \frac{1}{4}\frac{\rho_{\alpha\alpha}(k+p)}{i}\frac{\rho_{00}(p)}{i} \right.$$
$$\left. + F_{00}(k+p)F_{\alpha\alpha}(p) - \frac{1}{4}\frac{\rho_{00}(k+p)}{i}\frac{\rho_{\alpha\alpha}(p)}{i} \right], \tag{96}$$

$$\Pi_{T,\rho}(k) = \Pi_{L,\rho}(k) = -(ed_e)^2 \left(k^0\right)^2 \sum_{\alpha=-1,1} \int_p \left[ \rho_{\alpha\alpha}(k+p)F_{00}(p) - F_{\alpha\alpha}(k+p)\rho_{00}(p) \right.$$
$$\left. + \rho_{00}(k+p)F_{\alpha\alpha}(p) - F_{00}(k+p)\rho_{\alpha\alpha}(p) \right], \tag{97}$$

where we have omitted the label of the center-of-mass cordinate $X$ in Green's functions and self-energy. We find that the $\Pi_{T,F}(k) = \Pi_{L,F}(k)$ are symmetric ($\Pi_{T,F}(-k) = \Pi_{T,F}(k)$) under $k \to -k$ and that $\Pi_{T,\rho} = \Pi_{L,\rho}$ are anti-symmetric ($\Pi_{T,\rho}(-k) = -\Pi_{T,\rho}(k)$) under $k \to -k$, for any Green's functions for dipole fields. When we prepare initial conditions with symmetric $F_{T,L}$ and anti-symmetric $\rho_{T,L}$ for photons, we can derive symmetric $F_{T,L}$ and anti-symmetric $\rho_{T,L}$ at any time. In addition, since $\Pi(k)$'s are proportional to $(k^0)^2$, there is no mass gap for incoherent photons for the leading-order self-energy in the coupling expansion. The velocity of gapless modes of incoherent photons will decrease when we increase the density of dipoles in this theory.

Finally, we show the energy density $E_{tot}$. In the spatially homogeneous system in the $2+1$ dimensions, we can derive $\frac{\partial E_{tot}}{\partial X^0} = 0$ with the energy density given by,

$$E_{tot} \equiv \frac{1}{2I}\sum_{\alpha=-1,1}\bar{\psi}_\alpha^*\bar{\psi}_\alpha + \frac{1}{2}(\partial_0 A_i)^2 + \int_p p^0 \left( F_{00} + \sum_{\alpha=-1,1} F_{\alpha\alpha} \right) + \frac{1}{2}\int_p \left(p^0\right)^2 (F_T + F_L)$$
$$+ 2(ed_e)^2 \mathbf{E}^2 \sum_{\alpha=-1,1}\int_p \left( F_{00}(p)\mathrm{Re}\Delta_{g,\alpha\alpha,R}(p+\alpha\partial\zeta) + \mathrm{Re}\Delta_{00,R}(p)\Delta_{g,\alpha\alpha,F}(p+\alpha\partial\zeta) \right)$$
$$- \int_p (\mathrm{Re}\Sigma_{00,R}F_{00} + \mathrm{Re}\Delta_{00,R}\Sigma_{00,F}) - \sum_{\alpha=-1,1}\int_p (\mathrm{Re}\Sigma_{\alpha\alpha,R}F_{\alpha\alpha} + \mathrm{Re}\Delta_{\alpha\alpha,R}\Sigma_{\alpha\alpha,F})$$
$$- \frac{1}{2}\int_p (\mathrm{Re}\Pi_{R,T}F_T + \mathrm{Re}D_{R,T}\Pi_{F,T} + \mathrm{Re}\Pi_{R,L}F_L + \mathrm{Re}D_{R,L}\Pi_{F,L}), \tag{98}$$

where we have used the KB equations in this section, the Klein–Gordon Equation (20) and the Schödinger-like Equations (23) and (24) in Section 2. The first term represents the contribution of nonzero angular momenta for coherent dipole fields. The second term represents the contribution by electric fields $E_i = \partial_0 A_i$. The third and the fourth terms represent the contribution by quantum fluctuations for dipoles and photons, respectively. When the temperature is nonzero $T \neq 0$ at equilibrium states and the spectral width in the spectral functions is small enough, statistical functions which are proportional to the Bose–Einstein distributions $\frac{1}{e^{p^0/T}-1}$ give temperature-dependent terms $mT^2$ for dipole fields and $\propto T^3$ for photon fields in $2+1$ dimensions. The fifth term represents the potential energy in processes in Figure 2a. The sixth, seventh and eighth terms represent the potential energy in processes in Figure 2b. The coefficients in the sixth and seventh terms are not $\frac{1}{3}$ but 1. While the factor 1 might look like a contradiction with the preceding research in [68,69] which suggest that the factor $\frac{1}{3}$ appears in the interaction with 3-point-vertex, the factor 1 appears due to time derivative $(\partial^0)^2$ in self-energy for dipole fields and photon fields.

## 5. Dynamics of Coherent Fields

In this section, we show that our Lagrangian describes the super-radiance phenomena in time evolution equations of coherent fields. We shall assume that all the coherent fields are independent of $x^1$ (dependent on $x^0$ and $x^2$). We also assume the symmetry for $\alpha = -1$ and $\alpha = 1$, namely $\bar{\psi}_1^{(*)} = \bar{\psi}_{-1}^{(*)}$, $\Delta_{01} = \Delta_{0-1}$, and $\Delta_{10} = \Delta_{-10}$. We set initial conditions $E_2 = 0$ and $\partial_0 E_2 = 0$ at $x^0 = 0$.

We define $Z \equiv 2|\bar{\psi}_1|^2 - |\bar{\psi}_0|^2$. It is possible to derive the following equations from time evolution Equations (20), (23) and (24) with their complex conjugates for background coherent fields in Section 2.

$$\partial_0 Z = i4ed_e E_1 (\bar{\psi}_1^* \bar{\psi}_0 - \bar{\psi}_0^* \bar{\psi}_1), \tag{99}$$

$$\partial_0 (\bar{\psi}_1^* \bar{\psi}_0) = \frac{i}{2I} \bar{\psi}_1^* \bar{\psi}_0 + ied_e E_1 Z \tag{100}$$

$$\left[(\partial_0)^2 - (\partial_2)^2\right] E_1 = -2ed_e (\partial_0)^2 \left[\bar{\psi}_1^* \bar{\psi}_0 + \bar{\psi}_0^* \bar{\psi}_1 + \Delta_{01}(x,x) + \Delta_{10}(x,x)\right]. \tag{101}$$

We have used moderately varying spatial dependence $|\nabla_i^2 \bar{\psi}_{-1,0,1}/m| \ll |\partial_0 \bar{\psi}_{-1,0,1}|$. We derive aspects of the super-radiance and the Higgs mechanism in the above three equations.

## 5.1. Super-Radiance

In this section, we show the super-radiance in time evolution equations for coherent fields with the rotating wave approximations neglecting non-resonant terms and quantum fluctuations. We have used the derivations in [70,71] for background coherent fields.

We shall consider only $k^0 = \frac{1}{2I}$ in this section and we expand the electric field $E_1$ and the transition rate $\bar{\psi}_0 \bar{\psi}_1^*$ as,

$$E_1(x^0, x^2) = \frac{1}{2} \epsilon(x^0, x^2) e^{-i(k^0 x^0 - k^0 x^2)} + \frac{1}{2} \epsilon^*(x^0, x^2) e^{i(k^0 x^0 - k^0 x^2)}, \tag{102}$$

$$\bar{\psi}_1 \bar{\psi}_0^* = \frac{1}{2} R(x^0, x^2) e^{-i(k^0 x^0 - k^0 x^2)}, \tag{103}$$

We consider the following case,

$$|\partial_0 \epsilon| \ll |k^0 \epsilon|, \quad |\partial_0 R| \ll |k^0 R|,$$
$$|\partial_2 \epsilon| \ll |k^0 \epsilon|. \tag{104}$$

Neglect non-resonant terms like $e^{\pm 2ik^0 x^0}$ and quantum fluctuations (Green's functions $\Delta_{01}$ and $\Delta_{10}$) (the rotating wave approximation). Then from Equations (99)–(101), we arrive at the Maxwell–Bloch equations,

$$\frac{\partial \epsilon}{\partial x^0} + \frac{\partial \epsilon}{\partial x^2} = ied_e k^0 R, \tag{105}$$

$$\frac{\partial Z}{\partial x^0} = ied_e(\epsilon R^* - \epsilon^* R), \tag{106}$$

$$\frac{\partial R}{\partial x^0} = -ied_e \epsilon Z. \tag{107}$$

We assume that $\epsilon$, $Z$ and $R$ are independent of the spatial coordinate of the $x^2$ direction. We shall change $\epsilon \to i\epsilon$ in the above equations and assume real functions $R = R^*$ and $\epsilon = \epsilon^*$. Then we can write,

$$\frac{\partial \epsilon}{\partial x^0} = ed_e k^0 R, \tag{108}$$

$$\frac{\partial Z}{\partial x^0} = -2ed_e \epsilon R, \tag{109}$$

$$\frac{\partial R}{\partial x^0} = ed_e \epsilon Z. \tag{110}$$

We find the conservation law with the definition $B^2 \equiv 2R^2 + Z^2$,

$$\frac{\partial}{\partial x^0} B^2 = \frac{\partial}{\partial x^0} \left(2R^2 + Z^2\right) = 0. \tag{111}$$

The relation $\frac{\partial B}{\partial x^0} = 0$ represents the probability conservation since we can rewrite $B^2 = (2|\bar{\psi}_1|^2 + |\bar{\psi}_0|^2)^2$ by Equation (103) and $Z \equiv 2|\bar{\psi}_1|^2 - |\bar{\psi}_0|^2$. We also find the following conservation law,

$$\frac{\partial}{\partial x^0}\left[\frac{1}{2}\epsilon^2 + \frac{1}{2}k^0 Z\right] = 0, \tag{112}$$

which represents the energy conservation. By this relation, we might be able to estimate the maximum energy density of electric fields,

$$\left(\frac{1}{2}\epsilon^2\right)_{\max} = -\frac{1}{2}k^0 Z_{\min} = \frac{1}{2}k^0 B, \tag{113}$$

in case there is no external energy supply. We derive the following solutions in Equations (108)–(110),

$$R(x^0) = \frac{1}{\sqrt{2}} B \sin\theta(x^0), \quad Z(x^0) = B \cos\theta(x^0), \tag{114}$$

$$\theta(x^0) = \theta_0 + \sqrt{2}ed_e \int_0^{x^0} dx'^0 \epsilon(x'^0), \tag{115}$$

with $\frac{\partial \theta}{\partial x^0} = \sqrt{2}ed_e \epsilon$ and the constant $B$ in a similar way to [71]. The $\theta(x^0)$ swings around the position $\theta = \pi$ with the frequency $\Omega = ed_e\sqrt{k^0 B}$ in case we start with initial conditions at around $\theta_0 \sim \pi$ ($|\bar{\psi}_1|^2 = 0$), since we can rewrite Equation (108) as

$$\frac{\partial^2 \theta(x^0)}{\partial (x^0)^2} = (ed_e)^2 k^0 B \sin\theta(x^0). \tag{116}$$

The $B$ is the order of the number density of dipoles.

We introduce the damping term $\frac{1}{L}\epsilon$ for the release of radiation and the propagation length $L$ in Equation (108). We can write,

$$\frac{\partial \epsilon}{\partial x^0} + \frac{1}{L}\epsilon = \frac{ed_e k^0}{\sqrt{2}} B \sin\theta(x^0). \tag{117}$$

In $\kappa = \frac{1}{L} \gg$ time derivative, we can neglect the first term in the above equations, then

$$\frac{\partial \theta}{\partial x^0} = \frac{(ed_e)^2 k^0 B}{\kappa} \sin\theta(x^0). \tag{118}$$

The solution is,

$$\theta(x^0) = 2\tan^{-1}\left[\exp\left(\frac{(ed_e)^2 k^0 B x^0}{\kappa}\right)\tan\frac{\theta_0}{2}\right], \tag{119}$$

and,

$$\epsilon = \frac{1}{\sqrt{2}ed_e \tau_R} \times \left[\cosh\left(\frac{x^0 - \tau_0}{\tau_R}\right)\right]^{-1} \tag{120}$$

with $\tau_R = \frac{\kappa}{(ed_e)^2 k^0 B}$ and $\tau_0 = -\tau_R \ln(\tan\frac{\theta_0}{2})$. The $\tau_R \propto 1/B \sim 1/N$ with the number of dipoles $N$ represents the relaxation time of electric fields in the super-radiance. When $N$ dipoles decay within time scales $1/N$, the intensity of electric fields becomes the order $N^2$ (super-radiant decay with correlation among dipoles), not $N$ (spontaneous decay without correlation among dipoles).

## 5.2. Higgs Mechanism and Tachyonic Instability

In this section, we rewrite time evolution equations for coherent fields with only real functions. We assume the spatially homogeneous case. We do not adopt the rotating wave approximation in this section. We show how coherent electric fields $E_1$ are affected by $Z = 2|\bar{\psi}_1|^2 - |\bar{\psi}_0|^2$.

In Equation (101), the second derivatives of coherent fields on the right-hand side are written by,

$$\frac{ed_e}{2I^2}\left(\bar{\psi}_1^*\bar{\psi} + \bar{\psi}_0^*\bar{\psi}_1\right) + \frac{2(ed_e)^2 Z}{I}E_1,$$

where we have used Equation (100). As a result, we arrive at,

$$\left[(\partial_0)^2 - (\partial_2)^2 - \frac{2(ed_e)^2 Z}{I}\right]E_1 = \frac{\mu_1}{4I^2} + \frac{2(ed_e)^2 E_1}{I}\int_p \left(2F_{11}(X,p) - F_{00}(X,p) - \Delta_{g,11,F}(X,p)\right)$$

$$+ \frac{(ed_e)^2}{I^2}E_1\int_p \left(\text{Re}\Delta_{g,11,R}(X,p)F_{00}(X,p) + \Delta_{g,11,F}(X,p)\text{Re}\Delta_{00,R}(X,p)\right)$$

$$+ \frac{(ed_e)^2}{2I^2}\frac{\partial E_1}{\partial X^0}\int_p \left(\frac{\partial F_{00}}{\partial p^0}\frac{\Delta_{g,11,\rho}}{i} + \frac{\rho_{00}}{i}\frac{\partial \Delta_{g,11,F}}{\partial p^0}\right) + \frac{(ed_e)^2}{4I^2}E_1\frac{\partial}{\partial X^0}\int_p \left(\frac{\partial F_{00}}{\partial p^0}\frac{\Delta_{g,11,\rho}}{i} + \frac{\rho_{00}}{i}\frac{\partial \Delta_{g,11,F}}{\partial p^0}\right), \quad (121)$$

with the $x^1$ direction of the dipole moment (density) given by $\mu_1 = 2ed_e\left(\bar{\psi}_1^*\bar{\psi}_0 + \bar{\psi}_0^*\bar{\psi}_1\right)$, $F_{11}(X,p) = \frac{\Delta_{11}^{21}(X,p) + \Delta_{11}^{12}(X,p)}{2}$, $F_{00}(X,p) = \frac{\Delta_{00}^{21}(X,p) + \Delta_{00}^{12}(X,p)}{2}$ and $\Delta_{g,11,F}(X,p) = \frac{\Delta_{g,11}^{21}(X,p) + \Delta_{g,11}^{12}(X,p)}{2}$. In the Appendix A we have shown the detailed derivation for the second, third, fourth and fifth terms in the above equations. We have assumed the self-energy $\Sigma_{00} = \Sigma_{11} = 0$ in deriving the time derivatives of $\Delta_{10}$ and $\Delta_{01}$ in Equation (101). Even if we include contributions of self-energy in Equation (121), they are higher order $O\left((ed_e)^4\right)$ in the coupling expansion. We have neglected higher order terms in the gradient expansion for quantum fluctuations. In Equation (121), we leave the $-(\partial_2)^2 E_1$ term on the left-hand side in the above equation to compare with the sign of $-\frac{2(ed_e)^2 Z}{I}E_1$ term. We find the Higgs mechanism with the mass squared $-\frac{2(ed_e)^2 Z}{I}$ in the case of the normal population $Z = 2|\bar{\psi}_1|^2 - |\bar{\psi}_0|^2 < 0$. On the other hand, the tachyonic instability appears in the inverted population $Z > 0$ in the above equation. Then the electric field $E_1$ will increase exponentially until $Z$ becomes negative. In Equation (121), the second term on the right-hand side is proportional to $2F_{11} - F_{00} - \Delta_{g,11,F}$. Near equilibrium states, we might find $F_{00} > 2F_{11} - \Delta_{g,11,F}$, where statistical functions $F_{11}$, $F_{00}$ and $\Delta_{g,11,F}$ are proportional to the Bose–Einstein distribution $\frac{1}{e^{p^0/T} - 1}$ plus $\frac{1}{2}$ (with the Kadanoff–Baym ansatz) with different dispersion relations $p^0 \sim \frac{p^2}{2m}$ for $F_{00}$ and $p^0 \sim \frac{p^2}{2m} + \frac{1}{2I}$ for $F_{11}$ and $\Delta_{g,11,F}$, due to the energy difference $\frac{1}{2I} - \frac{0}{2I}$ between the ground state and first excited states. So the $2F_{11} - F_{00} - \Delta_{g,11,F}$ in the second term is negative near the equilibrium states, which might mean no tachyonic unstable terms appear from quantum fluctuations near equilibrium states. The contributions of quantum fluctuations on the right-hand side written by statistical functions (second, third, fourth and fifth terms) vanish at zero temperature $T = 0$. Quantum fluctuations represent finite temperature effects at equilibrium states, although we need not restrict ourselves to only the equilibrium case. We have shown general contributions of quantum fluctuations in both equilibrium and non-equilibrium case in this paper.

Finally we shall consider remaining equations for coherent dipole fields. By using Equations (99) and (100) and the definitions of real functions $\mu_1 = 2ed_e(\bar{\psi}_1^*\bar{\psi}_0 + \bar{\psi}_0^*\bar{\psi}_1)$, $P = ied_e(\bar{\psi}_1^*\bar{\psi}_0 - \bar{\psi}_0^*\bar{\psi}_1)$ and $Z = 2|\bar{\psi}_1|^2 - |\bar{\psi}_0|^2$, we can also derive,

$$\partial_0 Z = 4E_1 P, \qquad (122)$$

$$\partial_0 \mu_1 = \frac{P}{I}, \qquad (123)$$

$$\partial_0 P = -\frac{\mu_1}{4I} - 2(ed_e)^2 E_1 Z. \qquad (124)$$

We can show $\partial_0(2|\bar{\psi}_1|^2 + |\bar{\psi}_0|^2) = 0$ by using these three equations. In these equations with initial conditions $E_1 > 0$, $Z > 0$ (inverted population), $P = 0$ and $\mu_1 = 0$, the $P$ and the $\mu_1$ decrease at around the initial time and $Z$ starts to decrease due to $E_1 P < 0$. In initial conditions $E_1 > 0$, $Z < 0$ (normal population), $P = 0$ and $\mu_1 = 0$, the $P$ and the $\mu_1$ increase at around the initial time and $Z$ starts to increase due to $E_1 P > 0$. The absolute values of $Z$ decrease at around the initial time. We find that there is no term of quantum fluctuations in Equations (122)–(124).

We can solve Equations (121)–(124) with real functions in this section and the Kadanoff–Baym equations with real statistical functions and pure imaginary spectral functions in Section 4, simultaneously.

## 6. Discussion

In this paper, we have derived time evolution equations, namely the Klein–Gordon equations for coherent photon fields, the Schrödinger-like equations for coherent electric dipole fields and the Kadanoff–Baym equations for quantum fluctuations, starting with the Lagrangian in quantum electrodynamics with electric dipoles in $2+1$ dimensions. We have adopted the two-particle-irreducible effective action technique with the leading-order self-energy of the coupling expansion. We find that electric dipoles change their angular momenta due to coherent electric fields $E_1 \pm i\alpha E_2$ with $\alpha = \pm 1$. They also change momenta and angular momenta by scattering with incoherent photons. The proof of H-theorem is possible for these processes as shown in Section 3. Our analysis provides the dynamics of both the order parameters with coherent fields and quantum fluctuations for incoherent particles.

In Section 2, we adopt two-energy level approximation for the angular momenta of dipoles. Then, we find that the $i\Delta_0^{-1}$ is written by $3 \times 3$ matrix with zero $(-1, 1)$ and $(1, -1)$ components. The form of the matrix is similar to $3 \times 3$ matrix in the analysis in open systems, the central region, left and right reservoirs as in [59,61–63]. Hence we can simplify the Kadanoff–Baym equations for dipole fields in an isolated system with the same procedures as those in open systems. The difference between QED with dipoles and $\phi^4$ theory in open systems is that the coherent electric field changes the momenta of dipoles when the phase $\alpha\zeta$ in $E_1 \pm i\alpha E_2$ with $\alpha = \pm 1$ is dependent on space–time. The space dependence of coherent electric fields might disappear in the time evolution due to the instability by the lower entropy of the system, then electric fields will change angular momenta of dipoles but not change momenta $p$ due to $\partial\zeta = 0$. We can also trace the dynamics with $\partial\zeta = 0$. By setting the initial conditions with the symmetry $\alpha \to -\alpha$, namely $\bar{\psi}_\alpha^{(*)} = \bar{\psi}_{-\alpha}^{(*)}$, $\Delta_{\alpha 0} = \Delta_{-\alpha 0}$ and $\Delta_{0\alpha} = \Delta_{0-\alpha}$, with initial conditions $E_2 = 0$ and $\partial_0 E_2 = 0$ in spatially homogeneous systems in $\partial^\nu F_{\nu 2} = J_2$ in Equation (20), we can show $E_2 = 0$ at any time. Then we can use $\partial\zeta = 0$. This condition simplifies numerical simulations in the Kadanoff–Baym equations since we need not estimate the momentum shift $p \to p \pm \alpha\partial\zeta$ in the finite-size lattice for the momentum space. As a result, the simulations for Kadanoff–Baym equations for dipoles and photons will be similar to those in QED with charged bosons in [72].

In Section 3, we have introduced a kinetic entropy current and shown the H-theorem in the leading-order of the coupling expansion with $ed_e$. This entropy approaches the Boltzmann entropy in the limit of zero spectral width as in [58]. The mode-coupling processes between dipoles and photons produce entropy. When there are deviations between $(00)$ and $(\alpha\alpha)$ components of distribution functions, entropy production occurs. Entropy production stops when the Bose–Einstein distribution is realized in the dynamics of Kadanoff–Baym equations.

We can also derive the energy shifts in dispersion relations due to nonzero electric fields by using the retarded Green's functions in Section 3. The 0th order equations for retarded Green's functions are given by,

$$\left(p^0 - \frac{\mathbf{p}^2}{2m} + 2(ed_e)^2 E_1^2 \Delta_{g,11,R}\right) \Delta_{00,R} = -1, \tag{125}$$

$$\left(p^0 - \frac{\mathbf{p}^2}{2m} - \frac{1}{2I}\right) \Delta_{11,R} + (ed_e)^2 E_1^2 \Delta_{00,R} \Delta_{g,11,R} = -1, \tag{126}$$

with $\Delta_{g,11,R} = \frac{-1}{p^0 - \frac{\mathbf{p}^2}{2m} - \frac{1}{2I}}$. Multiply $p^0 - \frac{\mathbf{p}^2}{2m} - \frac{1}{2I}$, take the imaginary parts in the above equations and remember the imaginary parts of retarded Green's functions are the spectral functions, then we find,

$$W \begin{bmatrix} \rho_{00} \\ \rho_{11} \end{bmatrix} = 0, \text{ with,}$$

$$W = \begin{bmatrix} \left(p^0 - \frac{\mathbf{p}^2}{2m} - \frac{1}{2I}\right)\left(p^0 - \frac{\mathbf{p}^2}{2m}\right) - 2(ed_e)^2 E_1^2 & 0 \\ -(ed_e)^2 E_1^2 & \left(p^0 - \frac{\mathbf{p}^2}{2m} - \frac{1}{2I}\right)^2 \end{bmatrix} \tag{127}$$

By setting determinant $|W|$ to be zero, we find the following solutions for dispersion relations,

$$p^0 = \frac{\mathbf{p}^2}{2m} + \frac{1}{4I} \pm \frac{1}{2}\sqrt{\frac{1}{4I^2} + 8(ed_e)^2 E_1^2}. \tag{128}$$

Here we assumed the symmetry for $\alpha = \pm 1$ for Green's functions and zero self-energy $\Sigma_{00} = \Sigma_{11} = 0$. We find how electric fields shift two energy levels 0 and $\frac{1}{2I}$. The above energy shift is similar to the energy shift given in [27] in 3 + 1 dimensions due to nonzero electric fields.

In Section 5.1, we have derived the super-radiance from time evolution equations for coherent fields. We find that it is possible to derive the Maxwell–Bloch equations from our Lagrangian with the probability conservation law and the energy conservation law. Super-radiant decay with intensity of the order $\propto N^2$ (N: The number of dipoles) appears in a similar way to [70,71]. It is possible to derive the maximum energy of electric fields by use of Equation (113). We know that the moment of inertia of water molecule is $I = 2m_H R^2$ with $m_H = 940$ MeV with $R = 0.96 \times 10^{-10}$ m. Hence the $k^0 = \frac{1}{2I} = 1.1 \times 10^{-3}$ eV. Since $B = \frac{N}{V} = 3.3 \times 10^{28}$ /m$^3$ for liquid water, we find

$$\frac{1}{2}\epsilon_{max}^2 = \frac{1}{2}k^0 B = 1.8 \times 10^{25} \text{ eV/m}^3. \tag{129}$$

When we multiply the volume of all microtubules (MTs) in a brain,

$$V_{MT} = \pi \times 15\text{nm}^2 \times 1000\text{nm} \times 2000 \text{ MTs/neuron} \times 10^{11} \text{ neurons/brain} = 1.4 \times 10^{-7} \text{ m}^3, \tag{130}$$

we can arrive at,

$$\frac{1}{2}\epsilon_{max}^2 V_{MT} = 0.41 \text{ J} = 0.1 \text{ cal.} \tag{131}$$

If we maintain our brain 100 s without energy supply, we need at least $0.1 \times 10^{-2}$ cal/s or 86 cal/day to maintain the ordered states of memory. We can compare 86 cal/day with 4000 cal/day = 2000 kcal/day × 0.2 (energy consumption rate of brain) × 0.01 (energy rate to maintain the ordered system). The 86 cal/day is within the 4000 cal/day, which is consistent with our experiences. In this derivation, we have used coefficients in 2 + 1 dimensions and the number density of water molecules in 3 + 1 dimensions.

In Section 5.2, we have derived time evolution equations for electric field $E_1$. The Higgs mechanism appears in this equation in normal population $Z < 0$. As a result, the dynamical mass generation occurs with the maximum mass $\Omega_{\text{Higgs}} = 2ed_e\sqrt{k^0 B} = 30k^0$ where the number density of dipoles is $B = 2|\bar{\psi}_1|^2 + |\bar{\psi}_0|^2 = \frac{N}{V}$. The period is $2\pi/\Omega_{\text{Higgs}} = 1.3 \times 10^{-13}$ s. In normal population $Z < 0$, the Meissner effect appears with the penetrating length $1/\Omega_{\text{Higgs}} = 6.3$ μm. On the other hand, the tachyonic instability occurs in inverted population $Z > 0$. The electric field $E_1$ increases exponentially with $\exp(\Omega X^0)$ (with $\Omega \leq \Omega_{\max}$) where the time scale is $1/\Omega_{\max} = 2.1 \times 10^{-14}$ s with $\Omega_{\max} = \Omega_{\text{Higgs}}$. Due to energy conservation, since $Z$ decreases as the absolute value of the electric field increases, tachyonic instability stops in $Z < 0$.

We have prepared for numerical simulations with time evolution equations, namely the Schödinger-like equations for coherent electric dipole fields, the Klein–Gordon equations for coherent electric fields and the Kadanoff–Baym equations for quantum fluctuations. Our simulations might describe the dynamics towards equilibrium states for quantum fluctuations and the dynamics of super-radiant states for coherent fields. Our analysis is also extended to simulations in open systems by preparing the left and the right reservoirs like those in [59] or networks [73].

## 7. Conclusions

We have derived the Schrödinger equations for coherent electric dipole fields, the Klein–Gordon equations for coherent electric fields and the Kadanoff–Baym equations for quantum fluctuations in QED with electric dipoles in $2 + 1$ dimensions. It is possible to derive equilibration for quantum fluctuations and super-radiance for background coherent fields simultaneously. Total energy consumption to maintain super-radiance in microtubules is consistent with energy consumption in our experiences. We can describe dynamical information transfer with super-radiance via microtubules without violation of the second law in thermodynamics. We have also derived the Higgs mechanism in normal population and the tachyonic instability in inverted population. These dynamical properties might be significant to form and maintain coherent domains composed of dipoles and photons. We are ready to describe memory formation processes towards equilibrium states in $2 + 1$ dimensions with equations in this paper. Furthermore, our approach might pave the way to understand the dynamical thinking processes with memory recalling in QBD by investigating the case in open systems with the Kadanoff–Baym equations. This work will be extended to the $3 + 1$ dimensional analysis to describe memory formation processes in numerical simulations. We should derive the Schödinger-like equations, the Klein–Gordon equations and the Kadanoff–Baym equations by starting with the single Lagrangian in QED with electric dipoles in $3 + 1$ dimensions in the future study. These equations in $3 + 1$ dimensions will describe more realistic and practical dynamics in QBD.

**Author Contributions:** Conceptualization, A.N, S.T. and J.A.T.; methodology, A.N.; software, A.N.; validation, A.N., S.T. and J.A.T.; formal analysis, A.N.; investigation, A.N.; resources, S.T. and J.A.T.; data curation, A.N.; writing-original draft preparation, A.N.; writing-review and editing, A.N., S.T. and J.A.T.; visualization, A.N.; supervision, S.T. and J.A.T.; project administration, S.T. and J.A.T.; funding acquisition, S.T. and J.A.T.

**Funding:** This work was supported by JSPS KAKENHI Grant Number JP17H06353.

**Acknowledgments:** J.A.T. is grateful for research support received from NSERC (Canada).

**Conflicts of Interest:** The authors declare no conflict of interest.

## Appendix A. Quantum Fluctuations in the Klein–Gordon Equations

In this section, we shall derive the second, third, fourth and fifth terms involving quantum fluctuations on the right-hand side in Equation (121) in spatially homogeneous systems. They correspond to the following term,

$$-2ed_e(\partial_0)^2 \left[\Delta_{10}(x,x) + \Delta_{01}(x,x)\right],$$

in Equation (101) with the symmetry $\Delta_{10} = \Delta_{-10}$ and $\Delta_{01} = \Delta_{0-1}$. It appears in taking the time derivative of $J_1$ (given by Equation (21)) in Equation (20). Here $\Delta_{10}(x,x)$ and $\Delta_{01}(x,x)$ can be rewritten by,

$$\Delta_{10}(x,x) = -\frac{ed_e}{i}\int_w \Delta_{g,11}(x,w)E_1(w)\Delta_{00}(w,x), \tag{A1}$$

$$\Delta_{01}(x,x) = -\frac{ed_e}{i}\int_w \Delta_{00}(x,w)E_1(w)\Delta_{g,11}(w,x), \tag{A2}$$

where we have used Equations (31) and (34) by setting $E_2 = 0$. We rewrite second time derivatives of $\Delta_{10}(x,x)$ and $\Delta_{01}(x,x)$.

We shall rewrite Equation (30) without self-energy $\Sigma_{\alpha\alpha}$ as,

$$\left[i\frac{\partial}{\partial x^0} + \frac{\nabla_i^2}{2m} - \frac{1}{2I}\right]\Delta_{g,11}(x,w) = i\delta_C(x-w), \tag{A3}$$

$$\left[-i\frac{\partial}{\partial x^0} + \frac{\nabla_i^2}{2m} - \frac{1}{2I}\right]\Delta_{g,11}(w,x) = i\delta_C(w-x), \tag{A4}$$

where we have multiplied $\Delta_{g,11}$ from the right and left of Equation (30). By using the above equations and Equations (32) and (33) with Equations (A1) and (A2) and $\Delta_{0,00}^{-1}(x,y) = \left(i\frac{\partial}{\partial x^0} + \frac{\nabla_i^2}{2m}\right)\delta_C(x-y)$, we can show

$$\frac{\partial}{\partial x^0}\Delta_{10}(x,x) = ed_e\left[\left(-\frac{\nabla_i^2}{2m} + \frac{1}{2I}\right)\Delta_{g,11} + i\delta_C\right]E_1\Delta_{00}$$

$$+ \Delta_{g,11}E_1\frac{\nabla_i^2}{2m}\Delta_{00} + 2\Delta_{g,11}ed_eE_1\Delta_{01}E_1 - \Delta_{g,11}E_1i\delta_C\right]$$

$$= ed_e\left[\left(\frac{1}{2I}\Delta_{g,11} + i\delta_C\right)E_1\Delta_{00} + 2\Delta_{g,11}ed_eE_1\Delta_{01}E_1 - \Delta_{g,11}E_1i\delta_C\right], \tag{A5}$$

$$\frac{\partial}{\partial x^0}\Delta_{01}(x,x) = ed_e\left[(-2ed_eE_1\Delta_{10} + i\delta_C)E_1\Delta_{g,11} + \Delta_{00}E_1\left(-\frac{1}{2I}\Delta_{g,11}i\delta_C\right)\right], \tag{A6}$$

where $\delta_C$ represents the delta function in the closed-time path. Here the terms proportional to $\nabla_i^2$ are cancelled in spatially homogeneous systems. By use of the above two equations, we can show

$$\frac{\partial}{\partial x^0}(\Delta_{10} + \Delta_{01}) = \frac{1}{2iI}(\Delta_{10} - \Delta_{01}), \tag{A7}$$

and,

$$\frac{\partial^2}{\partial(x^0)^2}(\Delta_{10} + \Delta_{01}) = \frac{ed_e}{2iI}\left[\left(\frac{\Delta_{g,11}}{2I} + i\delta_C\right)E_1\Delta_{00} + 2\Delta_{g,11}ed_eE_1\Delta_{01}E_1 - \Delta_{g,11}E_1i\delta_C\right.$$

$$\left. - (-2ed_eE_1\Delta_{10} + i\delta_C)E_1\Delta_{g,11} - \Delta_{00}E_1\left(-\frac{1}{2I}\Delta_{g,11} - i\delta_C\right)\right]$$

$$= \frac{ed_e}{2iI}\left[2iE_1(\Delta_{00} - \Delta_{g,11}) + \frac{1}{2I}(\Delta_{g,11}E_1\Delta_{00} + \Delta_{00}E_1\Delta_{g,11})\right.$$

$$\left. + 2ed_e(\Delta_{g,11}E_1\Delta_{01}E_1 + E_1\Delta_{10}E_1\Delta_{g,11})\right]. \tag{A8}$$

Since we can rewrite Equations (35) or (36) by multiplying $i\Delta_{g,11}$ as,

$$\begin{aligned}i\Delta_{g,11} - i\Delta_{11} &= ed_e \Delta_{g,11} E_1 \Delta_{01} \\ &= ed_e \Delta_{10} E_1 \Delta_{g,11},\end{aligned} \tag{A9}$$

we arrive at,

$$\frac{\partial^2}{\partial (x^0)^2}(\Delta_{10} + \Delta_{01}) = -\frac{1}{4l^2}(\Delta_{10} + \Delta_{01}) + \frac{ed_e E_1}{l}(\Delta_{00} - 2\Delta_{11} + \Delta_{g,11}), \tag{A10}$$

where we have used Equations (A1) and (A2).

Finally by rewriting the statistical parts (subscript 'F') of $\Delta_{10} + \Delta_{01}$ with Equations (A1) and (A2), and using $E_1(w) = E_1(x) + (w^0 - x^0)\partial_0 E_1(x)$ in,

$$\left[\int dw \left[\Delta_{g,11}(x,w)E_1(w)\Delta_{00}(w,x) + \Delta_{00}(x,w)E_1(w)\Delta_{g,11}(w,x)\right]\right]_F,$$

and the relation in the first order in the gradient expansion,

$$\left[\int dw \Delta_{g,11}(x,w)\Delta_{00}(w,x)\right]_F = \int_p \left(\frac{\Delta_{g,11,R}(x,p)}{i}F_{00}(x,p) + \Delta_{g,11,F}\frac{\Delta_{00,A}}{i}\right.$$
$$\left. + \frac{i}{2}\left\{\frac{\Delta_{g,11,R}(x,p)}{i}, F_{00}(x,p)\right\} + \frac{i}{2}\left\{\Delta_{g,11,F}, \frac{\Delta_{00,A}}{i}\right\}\right), \tag{A11}$$

with the advanced (subscript 'A') $\Delta_{00,A} = i(\Delta_{00}^{11} - \Delta_{00}^{21}) = \text{Re}\Delta_{00,R} - \frac{\rho_{00}}{2}$ and the retarded $\Delta_{00,R} = i(\Delta_{00}^{11} - \Delta_{00}^{12}) = \text{Re}\Delta_{00,R} + \frac{\rho_{00}}{2}$, we can derive the third, fourth and fifth terms on the right-hand side in Equation (121).

## Reference

1. Day, J.J.; Sweatt, J.D. DNA methylation and memory formation. *Nat. Neurosci.* **2010**, *13*, 1319. [CrossRef] [PubMed]
2. Adolphs, R. The unsolved problems of neuroscience. *Trends Cogn. Sci.* **2015**, *19*, 173–175. [CrossRef] [PubMed]
3. Kukushkin, N.V.; Carew, T.J. Memory takes time. *Neuron* **2017**, *95*, 259–279. [CrossRef] [PubMed]
4. Jibu, M.; Yasue, K. *Quantum Brain Dynamics and Consciousness*; John Benjamins: Amsterdam, The Netherlands, 1995.
5. Vitiello, G. *My Double Unveiled: The Dissipative Quantum Model of Brain*; John Benjamins: Amsterdam, The Netherlands, 2001; Volume 32.
6. Sabbadini, S.A.; Vitiello, G. Entanglement and Phase-Mediated Correlations in Quantum Field Theory. Application to Brain-Mind States. *Appl. Sci.* **2019**, *9*, 3203. [CrossRef]
7. Lashley, K.S. *Brain Mechanisms and Intelligence: A Quantitative Study of Injuries to the Brain*; PsycBOOKS: Chicago, IL, USA, 1929.
8. Pribram, K.H. Languages of the brain: Experimental paradoxes and principles in neuropsychology. *Nerv. Ment. Dis.* **1973**, *157*, 69–70.
9. Pribram, K. *Brain and Perception: Holonomy and Structure in Figural Processing*; Lawrence Erlbaum Associates: Hillsdale, NJ, USA, 1991.
10. Umezawa, H. *Advanced Field Theory: Micro, Macro, and Thermal Physics*; AIP: College Park, MD, USA, 1995.
11. Nambu, Y.; Lasinio, G.J. Dynamical Model of Elementary Particles Based on an Analogy with Superconductivity. I. *Phys. Rev.* **1961**, *112*, 345. [CrossRef]
12. Goldstone, J. Field theories with «Superconductor» solutions. *Il Nuovo Cim. (1955–1965)* **1961**, *19*, 154–164. [CrossRef]

13. Goldstone, J.; Salam, A.; Weinberg, S. Broken symmetries. *Phys. Rev.* **1962**, *127*, 965. [CrossRef]
14. Ricciardi, L.M.; Umezawa, H. Brain and physics of many-body problems. *Kybernetik* **1967**, *4*, 44–48. [CrossRef]
15. Stuart, C.; Takahashi, Y.; Umezawa, H. On the stability and non-local properties of memory. *J. Theor. Biol.* **1978**, *71*, 605–618. [CrossRef]
16. Stuart, C.; Takahashi, Y.; Umezawa, H. Mixed-system brain dynamics: Neural memory as a macroscopic ordered state. *Found. Phys.* **1979**, *9*, 301–327. [CrossRef]
17. Fröhlich, H. Bose condensation of strongly excited longitudinal electric modes. *Phys. Lett. A* **1968**, *26*, 402–403. [CrossRef]
18. Fröhlich, H. Long-range coherence and energy storage in biological systems. *Int. J. Quantum Chem.* **1968**, *2*, 641–649. [CrossRef]
19. Fröhlich, H. Long range coherence and the action of enzymes. *Nature* **1970**, *228*, 1093–1093. [CrossRef]
20. Fröhlich, H. Selective long range dispersion forces between large systems. *Phys. Lett. A* **1972**, *39*, 153–154. [CrossRef]
21. Fröhlich, H. Evidence for Bose condensation-like excitation of coherent modes in biological systems. *Phys. Lett. A* **1975**, *51*, 21–22. [CrossRef]
22. Fröhlich, H. Long-range coherence in biological systems. *La Riv. Del Nuovo Cim. (1971–1977)* **1977**, *7*, 399–418. [CrossRef]
23. Davydov, A.; Kislukha, N. Solitons in One-Dimensional Molecular Chains. *Phys. Status Solidi (B)* **1976**, *75*, 735–742. [CrossRef]
24. Tuszyński, J.; Paul, R.; Chatterjee, R.; Sreenivasan, S. Relationship between Fröhlich and Davydov models of biological order. *Phys. Rev. A* **1984**, *30*, 2666. [CrossRef]
25. Del Giudice, E.; Doglia, S.; Milani, M.; Vitiello, G. Spontaneous symmetry breakdown and boson condensation in biology. *Phys. Lett. A* **1983**, *95*, 508–510. [CrossRef]
26. Del Giudice, E.; Doglia, S.; Milani, M.; Vitiello, G. A quantum field theoretical approach to the collective behaviour of biological systems. *Nucl. Phys. B* **1985**, *251*, 375–400. [CrossRef]
27. Del Giudice, E.; Preparata, G.; Vitiello, G. Water as a free electric dipole laser. *Phys. Rev. Lett.* **1988**, *61*, 1085. [CrossRef] [PubMed]
28. Del Giudice, E.; Smith, C.; Vitiello, G. Magnetic Flux Quantization and Josephson Systems. *Phys. Scr.* **1989**, *40*, 786–791. [CrossRef]
29. Jibu, M.; Yasue, K. A physical picture of Umezawa's quantum brain dynamics. *Cybern. Syst. Res.* **1992**, *92*, 797–804.
30. Jibu, M.; Yasue, K. Intracellular quantum signal transfer in Umezawa's quantum brain dynamics. *Cybern. Syst.* **1993**, *24*, 1–7. [CrossRef]
31. Jibu, M.; Hagan, S.; Hameroff, S.R.; Pribram, K.H.; Yasue, K. Quantum optical coherence in cytoskeletal microtubules: Implications for brain function. *Biosystems* **1994**, *32*, 195–209. [CrossRef]
32. Jibu, M.; Yasue, K. What is mind?- Quantum field theory of evanescent photons in brain as quantum theory of consciousness. *INF* **1997**, *21*, 471–490.
33. Dicke, R.H. Coherence in spontaneous radiation processes. *Phys. Rev.* **1954**, *93*, 99. [CrossRef]
34. Gross, M.; Haroche, S. Superradiance: An essay on the theory of collective spontaneous emission. *Phys. Rep.* **1982**, *93*, 301–396. [CrossRef]
35. Preparata, G. Quantum field theory of superradiance. *Probl. Fundam. Mod. Phys.* **1990**, 303. [CrossRef]
36. Preparata, G. *QED Coherence in Matter*; World Scientific: Singapore, 1995.
37. Enz, C.P. On Preparata's theory of a superradiant phase transition. *Helv. Phys. Acta* **1997**, *70*, 141–153.
38. Vitiello, G. Dissipation and memory capacity in the quantum brain model. *Int. J. Mod. Phys.* **1995**, *9*, 973–989. [CrossRef]
39. Vitiello, G. Classical chaotic trajectories in quantum field theory. *Int. J. Mod. Phys. B* **2004**, *18*, 785–792. [CrossRef]
40. Zheng, J.M.; Pollack, G.H. Long-range forces extending from polymer-gel surfaces. *Phys. Rev. E* **2003**, *68*, 031408. [CrossRef] [PubMed]
41. Del Giudice, E.; Voeikov, V.; Tedeschi, A.; Vitiello, G. The origin and the special role of coherent water in living systems. *F. Cell* **2014**, 95–111. [CrossRef]
42. Tegmark, M. Importance of quantum decoherence in brain processes. *Phys. Rev. E* **2000**, *61*, 4194. [CrossRef]

43. Craddock, T.J.; Tuszynski, J.A.; Hameroff, S. Cytoskeletal signaling: Is memory encoded in microtubule lattices by CaMKII phosphorylation? *PLoS Comput. Biol.* **2012**, *8*, e1002421. [CrossRef]
44. Baym, G.; Kadanoff, L.P. Conservation laws and correlation functions. *Phys. Rev.* **1961**, *124*, 287. [CrossRef]
45. Kadanoff, L.P.; Baym, G. *Quantum Statistical Mechanics: Green's Function Methods in Equilibrium Problems*; WA Benjamin: Los Angeles, CA, USA, 1962.
46. Baym, G.; Kadanoff, L.P. Self-Consistent Approximations in Many-Body Systems. *Phys. Rev.* **1962**, *127*, 1391. [CrossRef]
47. Cornwall, J.M.; Jackiw, R.; Tomboulis, E. Effective action for composite operators. *Phys. Rev. D* **1974**, *10*, 2428. [CrossRef]
48. Niemi, A.J.; Semenoff, G.W. Finite-temperature quantum field theory in Minkowski space. *Ann. Phys.* **1984**, *152*, 105–129. [CrossRef]
49. Calzetta, E.; Hu, B.L. Nonequilibrium quantum fields: Closed-time-path effective action, Wigner function, and Boltzmann equation. *Phys. Rev. D* **1988**, *37*, 2878. [CrossRef] [PubMed]
50. Schwinger, J. Brownian motion of a quantum oscillator. *J. Math. Phys.* **1961**, *2*, 407–432. [CrossRef]
51. Keldysh, L.V. Diagram technique for nonequilibrium processes. *Sov. Phys. Jetp.* **1965**, *20*, 1018–1026.
52. Kluberg-Stern, H.; Zuber, J. Renormalization of non-Abelian gauge theories in a background-field gauge. I. Green's functions. *Phys. Rev. D* **1975**, *12*, 482. [CrossRef]
53. Abbott, L.F. The background field method beyond one loop. *Nucl. Phys. B* **1981**, *185*, 189–203. [CrossRef]
54. Abbott, L.F. Introduction to the background field method. *Acta Phys. Pol. B* **1981**, *13*, 33–50.
55. Wang, Q.; Redlich, K.; Stöcker, H.; Greiner, W. From the Dyson–Schwinger to the transport equation in the background field gauge of QCD. *Nucl. Phys. A* **2003**, *714*, 293–334. [CrossRef]
56. Ivanov, Y.B.; Knoll, J.; Voskresensky, D. Resonance transport and kinetic entropy. *Nucl. Phys. A* **2000**, *672*, 313–356. [CrossRef]
57. Kita, T. Entropy in nonequilibrium statistical mechanics. *J. Phys. Soc. Jpn.* **2006**, *75*, 114005–114005. [CrossRef]
58. Nishiyama, A. Entropy production in 2D $\lambda\phi4$ theory in the Kadanoff–Baym approach. *Nucl. Phys. A* **2010**, *832*, 289–313. [CrossRef]
59. Nishiyama, A.; Tuszynski, J.A. Non-Equilibrium $\phi4$ in open systems as a toy model of quantum field theory of the brain. *Ann. Phys.* **2018**, *398*, 214. [CrossRef]
60. Myöhänen, P.; Stan, A.; Stefanucci, G.; van Leeuwen, R. A many-body approach to quantum transport dynamics: Initial correlations and memory effects. *EPL (Europhys. Lett. )* **2008**, *84*, 67001. [CrossRef]
61. Myöhänen, P.; Stan, A.; Stefanucci, G.; Van Leeuwen, R. Kadanoff-Baym approach to quantum transport through interacting nanoscale systems: From the transient to the steady-state regime. *Phys. Rev. B* **2009**, *80*, 115107. [CrossRef]
62. Wang, J.S.; Agarwalla, B.K.; Li, H.; Thingna, J. Nonequilibrium Green's function method for quantum thermal transport. *Front. Phys.* **2014**, *9*, 673–697. [CrossRef]
63. Dražić, M.S.; Cerovski, V.; Zikic, R. Theory of time-dependent nonequilibrium transport through a single molecule in a nonorthogonal basis set. *Int. J. Quantum Chem.* **2017**, *117*, 57–73. [CrossRef]
64. Stratonovich, R.L. Gauge Invariant Generalization of Wigner Distribution. *Dok. Akad. Nauk SSSR* **1956**, *109*, 72–75.
65. Fujita, S. *Introduction to Non-Equilibrium Quantum Statistical Mechanics*; Krieger Pub Co: Malabar, FL, USA, 1966.
66. Groenewold, H.J. *On the Principles of Elementary Quantum Mechanics*; Springer: The Netherlands, 1946; Volume 12, pp. 1–56.
67. Moyal, J.E. Quantum mechanics as a statistical theory. *Math. Proc. Camb. Philos. Soc.* **1949**, *45*, 99–124. [CrossRef]
68. Knoll, J.; Ivanov, Y.B.; Voskresensky, D.N. Exact conservation laws of the gradient expanded Kadanoff–Baym equations. *Ann. Phys.* **2001**, *293*, 126–146. [CrossRef]
69. Ivanov, Y.B.; Knoll, J.; Voskresensky, D. Self-consistent approach to off-shell transport. *Phys. At. Nucl.* **2003**, *66*, 1902–1920. [CrossRef]
70. Bonifacio, R.; Preparata, G. Coherent spontaneous emission. *Phys. Rev. A* **1970**, *2*, 336. [CrossRef]
71. Benedict, M.G. *Super-Radiance: Multiatomic Coherent Emission*; Routledge: London, UK, 2018.

72. Nishiyama, A.; Tuszynski, J.A. Nonequilibrium quantum electrodynamics: Entropy production during equilibration. *Int. J. Mod. Phys. B* **2018**, *32*, 1850265. [CrossRef]
73. Nishiyama, A.; Tuszynski, J.A. Non-Equilibrium $\phi4$ theory for networks: Towards memory formations with quantum brain dynamics. *J. Phys. Commun.* **2019**, *3*, 055020. [CrossRef]

© 2019 by the authors. Licensee MDPI, Basel, Switzerland. This article is an open access article distributed under the terms and conditions of the Creative Commons Attribution (CC BY) license (http://creativecommons.org/licenses/by/4.0/).

*Article*

# Variational Autoencoder Reconstruction of Complex Many-Body Physics

Ilia A. Luchnikov [1,2], Alexander Ryzhov [1], Pieter-Jan Stas [3], Sergey N. Filippov [2,4,5] and Henni Ouerdane [1,*]

1. Center for Energy Science and Technology, Skolkovo Institute of Science and Technology, 3 Nobel Street, Skolkovo, Moscow Region 121205, Russia; Ilia.Luchnikov@skoltech.ru (I.A.L.); a.ryzhov@skoltech.ru (A.R.)
2. Moscow Institute of Physics and Technology, Institutskii Per. 9, Dolgoprudny, Moscow Region 141700, Russia; sergey.filippov@phystech.edu
3. Department of Applied Physics, Stanford University 348 Via Pueblo Mall, Stanford, CA 94305, USA; pjstas@stanford.edu
4. Valiev Institute of Physics and Technology of Russian Academy of Sciences, Nakhimovskii Pr. 34, Moscow 117218, Russia
5. Steklov Mathematical Institute of Russian Academy of Sciences, Gubkina St. 8, Moscow 119991, Russia
* Correspondence: h.ouerdane@skoltech.ru

Received: 9 October 2019; Accepted: 6 November 2019; Published: 7 November 2019

**Abstract:** Thermodynamics is a theory of principles that permits a basic description of the macroscopic properties of a rich variety of complex systems from traditional ones, such as crystalline solids, gases, liquids, and thermal machines, to more intricate systems such as living organisms and black holes to name a few. Physical quantities of interest, or equilibrium state variables, are linked together in equations of state to give information on the studied system, including phase transitions, as energy in the forms of work and heat, and/or matter are exchanged with its environment, thus generating entropy. A more accurate description requires different frameworks, namely, statistical mechanics and quantum physics to explore in depth the microscopic properties of physical systems and relate them to their macroscopic properties. These frameworks also allow to go beyond equilibrium situations. Given the notably increasing complexity of mathematical models to study realistic systems, and their coupling to their environment that constrains their dynamics, both analytical approaches and numerical methods that build on these models show limitations in scope or applicability. On the other hand, machine learning, i.e., data-driven, methods prove to be increasingly efficient for the study of complex quantum systems. Deep neural networks, in particular, have been successfully applied to many-body quantum dynamics simulations and to quantum matter phase characterization. In the present work, we show how to use a variational autoencoder (VAE)—a state-of-the-art tool in the field of deep learning for the simulation of probability distributions of complex systems. More precisely, we transform a quantum mechanical problem of many-body state reconstruction into a statistical problem, suitable for VAE, by using informationally complete positive operator-valued measure. We show, with the paradigmatic quantum Ising model in a transverse magnetic field, that the ground-state physics, such as, e.g., magnetization and other mean values of observables, of a whole class of quantum many-body systems can be reconstructed by using VAE learning of tomographic data for different parameters of the Hamiltonian, and even if the system undergoes a quantum phase transition. We also discuss challenges related to our approach as entropy calculations pose particular difficulties.

**Keywords:** complex systems thermodynamics; machine learning; quantum phase transition; Ising model; variational autoencoder

## 1. Introduction

The development of the dynamical theory of heat or classical equilibrium thermodynamics as we know it was possible only with empirical data collection, processing, and analysis, which led, through a phenomenological approach, to the definition of two fundamental physical concepts, the actual pillars of the theory: energy and entropy [1]. It is with these two concepts that the laws (or principles) of thermodynamics could be stated and the absolute temperature be given a first proper definition. Though energy remains as fully enigmatic as entropy from the ontological viewpoint, the latter concept is not completely understood from the physical viewpoint. This of course did not preclude the success of equilibrium thermodynamics as evidenced not only by the development of thermal sciences and engineering, but also because of its cognate fields that owe it, at least partly or as an indirect consequence, their birth, from quantum physics to information theory.

Early attempts to refine and give thermodynamics solid grounds started with the development of the kinetic theory of gases and of statistical physics, which in turn permitted studies of irreversible processes with the development of nonequilibrium thermodynamics [2–6] and later on finite-time thermodynamics [7–9], thus establishing closer ties between the concrete notion of irreversibility and the more abstract entropy, notably with Boltzmann's statistical definition [10] and Gibbs' ensemble theory [11]. Notwithstanding conceptual difficulties inherent to the foundations of statistical physics, such as, e.g., irreversibility and the ergodic hypothesis [12,13], entropy acquired a meaningful statistical character and the scope of its definitions could be extended beyond thermodynamics, thus paving the way to information theory, as information content became a physical quantity per se, i.e., something that can be measured [14]. Additionally, although quantum physics developed independently from thermodynamics, it extended the scope of statistical physics with the introduction of quantum statistics, led to the definition of the von Neumann entropy [15], and also introduced new problems related to small, i.e., mesoscopic and nanoscopic systems [16,17], down to nuclear matter [18], where the concepts of thermodynamic limit and ensuing standard definitions of thermodynamic quantities may be put at odds.

Quantum physics problems that overlap with thermodynamics are typically classified into different categories: ground state characterization [19], thermal state characterization at finite temperature [20], the so-called eigenstate thermalization hypothesis [21–25], calculation of the dynamics of either closed or open systems [26,27], state reconstruction from tomographic data [28], and quantum system control, which, given the complexity for its implementation, requires the development of new methods [29]. Among the rich variety of methods applicable to such problems, including, e.g., mean-field approach [30], slave particle approach [31], dynamical mean-field theory [32], nonperturbative methods based on functional integrals [33], we believe two large families of techniques are of particular interest for numerical studies of many-body systems when strong correlations must be accounted for: One is based on the quantum Monte Carlo (QMC) framework [34], which is powerful to overcome the curse of dimensionality by using the stochastic estimation of high-dimensional integrals; the other family encompasses methods that search solutions in the parametric set of functions, also called ansatz. The most used ansatzes are based on different tensor network architectures [35,36] as tensor network-based methods show state-of-the-art performance for the characterization of one-dimensional strongly correlated quantum systems. One can solve either the ground-state problem by using the variational matrix product state (MPS) ground state search [37] or a dynamical problem using a time-evolving block decimation (TEBD) algorithm [38]. Quantum criticality of one-dimensional systems also can be studied by using a more advanced architecture called multiscale entanglement renormalization ansatz (MERA) [39]. The application of tensor networks is not restricted to one-dimensional systems, and one can describe an open quantum dynamics [40], characterize the numerical complexity of an open quantum dynamics [41,42], perform tomography of non-Markovian quantum processes by using tensor networks [43,44], analyze properties of two dimensional quantum lattices by using projected entangled pair states (PEPS) [45], or solve classical statistical physics problems [46,47].

The cross-fertilization of quantum physics and thermodynamics has benefited much from the powerful quantum formalism and computational techniques; however, as thermodynamic concepts evolved from intuitive/phenomenological definitions to classical-mechanics constructs, extended with quantum physics and formalism when needed, thermodynamics, in spite of its undeniable theoretical and practical successes, never managed to fully mature into a genuine fundamental theory that firmly rests on strong basic postulates. On one hand, this led a growing number of physicists to consider thermodynamics as incomplete, and on the other, to think quantum theory as the underlying framework from which equilibrium and nonequilibrium thermodynamics emerge. Quantum thermodynamics [48,49] is a fairly recent field of play, where new ideas are tested while revisiting old problems related to cycles, engines, refrigerators, and entropy production, to name a few [50,51]. Further, quantum technology is a burgeoning field at the interface of physics and engineering, which seeks to develop devices able to harness quantum effects for computing and secure communication purposes [52,53]. The wide scale development of such a kind of systems, which irreversibly interact with an infinite environment, rests on the ability to properly simulate the open quantum dynamics of their many-body properties and analyze coherence and dissipation at the quantum level.

How fast quantum thermodynamics will progress is difficult to anticipate as there exist numerous unsolved problems, especially those related to the proper characterization of the physical processes, e.g., what qualifies as heat or work on ultrashort time and length scales, where averages become irrelevant is unclear, and how the laws of thermodynamics may be systematically adapted still may be debated. To mitigate risks of slow progress, one may resort to approaches that do not rely on models of systems, but rather on data, the idea being to gain actual knowledge and understanding from data irrespective of how complex the studied system is. Machine learning (ML) provides perfectly suited tools for that purpose [54]. ML has a rather long history that can be dated back with the works of Bayes (1763) on prior knowledge that can be used to calculate the probability of an event as formulated by Laplace (1812). Much later (1913), Markov chains were proposed as a tool to describe sequences of events, each being characterized by a probability of occurrence that depends on the actuality of the previous event only. The main milestone is in 1950, with Turing's machine that can learn [55], shortly followed in 1951 by the first neural network machine [56]. Thanks to the huge increase in computational power over the last two decades, ML is now used for a wide variety of problems [54], and quantum machine learning now shows extraordinary potential for faster and more efficient than ever treatment of complex quantum systems problems [57], one major challenge still residing in the development of the hardware capable to harness and transform this potentiality into actual tool.

With the recent success in the field of deep learning, tools other than those based on tensor networks work as well as an ansatz. Restricted Boltzmann machine has been successfully applied as an ansatz to a ground state search, dynamics calculation, and quantum tomography [58–60], as well as convolution neural network to the two-dimensional frustrated $J_1 - J_2$ model [61]. The deep autoregressive model was applied very efficiently and elegantly to a ground state search of many-body quantum system and to classical statistical physics as well [62,63]. It was also recently shown how ML can establish and classify with high accuracy the chaotic or regular behavior of quantum billiards models and XXZ spin chains [64]. Thus, it can be useful to transfer deep architectures from the field of deep learning to the area of many-body quantum systems. A variational autoencoder (VAE) was used for sampling from probability distributions of quantum states in [65]; in the present work, we show that state-of-the-art generative architecture called conditional VAE can be applied to describe the whole family of the ground states of a quantum many-body system. For that purpose, using quantum tomography (albeit in an approximate fashion as discussed below) and reconstruction tools developed in [66], we consider the paradigmatic Ising model in a transverse-field as an illustration of the usefulness and efficiency of our approach. The use of VAE in such a problem is justified by the simplicity of VAE training, as well as its expressibility [67].

The article is organized as follows. In Section 2, we give a brief recap of the physics of the Ising model in a transverse field. In Section 3, we develop our generative model in the framework of the tensor network. Section 4 is devoted to the variational autoencoder architecture. The results are shown and discussed in Section 5. The article ends with concluding remarks, followed a by a short series of appendices.

## 2. Transverse-Field Ising Model

Among the rich variety of condensed matter systems, magnetic materials are a source of many fruitful problems, whose studies and solutions inspired discussions and new models beyond their immediate scope. The Kondo effect (existence of a minimum of electrical resistivity at low temperature in metals due to the presence of magnetic impurities) is one such problem [68,69], as it provides an excellent basis for studies of quantum criticality and absolute zero-temperature phase transitions [70,71] and, also, on a more fundamental level, a concrete example of asymptotic freedom [69]. Assuming infinite on-site repulsion, the single-impurity Anderson model [68,72] was used to establish a correspondence between Hamiltonian language and path integral for the development of nonperturbative methods in quantum field theory [73,74]. One other important model is that of the Heisenberg Hamiltonian, defined for the study of ferromagnetic materials, and which, assuming a crystal subjected to an external magnetic field $B$, reads [75] as

$$H = -\sum_{\langle i,j \rangle} J_{ij} \hat{S}^i \hat{S}^j - h \cdot \sum_j \hat{S}^j \tag{1}$$

where, for ease of notations, we introduced $h = g\mu_B B$, with $g$ being the Landé factor and $\mu_B = e\hbar/2m_e$ being the Bohr magneton ($e$: elementary electric charge, and $m_e$: electron mass); $J_{ij}$ is a parameter that characterizes the nearest-neighbors exchange interaction between electron spins on the crystal sites $i$ and $j$ (the quantum spins $\hat{S}^i$ and $\hat{S}^j$ are vector operators whose components are proportional to the Pauli matrices). For simplicity, one may consider $J_{ij} \equiv J$ constant. If $J > 0$, then the system is ferromagnetic and if $J < 0$ the system is antiferromagnetic. Hereafter, we fix the electron's magnetic moment $g\mu_B = 1$.

Although Equation (1) has a fairly simple form, the exact calculation of the partition function is

$$Z = \text{Tr } e^{-\beta H} \tag{2}$$

where $\beta = 1/k_B T$ is the inverse thermal energy, which is possible on the analytical level with the mean-field approximation that simplifies the Hamiltonian (1), and also for one-dimensional systems, one difficulty of the Heisenberg Hamiltonian being that the three components of a spin vector operator do not commute. That said, Heisenberg's Hamiltonian is very useful to, e.g., study spin frustration [76], entanglement entropy [77], and also serve as a test case for density-matrix renormalization group algorithms [78]. Under zero field, Heisenberg's Hamiltonian is also a simplified form of the Hubbard model at half-filling, thus including ferromagnetism in the scope of strongly correlated systems studies.

A particular, but very important, approximation of Heisenberg's Hamiltonian, whose significance lies in physics, especially for the study of critical phenomena, cannot be underestimated: the so-called Ising model. In its initial formulation [79], Ising spins are $N$ classical variables, which may take $\pm 1$ as values and form a one-dimensional (1D) system characterized by free or periodic boundary conditions. The classical partition function $Z$ may be calculated analytically for the 1D Ising model, and quantities such as the average total magnetization are obtained directly [80]:

$$M = \frac{1}{\beta} \frac{\partial \ln Z}{\partial h} \tag{3}$$

In the present work, we consider a 1D quantum spin chain whose Hilbert space is given by $\mathcal{H} = \otimes_i^N \mathbb{C}^2$. The system is described by the transverse-field Ising (TFI) Hamiltonian [81]:

$$H = -J \sum_{\langle i,j \rangle} \sigma_z^i \sigma_z^{i+1} + h_x \sum_{i=1}^{N} \sigma_x^i. \tag{4}$$

where $\sigma_\alpha^i$ ($\alpha \equiv x, z$) is the Pauli matrix for the $\alpha$-component of the $i$-th spin in the chain, and $h_x$ is the magnetic field applied in the transverse direction $x$. In this case, the spins are no longer the classical Ising ones and the two terms that compose the Hamiltonian $H$ do not commute, therefore requiring a full quantum approach. An example of a real-world system that may be studied as a quantum Ising chain is cobalt niobate ($CoNb_2O_6$); in this case, the spins that undergo the phase transition as the transverse field varies are those of the $Co^{2+}$ ions [82]. The spin states are denoted $|+\rangle_i$ and $|-\rangle_i$ at ion site $i$. There are two possible ground states: when all $N$ spins are in the state $|+\rangle$ or in the state $|-\rangle$, i.e., when they are all aligned, which defines the ferromagnetic phase.

The phase transition from the ferromagnetic phase to the paramagnetic phase that we speak of now is of a quantum nature, and not of a thermal nature, as here it is driven only by the external magnetic field. More precisely, when the transverse field $h_x$ is applied with sufficient strength, the spins align along the $x$ direction, and the spin state at site $i$ is given as the superposition $(|+\rangle_i + |-\rangle_i)/\sqrt{2}$, which is nothing else but the eigenstate of the $x$-component of the spin. Therefore, in this particular case, there is no need to raise the temperature of the system initially in the ferromagnetic phase beyond the Curie temperature to make it a paramagnet: the many-body system remains in its ground state, but its properties have changed. Further, note that unlike for the ferromagnetic phase, the quantum paramagnetic phase has spin-inversion symmetry. An insightful discussion on quantum criticality can be found in Reference [83].

Now, we briefly comment on the quantity $\beta = 1/k_B T$ in the context of quantum phase transitions, which, strictly speaking, can only occur at temperature $T = 0$ K. In fact, close to the absolute zero, where $\beta \to \infty$, their signatures can be observed as quantum fluctuations dominate thermal fluctuations in the criticality region, where the quantum critical point lies. The imaginary time formalism [84], where $\exp(-\beta H)$ is interpreted as an evolution operator, and the partition function $Z$ as a path integral, provides a way to map a quantum problem onto a classical one with the introduction of the imaginary time $\beta$ resulting from a Wick rotation in the complex plane, thus yielding one extra dimension to the model. In classical thermodynamics, to observe a phase transition in a system requires that its size (i.e., the number of constituents $N$) tends to infinity so that the order parameter is non-analytic at the transition point; so, for the quantum transition, the thermodynamic limit entails the limit $\beta \to \infty$ also: the 1D TFI model is mapped onto an equivalent 2D classical Ising model [85]. The imaginary time formalism permits implementation of classical Monte Carlo simulations to study quantum systems. Further discussion, including the sign problem for the quantum spin-1/2 system, is available in Reference [4].

We have chosen the transverse-field Ising model as an illustrative case for our study for several reasons. First, as this system is 1-dimensional, we can apply an MPS variational ground state solver [37], and therefore obtain the ground state solution in MPS representation. We can then perform fast and exact sampling for generation of large data sets for the training of the VAE. Next, this model can be solved analytically, which allows us to adequately benchmark our results. Finally, this model shows a nontrivial behavior around the quantum phase transition point at $h_x = 1$, and thus constitutes an interesting example to apply a VAE.

## 3. Generative Model as a Quantum State

Many-body quantum physics is rich in high-dimensional problems. Often, however, with increasing dimensionality, these become extremely difficult or impossible to solve. One solving method is through the reformulation of the quantum mechanical problem as a statistical problem, when possible. This way, machine learning can be used to effectively solve such a problem, as machine learning is a tool for the solving of high-dimensional statistical problems [86]. Probabilistic interpretation allows for using powerful sampling-based methods that work efficiently with high dimensional data.

An example of the reformulation of a quantum problem as a statistical problem is with informationally complete (IC) positive-operator valued measures (POVMs) [87]. POVMs describe the most general measurements of a quantum system. Each particular POVM is defined by a set of positive semidefinite operators $M^\alpha$, with the normalization condition $\sum_\alpha M^\alpha = \mathbb{1}$, where $\mathbb{1}$ is the identity operator. The fact that the POVM is informationally complete means that using measurement outcomes one can reconstruct the state of a system with arbitrary accuracy.

The probability of measurement outcome for a quantum system with the density operator $\rho$ is governed by Born's rule: $P[\alpha] = \text{Tr}(\varrho M^\alpha)$, where $\{M^\alpha\}$ is a particular POVM and $\alpha$ is an outcome result. In other words, any density matrix can be mapped on a mass function, although not all mass functions can be mapped on a density matrix [88,89]. Some mass functions lead to non-positive semidefinite "density matrices", which is not physically allowed. As such, quantum theory is a constrained version of probability theory. For a many-body system, these constraints can be very complicated, and direct consideration of quantum theory as a constrained probability theory is not fruitful. However, if one can access the samples of the IC POVM induced mass function, which is by definition physically allowed, this mass function can be reconstructed using generative modeling [66,67]. Samples can be obtained either by performing generalized measurements over the quantum system or by in silico simulation.

In the present work, we simulate measurements of the ground state of a spin chain with the TFI Hamiltonian, Equation (4). As a local (one spin) IC POVM, we use the so-called symmetric IC POVM for qubits (tetrahedral) POVM [90]:

$$M^\alpha_{\text{tetra}} = \frac{1}{4}(\mathbb{1} + s^\alpha \sigma), \ \alpha \in (0,1,2,3), \ \sigma = (\sigma_x, \sigma_y, \sigma_z),$$

$$s^0 = (0,0,1), \ s^1 = \left(\frac{2\sqrt{2}}{3}, 0, -\frac{1}{3}\right), \ s^2 = \left(-\frac{\sqrt{2}}{3}, \sqrt{\frac{2}{3}}, -\frac{1}{3}\right), \ s^3 = \left(-\frac{\sqrt{2}}{3}, -\sqrt{\frac{2}{3}}, -\frac{1}{3}\right). \quad (5)$$

Note that the many-spin generalization of local IC POVM can easily be obtained by considering the tensor product of local ones:

$$M^{\alpha_1,\ldots,\alpha_N}_{\text{tetra}} = M^{\alpha_1}_{\text{tetra}} \otimes M^{\alpha_2}_{\text{tetra}} \otimes \cdots \otimes M^{\alpha_N}_{\text{tetra}}. \quad (6)$$

To simulate measurements outcome under the IC POVM described above, we implement the following numerical scheme: First, we run a variational MPS ground state solver to obtain the ground state of the TFI model in the MPS form:

$$\Omega_{i_1,i_2,\ldots,i_N} = \sum_{\beta_1,\beta_2,\ldots,\beta_{N-1}} A^1_{i_1\beta_1} A^2_{\beta_1 i_2 \beta_2} \cdots A^N_{\beta_{N-1} i_N} \quad (7)$$

where we use the tensor notation instead of the bra-ket notation for further simplicity, and we obtain the MPS representation of IC POVM induced mass function:

$$P[\alpha_1, \alpha_2, \ldots, \alpha_N] = \sum_{\delta_1, \delta_2, \ldots, \delta_{N-1}} \pi_{\alpha_1 \delta_1} \pi_{\delta_1 \alpha_2 \delta_2} \cdots \pi_{\delta_{N-1} \alpha_N},$$

$$\pi_{\delta_{n-1} \alpha_n \delta_n} = \pi_{\underbrace{\beta_{n-1}\beta'_{n-1}}_{\text{multi-index } \delta_{n-1}} \alpha_n \underbrace{\beta_n \beta'_n}_{\text{multi-index } \delta_n}} = [M_{\text{tetra}}]^{\alpha_n}_{ij} A^n_{\beta_{n-1} j \beta_n} [A^n]^*_{\beta'_{n-1} i \beta'_n} \quad (8)$$

whose diagrammatic representation [35] is shown in Figure 1. Next, we produce a set of samples of size $M$: $\{\alpha^i_1, \alpha^i_2, \ldots, \alpha^i_N\}^M_{i=1}$ from the given probability. The sampling can be efficiently implemented as shown in Appendix B. We call this set of samples (outcome measurements) a data set, which may then be used to train a generative model $p[\alpha_1, \alpha_2, \ldots, \alpha_N | \theta]$ to emulate the true mass function $P[\alpha_1, \alpha_2, \ldots, \alpha_N]$. Here, $\theta$ is the set of parameters of the generative model, which is trained by maximizing the logarithmic likelihood $\mathcal{L}(\theta) = \sum_{i=1}^{M} \log p[\alpha_1^i, \alpha_2^i, \ldots, \alpha_N^i | \theta]$ with respect to the parameters $\theta$ [91]. The trained generative model fully characterizes a quantum state. The density matrix is obtained by applying an inverse transformation to the mass function [92]:

$$\varrho = \sum_{\alpha_1, \alpha_2, \ldots, \alpha_N} p[\alpha_1, \alpha_2, \ldots, \alpha_N | \theta] [M_{\text{tetra}}^{\alpha_1}]^{-1} \otimes [M_{\text{tetra}}^{\alpha_2}]^{-1} \otimes \cdots \otimes [M_{\text{tetra}}^{\alpha_N}]^{-1},$$

$$[M_{\text{tetra}}^{\alpha}]^{-1} = \sum_{\alpha'} T_{\alpha\alpha'}^{-1} M_{\text{tetra}}^{\alpha'}, \qquad (9)$$

$$T_{\alpha\alpha'} = \text{Tr}\left(M_{\text{tetra}}^{\alpha} M_{\text{tetra}}^{\alpha'}\right),$$

the diagrammatic representation of which is given in Figure 2. Note that the summation included in the density matrix representation is numerically intractable, but we can estimate it using samplings from the generative model.

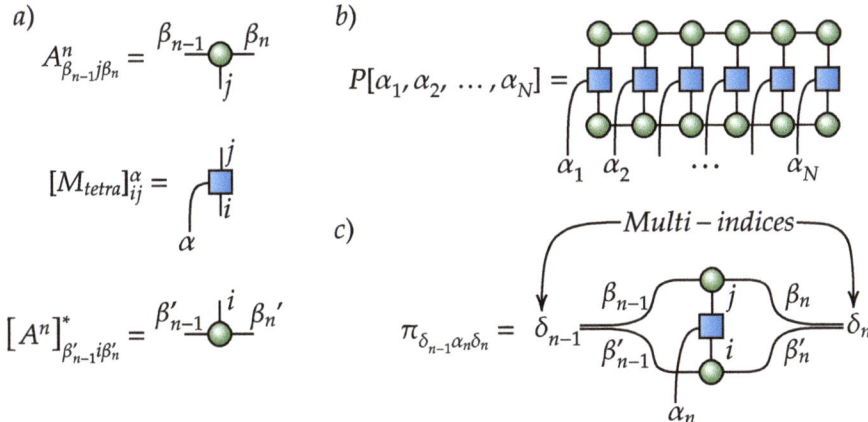

**Figure 1.** Tensor diagrams for (**a**) building blocks, (**b**) matrix product state (MPS) representation of measurement outcome probability, and (**c**) its subtensor.

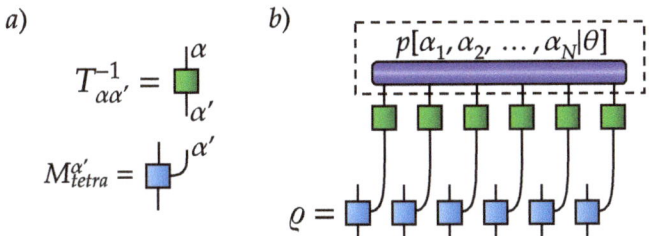

**Figure 2.** Tensor diagrams for (**a**) building blocks and (**b**) inverse transformation from a mass function to a density matrix.

Our goal is to use a generative model as an effective representation of quantum states to calculate the mean values of observables such as, e.g., two-point and higher-order correlation functions. An explicit expression of the two-point correlation function obtained by sampling from the trained generative model is shown in Figure 3. To obtain the ground state of the TFI model, we use a variational MPS ground state search, and we pick the bond dimension of MPS equal to 25 and perform 5 DMRG sweeps to get an approximate ground state in the MPS form. We use the variational MPS solver provided by the mpnum toolbox [93].

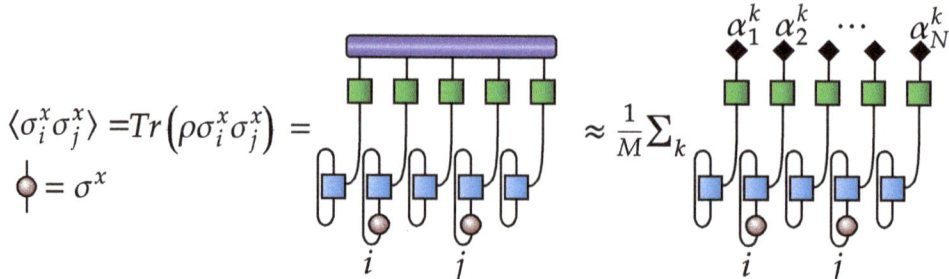

**Figure 3.** Tensor diagrams representing calculation of two-point correlation function.

## 4. Variational Autoencoder Architecture

In our work, we use a conditional VAE [94] to represent quantum states. A conditional VAE is a generative model expressed by the following probability distribution,

$$p[x|\theta, h] = \int p[x|z, \theta, h] p[z] dz, \tag{10}$$

where $x$ is the data we want to simulate; $\theta$ represents the VAE parameters, which can be tuned to get the desired probability distribution over $x$; $h$ is the condition; and $z$ is a vector of latent variables. In our case, $x$ is the quantum measurement outcome in one-hot notation. A collection of measurement outcomes is a matrix of size $N \times 4$, where $N$ is the number of particles in the chain and 4 is the number of possible outcomes of the tetrahedral IC POVM, which is either [1000], [0100], [0010], or [0001]. $h$ is the external magnetic field. The probability distribution $p[x|z, \theta, h]$ can thus be written as

$$p[x|z, \theta, h] = \prod_{i=1}^{N} \prod_{j=1}^{4} \pi_{ij}(z, h, \theta)^{x_{ij}}, \tag{11}$$

where $\pi_{ij}(z, h, \theta)$ is the neural network in our architecture, and, more precisely, $\pi_{ij}$ is the probability of the $j^{th}$ outcome of the POVM for the $i^{th}$ spin with $\sum_{j=1}^{N} \pi_{ij} = 1$ and $\pi_{ij} \geq 0$. The quantity $p[z]$ is the prior distribution over latent variables, which is simply given by $\mathcal{N}(0, I) = \frac{1}{\sqrt{2\pi}^N} \exp\left\{-\frac{1}{2} z^T z\right\}$, with $I$ being the identical covariance matrix. We take the number of latent variables equal to the number of spins, $N$. Essentially, we want to optimize our VAE so that its probability matches the probability of the quantum measurement outcomes as closely as possible. This can be done using the well-known maximum likelihood estimation:

$$\theta_{MLE} = \underset{\theta}{\operatorname{argmax}} \sum_{i=1}^{M} \log(p[x_i|\theta, h]), \tag{12}$$

where $\{x_i\}_{i=1}^{M}$ is the data set of outcome measurements. We cannot simply maximize this function using, for example, a gradient descent method, due to the presence of hidden variables in the structure of this function. However, we can overcome this problem by using the Evidence Lower Bound (ELBO) [95] and the reparametrization trick shown in [96]. The detailed description of the procedure is given in the Appendix A.

Once trained, the VAE is a simple and efficient way to produce new samples from its probability distribution. It can be done in three steps. First, we produce a sample from the prior distribution $p[z] = \mathcal{N}(0, I)$. Next, we feed this sample and the external magnetic field value into the neural network decoder $\pi_{ij}(z, \theta, h)$, which returns the matrix of probabilities. Finally, we sample from the matrix of probability $\pi_{ij}(z, \theta, h)$ to generate "fake" outcome measurements. A visual representation of the sampling method is shown in Figure 4.

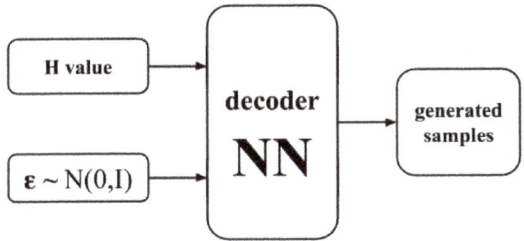

**Figure 4.** Sampling scheme with the trained variational autoencoder (VAE).

In many problems, gradients of observables with respect to different model parameters yield quantities of interest. For example, one may consider the magnetic differential susceptibility tensor $\chi_{ij} = \partial \mu_i / \partial h_j$. It can be done efficiently by using backpropagation through the VAE architecture but, as samples from the VAE are discrete, a straightforward backpropagation is impossible. In recent papers [97–99], a method called the Gumbel-softmax was introduced to overcome this difficulty through continuous relaxation. The spirit, and therefore the physical meaning of the method, may be understood with a short discussion of the so-called simulated annealing technique, which is often used to solve discrete optimization problems. Broadly speaking, the simulated annealing rests on the introduction of a parameter that acts as an artificial "temperature", which varies continuously to modify the state of the system in search of a global optimum. Starting from a given state, for some values of the temperature, if the system mostly explores the neighboring states, moving among them and possibly in the vicinity of the "better" ones, i.e., with lower energy, it may get and remain close to a local optimum, or local energy minimum in the thermodynamic language; however, to avoid remaining in a locally optimal region, "bad" moves leading to worse (i.e., higher energy) states are useful to explore the temperature space more completely improving the chance to find a global optimum or at least to be near it. To each move an energy variation, $\Delta E$, is associated; it is the continuous character of the fictitious temperature that makes the discrete problem continuous as the probability $\exp(-\Delta E)/k_B T$ of acceptance of a state is continuous. Although this approach has been known for a long time [100], it remains topical and under active development [101,102]. The method of continuous relaxation we use also exploits such an artificial temperature to make discrete samples continuous.

The Gumbel-softmax trick, consists of three steps:

1. We calculate the matrix of log probabilities, taking element-wise logarithm of decoder network output: $\log \Pi = \begin{bmatrix} \log \pi_{11} & \log \pi_{12} \ldots \log \pi_{1N} \\ \log \pi_{21} & \log \pi_{22} \ldots \log \pi_{2N} \\ \log \pi_{31} & \log \pi_{32} \ldots \log \pi_{3N} \\ \log \pi_{41} & \log \pi_{42} \ldots \log \pi_{4N} \end{bmatrix}$,
2. We generate a matrix of samples from the standard Gumbel distribution $G$ and sum it up element-wise with the matrix of log probabilities $\log \Pi$: $Z = \log \Pi + G$,
3. Finally, we take the softmax function of the result from the previous step: $x_{\text{soft}}^{\text{fake}}(T) = \text{softmax}(Z/T)$, where $T$ is a temperature of softmax. The softmax functions is defined by the expression $\text{softmax}(x_{ij}) = \frac{\exp(x_{ij})}{\sum_i \exp(x_{ij})}$.

The quantity $x_{\text{soft}}^{\text{fake}}(T)$ has a number of remarkable properties: first, it becomes an exact one-hot sample when $T \to 0$; second, we can backpropagate through soft samples for any T> 0. The method is validated in the next section.

Before we proceed to the presentation and discussion of our results, and to better see the added value of the VAE, it is instructive to compare MPS and VAE (NN) in terms of expressibility, i.e., "estimation of MPS states via incomplete local measurements" vs "VAE reconstruction". As the state of the system is assumed to be unknown, and some measurement outcomes are only known for different magnetic fields, these outcomes are too few for exact tomography. Further, it is known that for a given bond dimension $d$, the entangled entropy cannot be larger than $\log(d)$; in other words, the bond dimension of MPS places an upper bound on the entangled entropy. Thus, the MPS representation describes well only quantum states with low entangled entropy, i.e., quantum states which satisfy the area law [103,104]. The situation with neural network quantum states (NQS) is different: there is no such a restriction for NQS. Moreover, the existence of NQS with volume-law entanglement [105] shows a promising development of new, and possibly powerful, NN-based approaches to representing many-body quantum systems.

## 5. Results

Here, we show that the VAE trained on a set of preliminary measurements is capable to describe the physics of the whole family of TFI models. We validate our results by comparing VAE-based calculations with numerically exact calculations performed by variational MPS algorithm [35]. Additionally, to assess the capabilities of the VAE, we consider a spin chain with 32 spins. We calculate the MPS representation of the ground state and extract information from it by performing measurements over the state. The external field in the $x$-direction is varied from 0 to 2 with a step of 0.1. The VAE is trained on a data set (TFI measurement outcomes) consisting of 10.5 million samples in total: 21 external fields $h_x$ with 500,000 samples per field.

To evaluate the VAE performance, we simply compare directly the numerically exact correlation functions with those reconstructed with our VAE. Those of $n = 1, \ldots, 32$, $\langle \sigma_z^1 \sigma_z^n \rangle$, and $\langle \sigma_x^1 \sigma_x^n \rangle$ are shown in Figures 5 and 6, respectively, and we compare the numerically exact and the VAE-based average magnetizations along $x$, given by $\langle \sigma_x^n \rangle$ for each position of the spin along the chain, in Figure 7. We see that the VAE captures well the physics of the one- and two-point correlation functions. Figure 8 shows the total magnetizations, $\mu_x$ and $\mu_z$, in the $x$ and $z$ directions, respectively, with $\mu_i = \frac{1}{N}\sum_{j=1}^{N} \langle \sigma_i^j \rangle$, and we see that the VAE is a tool well-suited for the description of the quantum phase transition and also finite-size effects: whereas for the infinite TFI chain, i.e., in the thermodynamic limit, the phase transition is observed at $h_x = 1$, and the finite size of the system yields a shift of the critical point at $h_x \approx 0.9$. Also note that in the $T \to 0$ limit, the magnetization $M$ defined in Equation (3) coincides exactly with the magnetization $\mu$ defined above.

A backpropagation algorithm combined with the Gumbel-softmax trick may be used to evaluate the derivative of an output over an input. We use this approach to calculate some elements of a magnetic differential susceptibility tensor $\chi_{ij} = \partial \mu_i / \partial h_j$, in particular, $\chi_{xx}$ and $\chi_{zx}$ shown in Figure 9. The backpropagation-based magnetic differential susceptibility agrees well with the numerically calculated one (central differences). The main advantage of the backpropagation-based calculation is its numerical efficiency. The VAE may thus be trained with an arbitrary set of external parameters, i.e., not only $h_x$, but also $h_y$ and $h_z$, and yield the full differential susceptibility tensor.

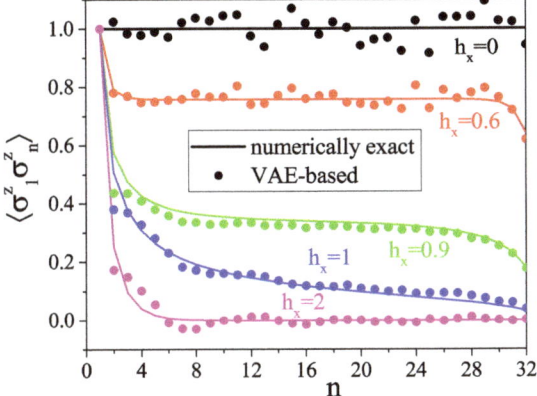

**Figure 5.** Two-point correlation function $\langle \sigma_1^z \sigma_n^z \rangle$ for different values of external magnetic field $h_x$.

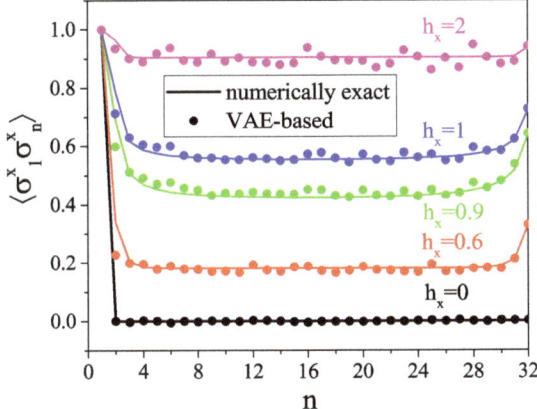

**Figure 6.** Two-point correlation function $\langle \sigma_1^x \sigma_n^x \rangle$ for different values of external magnetic field $h_x$.

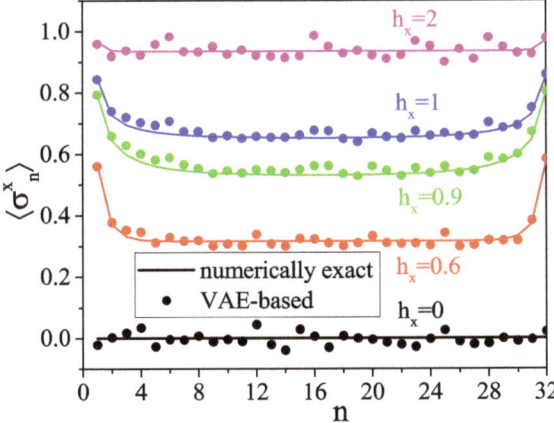

**Figure 7.** Average magnetization per site along $x$ for different values of external magnetic field $h_x$.

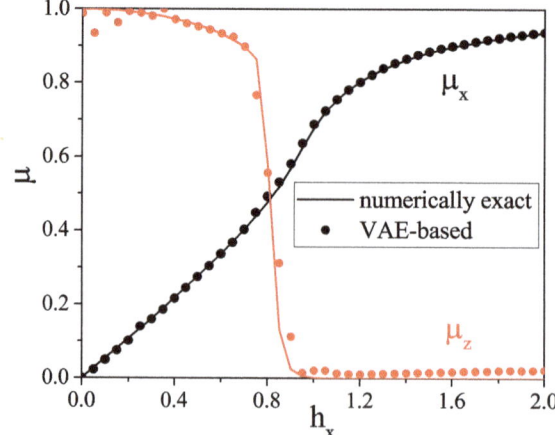

**Figure 8.** Total magnetization along $x$ and $z$ axes for different values of external magnetic field $h_x$. The location of the critical region is slightly shifted towards smaller values of $h_x$ due to the finite size of the chain.

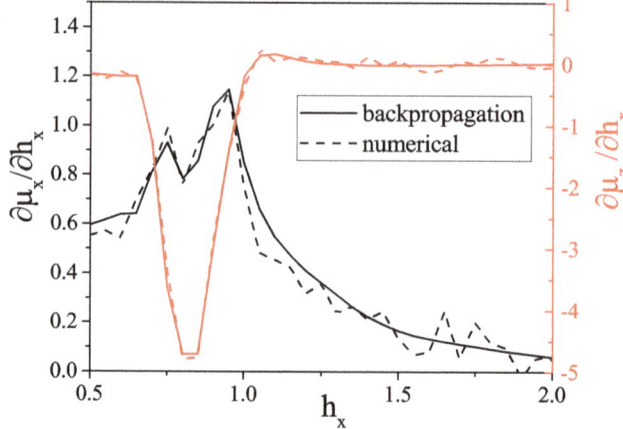

**Figure 9.** Backpropagation-based and numerical-based (central differences) values of $\chi_{xx}$ and $\chi_{zx}$ for different values of external magnetic field $h_x$. Both derivatives slightly fluctuate due to VAE error.

At this stage, we could conclude that the VAE is capable to describe the physics of one- and two-point correlation functions, and therefore the TFI physics. However, notwithstanding the ability of the VAE to yield correlation functions that fit well numerically-exact correlation functions, this is not yet a full proof that it represents quantum states well. To address this point, we consider a small spin chain (five spins with TFI Hamiltonian and an external magnetic field $h_x = 0.9$) for which we calculate both the exact mass function and that estimated from VAE samples. Figure 10 shows that the VAE result again fits the numerically exact mass function with high accuracy. Further, we calculate the Bhattacharyya coefficient [106]: $BC(p_{vae}, p_{exact}) = \sum_\alpha p_{exact}[\alpha] \sqrt{\frac{p_{vae}[\alpha]}{p_{exact}[\alpha]}}$ as a function of the external magnetic field $h_x$. Results reported in Figure 11 show that $BC(p_{vae}, p_{exact}) > 0.99$ over the whole $h_x$ range, which thus proves that the VAE represents a quantum state well, at least for small spin chains.

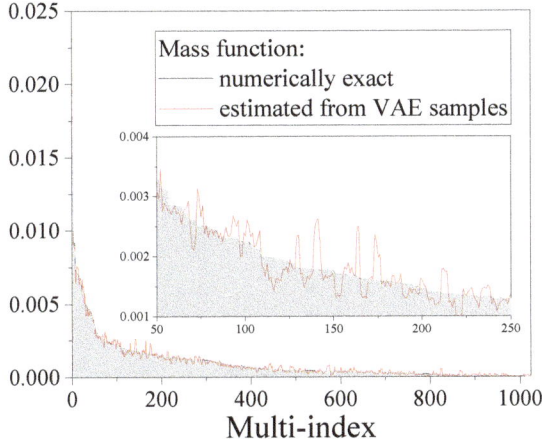

**Figure 10.** Comparison of two positive-operator valued measure (POVM)-induced mass functions ($P[\alpha] = \mathrm{Tr}(\rho M^\alpha)$) for a chain of size 5: numerically exact mass function and reconstructed from VAE samples mass function. A sequence of indices $\alpha$ has been transformed into a single multi-index. Indices have been ordered to put numerically exact probability in descending order. A good agreement between the mass functions is observed.

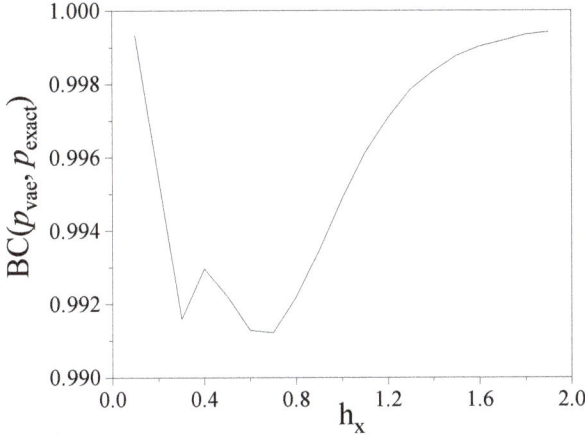

**Figure 11.** Dependence of the classical fidelity on the external magnetic field. A high predictive accuracy is demonstrated for the whole set of fields.

The structure of the entanglement is an another interesting subject that we would like to validate. The essence of entanglement between two parts of the chain, which is split into $n$ left spins and $N - n$ right spins, can be described by the Rényi entropy of the left part of this chain: $S_\alpha = \frac{1}{1-\alpha} \log \mathrm{Tr}\rho_n^\alpha$, where $\rho_n$ is the density matrix of the first $n$ spins in the chain. We estimate the Rényi entropy of order 2: $S_2 = -\log(\mathrm{Tr}\rho^2)$, as it can be efficiently calculated from the matrix product representation of the density matrix and from the VAE samples. However, as sample-based estimation of the entangled entropy has a variance that grows exponentially with the number of spins, we consider a small spin chain of size 10. A direct comparison between the numerically exact and the VAE-based entangled entropies is shown for different values of $n$ in Figure 12. For this particular case, the VAE clearly overestimates the entangled entropy. This undesirable effect is indeed observed for all sizes of spin chains, and even for the spin chain of size 5, for which we have an excellent agreement between the numerically exact mass function and the VAE-based result. The entropy $S_2$ is sensitive to small errors in the mass function, but it also appears that the primary method of state reconstruction used in the present work has the following shortcomings.

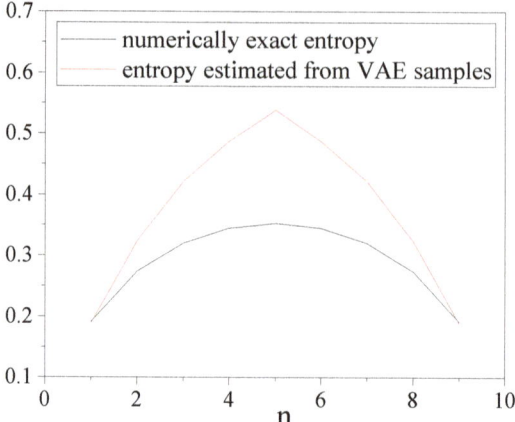

**Figure 12.** Comparison of the numerically exact Rényi entropy and that reconstructed from the VAE samples for different values of $n$.

1. If one reconstructs a pure state, the VAE smooths the spectrum of the density matrix and approximates the pure state by a slightly mixed state, as illustrated with a simple example in Figure 13.
2. The VAE does not account the positivity constraints, which yields negative eigenvalues for the density matrix. These negative eigenvalues even appear in the spectrum of the reduced density matrix, as shown in Figure 13.

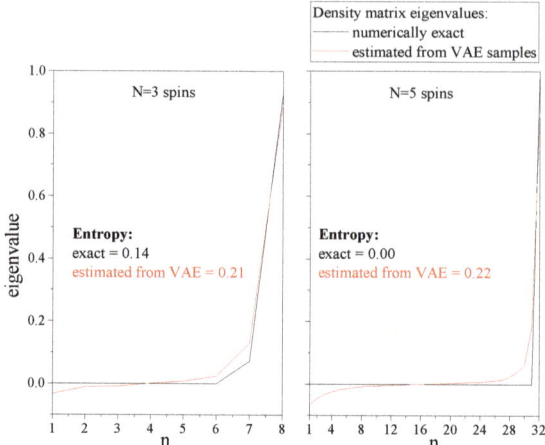

**Figure 13.** Comparison of numerically exact spectra of density matrices and VAE-estimated spectra. The ground state spectra of the spin chain of size 5 with an external magnetic field $h = 0.9$ is shown on the right panel, and the spectra of the reduced density matrix (last 3 spins) are shown on the left panel.

These drawbacks hinder a robust description of the entanglement structure. In addition to the mismatch between the Rényi entropies ($S_2$), the entropy of a reduced density matrix can be larger than the entropy of the whole density matrix, which is erroneous. This particular issue, now identified, may be resolved by introduction of a particular regularization term into the VAE loss. This is the object of future work.

Finally, it is also instructive to comment on the memory costs of the use of either MPS or VAE, which is somehow a tricky question, as it is unclear for any NN-based architecture what numbers of layers and neurons per layer are needed because there is no criterion for NN, whereas for the MPS and tensor networks, there is one. Thus, a direct comparison of NN architectures and tensor networks (MPS, etc.) is certainly a difficult task, and in our opinion, likely an impossible one. At this stage, we may say the following. For a given spin chain of size $N$ and maximal entangled entropy between subchains $S = -\text{Tr}\rho \log \rho$, the MPS requires to store approximately $2N \exp(2S)$ complex numbers; this follows from the fact that one then considers $N$ subtensors of size $\exp(S) \times 2 \times \exp(S)$, where $\exp(S)$ is the typical (approximate) size of bond dimension. For a VAE, although it seems that there are no entropic restrictions, the proper quantitative characterization of the "neural network" complexity of a quantum state still is an open question (for tensor networks, it is the entangled entropy). A VAE contains two neural networks: encoder and decoder. To store a feed-forward neural network, one has to store $\sum_i l_{i-1} \times l_i + l_i$ real numbers, with $l_i$ being the number of neurons in the layer number $i$. In general, one may conclude that the MPS is preferable for low entangled states, and the VAE is preferable for highly entangled states.

## 6. Conclusions

The thermodynamic study of complex many-body quantum systems still requires the development of new methods, including those that may stem from machine learning. The quantum Ising model, which is of particular importance for practical purposes [107,108], provides a rich framework to test these new methods that are also useful to obtain deeper physical insight into its nonequilibrium dynamics properties such as, e.g., quantum fluctuations propagation [109]. In the present work, we studied the ability of a VAE to reconstruct the physics of quantum many-body systems, using the transverse-field Ising model as a nontrivial example. We used the IC POVM to map the quantum problem onto a probabilistic domain and vice versa. We trained the VAE on a set of samples from the transformed quantum problem, and our numerical experiments show the following results.

- For a large system (32 spins), the VAE's reliability is verified by comparing one- and two-point correlation functions.
- For small system (five spins), the VAE's reliability is verified by direct comparison of mass functions.
- The VAE can capture a quantum phase transition.
- The response functions (magnetic differential susceptibility tensor) can be obtained using backpropagation through VAE.
- Despite the very good agreement between the VAE-based mass function and the true mass function, the VAE shows limited performance with the determination of the entangled entropy. This is point is the object of further development.

Our method can be extended to any other thermodynamic system by introduction of the temperature as an external parameter, thereby considering also thermal phase transitions. As one can calculate different thermodynamic quantities by applying backpropagation through VAE, a worthwhile and highly complex system to study would be water under its difference phases, so as to test recent new ideas and models [110,111].

Our code for our numerical experiments is available on the GitHub repository website [112].

**Author Contributions:** Conceptualization, I.A.L., S.N.F., and H.O.; methodology, I.A.L. and A.R.; software, I.A.L., A.R., and P.J.S.; validation, all authors; writing-original draft preparation, I.A.L., A.R., P.J.S., and H.O; writing-review and editing, S.N.F. and H.O.

**Funding:** This research was supported by the Russian Foundation for Basic Research grants under the Project No. 18-37-00282 and the Project No. 18-37-20073. This research was also partially supported by the Skoltech NGP Program (Skoltech-MIT joint project).

**Acknowledgments:** The authors thank Stepan Vintskevich for fruitful discussions. The authors also thank Google Colaboratory for providing access to GPU for the acceleration of computations.

**Conflicts of Interest:** The authors declare no conflicts of interest.

## Abbreviations

The following abbreviations are used in this manuscript.

| | |
|---|---|
| VAE | Variational Autoencoder |
| MPS | Matrix product state |
| TFI | Transverse-field Ising |
| IC | Informationally incomplete |
| POVM | Positive-operator valued measure |
| ELBO | Evidence lower bound |
| NN | Neural network |
| KL | Kullback–Leibler |
| DMRG | Density matrix renormalization group |

## Appendix A. VAE: Training and Implementation Details

When training our VAE, we find the arg maximum of the logarithmic likelihood $\mathcal{L}(\theta)$ w.r.t. its parameters $\theta$:

$$\theta_{MLE} = \underset{\theta}{\operatorname{argmax}} \mathcal{L}(\theta) = \underset{\theta}{\operatorname{argmax}} \log(p[x|\theta, h]), \tag{A1}$$

Equation (A1) cannot directly be evaluated, because of hidden variables in the structure of $p[x|\theta, h]$. We can, however, simplify this problem by introducing a distribution over hidden variables $z$. Remember that the probability distribution can be described as $p[x|\theta, h] = \int p[x|z, \theta, h] p[z] dz$, so that the expression for the log likelihood becomes

$$\mathcal{L}(\theta) = \log\left(\int p[x|z, \theta, h] p[z] dz\right). \tag{A2}$$

We can then use a mathematical trick that might seem counterintuitive at first glance, but ultimately becomes quite powerful. We multiply the function inside the integral by $\frac{q[z|x,\tilde{\theta},h]}{q[z|x,\tilde{\theta},h]} = 1$, where $q[z|x, \tilde{\theta}, h]$ is some arbitrary distribution that can be adjusted with $\tilde{\theta}$, so that

$$\mathcal{L}(\theta) = \log\left(\int p[x|z, \theta, h] p[z] dz\right) = \log\left(\int \frac{q[z|x, \tilde{\theta}, h]}{q[z|x, \tilde{\theta}, h]} p[x|z, \theta, h] p[z] dz\right)$$

$$= \log\left(\mathbb{E}_{q[z|x,\tilde{\theta},h]} p[x|z, \theta, h] \frac{p[z]}{q[z|x, \tilde{\theta}, h]}\right) \tag{A3}$$

where the quantity $\mathbb{E}_{f[x]}$ denotes the expectation value w.r.t some distribution $f[x]$. We can then use Jensen's inequality to show that

$$\log\left(\mathbb{E}_{q[z|x,\tilde{\theta},h]} p[x|z, \theta, h] \frac{p[z]}{q[z|x, \tilde{\theta}, h]}\right) \geq \mathbb{E}_{q[z|x,\tilde{\theta},h]} \log\left(p[x|z, \theta, h] \frac{p[z]}{q[z|x, \tilde{\theta}, h]}\right). \tag{A4}$$

where the rhs of this inequality is the lower bound of the log likelihood, as it will always be greater than or equal to the lower bound, and equality can always be achieved by a proper choice of $q$ if it is in a complex enough family.

Maximizing the lower bound is equivalent to maximizing the log likelihood. We can decompose this lower bound term into two terms:

$$\mathcal{L}(\theta) \geq \text{ELBO}(\theta, \tilde{\theta}) = \mathbb{E}_{q[z|x,\tilde{\theta},h]} \log(p[x|z,\theta,h]) - \int q[z|x,\tilde{\theta},h] \log \frac{q[z|x,\tilde{\theta},h]}{p[z]} dz \quad (A5)$$

Note that the second term is equivalent to the Kullback–Leibler divergence $KL(q[z|x,\tilde{\theta},h] \| p[z])$. In our case, we picked the particular distribution forms that reflect the structure of our problem:

$$p[x|z,\theta,h] = \prod_{i=1}^{N} \prod_{j=1}^{4} \pi_{ij}(z,\theta,h)^{x_{ij}},$$

$$q[z|x,\tilde{\theta},h] = \mathcal{N}(\mu_i(x,\tilde{\theta},h), \text{Diag}(\sigma_i^2(x,\tilde{\theta},h))), \quad (A6)$$

$$P[z] = \mathcal{N}(0, I)$$

where $\mu_i$ and $\sigma_i$ are given by the encoder neural network, and $\pi_{ij}$ is given by the decoder neural network, with $\sum_{j=1}^{4} \pi_{ij} = 1$ and $\pi_{ij} \geq 0$, which can be achieved by applying the softmax funtion to the output of the neural network. Now, we can use the reparametrization trick to change the variable in the integral $z = \sigma_j(x,\tilde{\theta},h)\varepsilon + \mu_j(x,\tilde{\theta},h)$, where $\varepsilon_j \sim \mathcal{N}(0, I)$, to simplify this expression to

$$\text{ELBO}(\theta, \tilde{\theta}) = \sum_{i=1}^{N} \sum_{j=1}^{4} x_{ij} \left\langle \log\left(\pi_{ij}(\sigma_i(x,\tilde{\theta},h)\varepsilon + \mu_i(x,\tilde{\theta},h), \theta, h)\right) \right\rangle_{\varepsilon_j \sim \mathcal{N}(0,I)}$$

$$- \sum_{i=1}^{N} \left( \log \sigma_i(x,\tilde{\theta},h) - \frac{\sigma_i^2(x,\tilde{\theta},h) + \mu_i^2(x,\tilde{\theta},h) - 1}{2} \right). \quad (A7)$$

The first term is the cross-entropy, which pushes the probability distribution to be as close as possible to the data. The second term is the regularizer, which forces the latent variable $z$ not to diverge too much from the normal distribution $\mathcal{N}(0, I)$, so that the VAE can be used to generate new data once it is trained. Note that both $x_{ij}$ and $\sigma_i$ must be positive. Instead of adding a constraint to the VAE, which would be difficult to do, we train the VAE for the variables $\Pi = \log \pi$ and $\xi = 2\log\sigma$. Equation (A7) then becomes

$$\text{ELBO}(\theta, \tilde{\theta}) = \sum_{i=1}^{N} \sum_{j=1}^{4} x_{ij} \left\langle \Pi_{ij}(e^{\xi_i(x,\tilde{\theta},h)/2}\varepsilon + \mu_i(x,\tilde{\theta},h), \theta, h) \right\rangle_{\varepsilon_j \sim \mathcal{N}(0,I)}$$

$$- \frac{1}{2} \sum_{i=1}^{N} \left( \xi_i(x,\tilde{\theta},h) - e^{\xi_i(x,\tilde{\theta},h)} - \mu_i^2(x,\tilde{\theta},h) + 1 \right). \quad (A8)$$

Now, $\text{ELBO}(\theta, \tilde{\theta})$ can be effectively optimized using gradient descent methods, averaging over $\varepsilon$ can be done by sampling. Generalizing to a data set of size $M$: $\{x^k\}_{k=1}^{M}$ can be easily done and is shown by

$$\text{ELBO}(\theta, \tilde{\theta}) = \sum_{k=1}^{M} \sum_{i=1}^{N} \sum_{j=1}^{4} x_{ij}^k \left\langle \Pi_{ij}(e^{\xi_i(x^k,\tilde{\theta},h)/2}\varepsilon + \mu_i(x^k,\tilde{\theta},h), \theta, h) \right\rangle_{\varepsilon_j \sim \mathcal{N}(0,I)}$$

$$- \frac{1}{2} \sum_{k=1}^{M} \sum_{i=1}^{N} \left( \xi_i(x^k,\tilde{\theta},h) - e^{\xi_i(x^k,\tilde{\theta},h)} - \mu_i^2(x^k,\tilde{\theta},h) + 1 \right). \quad (A9)$$

A visual representation of the VAE architecture is shown in Figure A1.

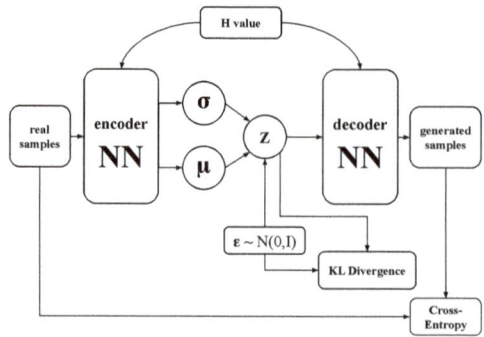

**Figure A1.** Architecture of the variational autoencoder.

To solve the optimization problem, we use Adam optimizer [113] with standard parameters (lr = 0.001, $\beta_1 = 0.9, \beta_2 = 0.999$). For the encoder and decoder, we use fully-connected neural networks with two hidden layers and 256 neurons on each. We train the VAE using batches of size 100,000 samples and for 750 epochs.

## Appendix B. Sampling from POVM-Induced Mass Function

The mass function induced by POVM $P[\alpha_1, \alpha_2, \ldots, \alpha_N]$ has a form of matrix product state. Thus, one can easily calculate any marginal mass function because a summation over any $\alpha$ can be done locally. Any conditional mass functions can be also calculated by using marginal mass functions. Thus, one can calculate chain decomposition of the whole mass function:

$$P[\alpha_1, \alpha_2, \ldots, \alpha_N] = P[\alpha_N] P[\alpha_{N-1} | \alpha_N] P[\alpha_{N-2} | \alpha_{N-1}, \alpha_N] \ldots P[\alpha_1 | \alpha_2, \ldots, \alpha_N] \quad (A10)$$

With this decomposition, one can produce a sample $\tilde{\alpha}_N$ from $P[\alpha_N]$ first, then a sample $\tilde{\alpha}_{N-1}$ from $P[\alpha_{N-1} | \tilde{\alpha}_N]$, and continue up to the end of the chain. The obtained set $\{\tilde{\alpha}_1, \tilde{\alpha}_2, \ldots, \tilde{\alpha}_N\}$ is a valid sample from the mass function.

## References

1. Muller, I. *A History of Thermodynamics*; Springer: Berlin/Heidelberg, Germany, 2007.
2. Onsager, L. Reciprocal Relations in Irreversible Processes. I. *Phys. Rev.* **1931**, *37*, 405–426. [CrossRef]
3. De Groot, S.R. *Thermodynamics of Irreversible Processes*; Interscience: New York, NY, USA, 1958.
4. Le Bellac, M.; Mortessagne, F.; Batrouni, G.G. *Equilibrium and Non-Equilibrium Statistical Thermodynamics*; Cambridge University Press: Cambridge, UK, 2006.
5. Apertet, Y.; Ouerdane, H.; Goupil, C.; Lecoeur, P. Revisiting Feynman's ratchet with thermoelectric transport theory. *Phys. Rev. E* **2014**, *90*, 012113. [CrossRef] [PubMed]
6. Goupil, C.; Ouerdane, H.; Herbert, E.; D'Angelo, Y.; Lecoeur, P. Closed-loop approach to thermodynamics. *Phys. Rev. E* **2016**, *94*, 032136. [CrossRef]
7. Andresen, B. Current trends in finite-time thermodynamics. *Angew. Chem.-Int. Edit.* **2011**, *50*, 2690–2704. [CrossRef] [PubMed]
8. Ouerdane, H.; Apertet, Y.; Goupil, C.; Lecoeur, P. Continuity and boundary conditions in thermodynamics: From Carnot's efficiency to efficiencies at maximum power. *Eur. Phys. J. Spec. Top.* **2015**, *224*, 839–864. [CrossRef]
9. Apertet, Y.; Ouerdane, H.; Goupil, C.; Lecoeur, P. True nature of the Curzon-Ahlborn efficiency. *Phys. Rev. E* **2017**, *96*, 022119. [CrossRef]
10. Boltzmann, L. Uber die beziehung dem zweiten Haubtsatze der mechanischen Warmetheorie und der Wahrscheinlichkeitsrechnung respektive den Satzen uber das Warmegleichgewicht. *Wiener Berichte* **1877**, *76*, 373–435.

11. Gibbs, J.W. *Elementary Principles in Statistical Mechanics*; Charles Scribner's Sons: New York, NY, USA, 1902.
12. Penrose, O. Foundations of statistical mechanics. *Rep. Prog. Phys.* **1979**, *42*, 1937–2006. [CrossRef]
13. Goldstein, S.; Lebowitz, J.L.; Zanghì, N. Gibbs and Boltzmann entropy in classical and quantum mechanics. *arXiv* **2019**, arXiv:1903.11870. Available online: https://arxiv.org/abs/1903.11870 (accessed on 6 November 2019).
14. Shannon, C.E. A mathematical theory of communication. *Bell Labs Tech. J.* **1948**, *27*, 379–423. [CrossRef]
15. von Neumann, J. *Mathematical Foundations of Quantum Mechanics. New Edition*; Princeton University Press: Princeton, NJ, USA, 2018.
16. Datta, S. *Electronic Transport in Mesoscopic Systems*; Cambridge University Press: Cambridge, UK, 1995.
17. Heikillä, T.T. *The Physics of Nanoelectronics*; Oxford University Press: Oxford, UK, 2013.
18. Chomaz, P.; Colonna, M.; Randrup, J. Nuclear spinodal fragmentation. *Phys. Rep.* **2004**, *389*, 263–440. [CrossRef]
19. Bressanini, D.; Morosi, G.; Mella, M. Robust wave function optimization procedures in quantum Monte Carlo methods. *J. Chem. Phys.* **2002**, *116*, 5345–5350. [CrossRef]
20. Feiguin, A.E.; White, S.R. Finite-temperature density matrix renormalization using an enlarged Hilbert space. *Phys. Rev. B* **2005**, *72*, 220401. [CrossRef]
21. Deutsch, J.M. Quantum statistical mechanics in a closed system. *Phys. Rev. A* **1991**, *43*, 2046–2049. [CrossRef]
22. Srednicki, M. Chaos and quantum thermalization. *Phys. Rev. E* **1994**, *50*, 888–901. [CrossRef]
23. Rigol, M.; Dunjko, V.; Olshanii, M. Thermalization and its mechanism for generic isolated quantum systems. *Nature* **2008**, *452*, 854–858. [CrossRef]
24. Dymarsky, A.; Lashkari, N.; Liu, H. Subsystem eigenstate thermalization hypothesis *Phys. Rev. E* **2018**, *97*, 012140. [CrossRef] [PubMed]
25. Dymarsky, A. Mechanism of macroscopic equilibration of isolated quantum systems *Phys. Rev. B* **2019**, *99*, 224302. [CrossRef]
26. Carleo, G.; Becca, F.; Schiró, M.; Fabrizio, M. Localization and glassy dynamics of many-body quantum systems. *Sci. Rep.* **2012**, *2*, 243. [CrossRef] [PubMed]
27. Chen, L.; Gelin, M.; Zhao, Y. Dynamics of the spin-boson model: A comparison of the multiple Davydov $D_1$, $D_{1.5}$, $D_2$ Ansätze. *Chem. Phys.* **2018**, *515*, 108–118. [CrossRef]
28. Lanyon, B.; Maier, C.; Holzäpfel, M.; Baumgratz, T.; Hempel, C.; Jurcevic, P.; Dhand, I.; Buyskikh, A.; Daley, A.; Cramer, M.; et al. Efficient tomography of a quantum many-body system. *Nat. Phys.* **2017**, *13*, 1158. [CrossRef]
29. Liao, H.J.; Liu, J.G.; Wang, L.; Xiang, T. Differentiable programming tensor networks. *Phys. Rev. X* **2019**, *9*, 031041. [CrossRef]
30. Fetter, A.L.; Walecka, J.D. *Quantum Theory of Many-Particle Systems*; Dover: New York, NY, USA, 2003.
31. Frésard, R.; Kroha, J.; Wölfle, P., The pseudoparticle approach to strongly correlated electron systems. In *Strongly Correlated Systems*; Avella, A., Mancini, F., Eds.; Springer: Berlin/Heidelberg, Germany, 2011; Volume 171.
32. Georges, A.; Kotliar, G.; Krauth, W.; Rozenberg, M.J. Dynamical mean-field theory of strongly correlated fermion systems and the limit of infinite dimensions. *Rev. Mod. Phys.* **1996**, *68*, 13–125. [CrossRef]
33. Negele, J.W.; Orland, H. *Quantum Many-Particle Systems*; Perseus Books: New York, NY, USA, 1998.
34. Foulkes, W.; Mitas, L.; Needs, R.; Rajagopal, G. Quantum Monte Carlo simulations of solids. *Rev. Mod. Phys.* **2001**, *73*, 33. [CrossRef]
35. Orús, R. A practical introduction to tensor networks: Matrix product states and projected entangled pair states. *Ann. Phys.* **2014**, *349*, 117–158. [CrossRef]
36. Orús, R. Tensor networks for complex quantum systems. *Nat. Rev. Phys.* **2019**, *1*, 538–550. [CrossRef]
37. Schollwöck, U. The density-matrix renormalization group in the age of matrix product states. *Ann. Phys.* **2011**, *326*, 96–192. [CrossRef]
38. Vidal, G. Efficient classical simulation of slightly entangled quantum computations. *Phys. Rev. Lett.* **2003**, *91*, 147902. [CrossRef]
39. Evenbly, G.; Vidal, G. Quantum criticality with the multiscale entanglement renormalization ansatz. In *Strongly Correlated Systems*; Springer: Berlin/Heidelberg, Germany, 2013; pp. 99–130.
40. Pollock, F.A.; Rodríguez-Rosario, C.; Frauenheim, T.; Paternostro, M.; Modi, K. Non-Markovian quantum processes: Complete framework and efficient characterization. *Phys. Rev. A* **2018**, *97*, 012127. [CrossRef]

41. Luchnikov, I.; Vintskevich, S.; Ouerdane, H.; Filippov, S. Simulation complexity of open quantum dynamics: Connection with tensor networks. *Phys. Rev. Lett.* **2019**, *122*, 160401. [CrossRef]
42. Taranto, P.; Pollock, F.A.; Modi, K. Memory strength and recoverability of non-Markovian quantum stochastic processes. *arXiv* **2019**, arXiv:1907.12583. Available online: https://arxiv.org/abs/1907.12583 (accessed on 6 November 2019).
43. Milz, S.; Pollock, F.A.; Modi, K. Reconstructing non-Markovian quantum dynamics with limited control. *Phys. Rev. A* **2018**, *98*, 012108. [CrossRef]
44. Luchnikov, I.A.; Vintskevich, S.V.; Grigoriev, D.A.; Filippov, S.N. Machine learning of Markovian embedding for non-Markovian quantum dynamics. *arXiv* **2019**, arXiv:1902.07019. Available online: https://arxiv.org/abs/1902.07019 (accessed on 6 November 2019).
45. Verstraete, F.; Murg, V.; Cirac, J.I. Matrix product states, projected entangled pair states, and variational renormalization group methods for quantum spin systems. *Adv. Phys.* **2008**, *57*, 143–224. [CrossRef]
46. Levin, M.; Nave, C.P. Tensor renormalization group approach to two-dimensional classical lattice models. *Phys. Rev. Lett.* **2007**, *99*, 120601. [CrossRef]
47. Evenbly, G.; Vidal, G. Tensor network renormalization. *Phys. Rev. Lett.* **2015**, *115*, 180405. [CrossRef] [PubMed]
48. Gemmer, J.; Michel, M. *Quantum Thermodynamics*; Springer: Berlin/Heidelberg, Germany, 2009.
49. Kosloff, R. Quantum thermodynamics and open-systems modeling. *J. Phys. Chem.* **2019**, *150*, 204105. [CrossRef]
50. Allahverdyan, A.E.; Johal, R.S.; Mahler, G. Work extremum principle: Structure and function of quantum heat engines. *Phys. Rev. E* **2008**, *77*, 041118. [CrossRef]
51. Thomas, G.; Johal, R.S. Coupled quantum Otto cycle. *Phys. Rev. E* **2011**, *83*, 031135. [CrossRef]
52. Makhlin, Y.; Schön, G; Shnirman, A. Quantum-state engineering with Josephson-junction devices. *Rev. Mod. Phys.* **2001**, *73*, 357–400. [CrossRef]
53. Navez, P.; Sowa, A.; Zagoskin, A. Entangling continuous variables with a qubit array. *Phys. Rev. B* **2019**, *100*, 144506. [CrossRef]
54. Bishop, C.M. *Pattern Recognition and Machine Learning*; Springer: Berlin/Heidelberg, Germany, 2006.
55. Turing, M.A. Computing machinery and intelligence. *Mind* **1950**, *59*, 433–460. [CrossRef]
56. Crevier, D. *AI: The Tumultuous Search for Artificial Intelligence*; BasicBooks: New York, NY, USA, 1993.
57. Biamonte, J.; Wittek, P.; Pancotti, N.; Rebentrost, P.; Wiebe, N.; Lloyd, S. Quantum machine learning. *Nature* **2017**, *549*, 195–202. [CrossRef] [PubMed]
58. Carleo, G.; Troyer, M. Solving the quantum many-body problem with artificial neural networks. *Science* **2017**, *355*, 602–606. [CrossRef] [PubMed]
59. Torlai, G.; Mazzola, G.; Carrasquilla, J.; Troyer, M.; Melko, R.; Carleo, G. Neural-network quantum state tomography. *Nat. Phys.* **2018**, *14*, 447. [CrossRef]
60. Tiunov, E.S.; Tiunova, V.V.; Ulanov, A.E.; Lvovsky, A.I.; Fedorov, A.K. Experimental quantum homodyne tomography via machine learning. *arXiv* **2019**, arXiv:1907.06589. Available online: https://arxiv.org/abs/1907.06589 (accessed on 6 November 2019).
61. Choo, K.; Neupert, T.; Carleo, G. Study of the two-dimensional frustrated J1-J2 model with neural network quantum states. *Phys. Rev. B* **2019**, *100*, 124125. [CrossRef]
62. Sharir, O.; Levine, Y.; Wies, N.; Carleo, G.; Shashua, A. Deep autoregressive models for the efficient variational simulation of many-body quantum systems. *arXiv* **2019**, arXiv:1902.04057. Available online: https://arxiv.org/abs/1902.04057 (accessed on 6 November 2019).
63. Wu, D.; Wang, L.; Zhang, P. Solving statistical mechanics using variational autoregressive networks. *Phys. Rev. Lett.* **2019**, *122*, 080602. [CrossRef]
64. Kharkov, Y.A.; Sotskov, V.E.; Karazeev, A.A.; Kiktenko, E.O.; Fedorov, A.K. Revealing quantum chaos with machine learning. *arXiv* **2019**, arXiv:1902.09216. Available online: https://arxiv.org/abs/1902.09216 (accessed on 6 November 2019).
65. Rocchetto, A.; Grant, E.; Strelchuk, S.; Carleo, G.; Severini, S. Learning hard quantum distributions with variational autoencoders. *npj Quantum Inf.* **2018**, *4*, 28. [CrossRef]
66. Carrasquilla, J.; Torlai, G.; Melko, R.G.; Aolita, L. Reconstructing quantum states with generative models. *Nat. Mach. Intell.* **2019**, *1*, 155. [CrossRef]

67. Generative Models for Physicists. Lecture note. Available online: http://wangleiphy.github.io/lectures/PILtutorial.pdf (accessed on 7 November 2019).
68. Hewson, A.C. *The Kondo Problem to Heavy Fermions*; Cambridge University Press: Cambridge, UK, 1993.
69. Coleman, P. Heavy fermions: Electrons at the edge of magnetism. In *Handbook of Magnetism and Advanced Magnetic Materials*; Kronmüller, H., Parkin, S., Eds.; John Wiley & Sons: Chichester, UK, 2007.
70. Sachdev, S. *Quantum Phase Transitions*; Cambridge University Press: Cambridge, UK, 2000.
71. Coleman, P.; Schofield, A. Quantum criticality. *Nature* **2000**, *433*, 226–229. [CrossRef] [PubMed]
72. Anderson, P.W. Localized Magnetic States in Metals. *Phys. Rev.* **1961**, *124*, 41–53. [CrossRef]
73. Frésard, R.; Ouerdane, H.; Kopp, T. Slave bosons in radial gauge: a bridge between path integral and Hamiltonian language. *Nucl. Phys. B* **2007**, *785*, 286–306. [CrossRef]
74. Frésard, R.; Ouerdane, H.; Kopp, T. Barnes slave-boson approach to the two-site single-impurity Anderson model with non-local interaction. *EPL* **2008**, *82*, 31001. [CrossRef]
75. Diu, B.; Guthmann, C.; Lederer, D.; Roulet, B. *Physique Statistique*; Éditions Hermann: Paris, France, 1996.
76. Mila, F., Frustrated spin systems. In *Many-Body Physics: From Kondo to Hubbard*; Pavarini, E., Koch, E., Coleman, P., Eds.; Verlag des Forschungszentrum Jülich: Kreis Düren, Rheinland, 2015.
77. Refael, G.; Moore, J.E. Entanglement Entropy of Random Quantum Critical Points in One Dimension. *Phys. Rev. Lett.* **2004**, *93*. [CrossRef]
78. Schollwóck, U. The density-matrix renormalization group. *Rev. Mod. Phys.* **2005**, *77*, 259–315. [CrossRef]
79. Ising, E. Beitrag zur Theorie des Ferromagnetismus. *Z. Phys.* **1925**, *31*, 253–258. [CrossRef]
80. Kramers, H.A.; Wannier, G.H. Statistics of the two-dimensional ferromagnet. Part I. *Phys. Rev.* **1941**, *60*, 252–262. [CrossRef]
81. Ovchinnikov, A.A.; Dmitriev, D.V.; Krivnov, V.Y.; Cheranovskii, V.O. Antiferromagnetic Ising chain in a mixed transverse and longitudinal magnetic field. *Phys. Rev. B* **2003**, *68*, 214406. [CrossRef]
82. Coldea, R.; Tennant, D.A.; Wheeler, E.M.; Wawrzynska, E.; Prabhakaran, D.; Telling, M.; Habicht, K.; Smeibidl, P.; K, K. Quantum criticality in an Ising chain: Experimental evidence for emergent $E_8$ symmetry. *Science* **2010**, *327*, 177–180. [CrossRef] [PubMed]
83. Sachdev, S.; Keimer, B. Quantum criticality. *Phys. Today* **2011**, *64*, 29–35. [CrossRef]
84. Matsubara, T. A new approach to quantum statistical mechanics. *Prog. Theor. Exp.* **1955**, *14*, 351–378. [CrossRef]
85. Kogut, J.B. An introduction to lattice gauge theory and spin systems. *Rev. Mod. Phys.* **1979**, *51*, 659–713. [CrossRef]
86. Krizhevsky, A.; Sutskever, I.; Hinton, G.E. Imagenet classification with deep convolutional neural networks. In Proceedings of the NIPS: Advances in Neural Information Processing Systems 25, Stateline, NV, USA, 3–8 December 2012; pp. 1097–1105.
87. Holevo, A.S. *Probabilistic and Statistical Aspects of Quantum Theory*; Springer: Berlin/Heidelberg, Germany, 2011; Volume 1.
88. Filippov, S.N.; Man'ko, V.I. Inverse spin-s portrait and representation of qudit states by single probability vectors. *J. Russ. Laser Res.* **2010**, *31*, 32–54. [CrossRef]
89. Appleby, M.; Fuchs, C.A.; Stacey, B.C.; Zhu, H. Introducing the Qplex: A novel arena for quantum theory. *Eur. Phys. J. D* **2017**, *71*, 197. [CrossRef]
90. Caves, C.M. Symmetric informationally complete POVMs - UNM Information Physics Group internal report (1999). Available online: http://info.phys.unm.edu/~caves/reports/infopovm.pdf (accessed on 7 November 2019).
91. Myung, I.J. Tutorial on maximum likelihood estimation. *J. Math. Psychol.* **2003**, *47*, 90–100. [CrossRef]
92. Filippov, S.N.; Man'ko, V.I. Symmetric informationally complete positive operator valued measure and probability representation of quantum mechanics. *J. Russ. Laser Res.* **2010**, *31*, 211–231. [CrossRef]
93. mpnum: A Matrix Product Representation Library for Python. Available online: https://mpnum.readthedocs.io/en/latest/ (accessed on 7 November 2019).
94. Sohn, K.; Lee, H.; Yan, X. Learning structured output representation using deep conditional generative models. In Proceedings of the NIPS: Advances in Neural Information Processing Systems 28, Montreal, QC, Canada, 7–12 December 2015; pp. 3483–3491.
95. Kingma, D.P.; Welling, M. Auto-encoding variational Bayes. *arXiv* **2013**, arXiv:1312.6114. Available online: https://arxiv.org/abs/1312.6114 (accessed on 6 November 2019).

96. Rezende, D.J.; Mohamed, S.; Wierstra, D. Stochastic backpropagation and approximate inference in deep generative models. In Proceedings of the 31st International Conference on Machine Learning (ICML), Beijing, China, 21–26 June 2014; Volume 32.
97. Jang, E.; Gu, S.; Poole, B. Categorical reparameterization with Gumbel-softmax. *arXiv* **2016**, arXiv:1611.01144. Available online: https://arxiv.org/abs/1611.01144 (accessed on 6 November 2019).
98. Kusner, M.J.; Hernández-Lobato, J.M. Gans for sequences of discrete elements with the Gumbel-softmax distribution. *arXiv* **2016**, arXiv:1611.04051. Available online: https://arxiv.org/abs/1611.04051 (accessed on 6 November 2019).
99. Maddison, C.J.; Mnih, A.; Teh, Y.W. The concrete distribution: A continuous relaxation of discrete random variables. *arXiv* **2016**, arXiv:1611.00712. Available online: https://arxiv.org/abs/1611.00712 (accessed on 6 November 2019).
100. Metropolis, N.; Rosenbluth, A.W.; Rosenbluth, M.N.; Teller, A.H.; Teller, E. Equation of State Calculations by Fast Computing Machines. *J. Chem. Phys.* **1953**, *21*, 1087–1092. [CrossRef]
101. Das, A.; Chakrabarti, B.K. Colloquium: Quantum annealing and analog quantum computation. *Rev. Mod. Phys.* **2008**, *80*, 1061–1081. [CrossRef]
102. Yavorsky, A.; Markovich, L.A.; Polyakov, E.A.; Rubtsov, A.N. Highly parallel algorithm for the Ising ground state searching problem. *arXiv* **2019**, arXiv:1907.05124. Available online: https://arxiv.org/abs/1907.05124 (accessed on 6 November 2019).
103. Verstraete, F.; Cirac, J.I. Matrix product states represent ground states faithfully. *Phys. Rev. B* **2006**, *73*, 094423. [CrossRef]
104. Eisert, J.; Cramer, M.; Plenio, M.B. Colloquium: Area laws for the entanglement entropy. *Rev. Mod. Phys.* **2010**, *82*, 277–306. [CrossRef]
105. Deng, D.-L.; Li, X.; Das Sarma, S. Quantum entanglement in neural network states. *Phys. Rev. X* **2017**, *7*, 021021. [CrossRef]
106. Bhattacharyya, A. On a measure of divergence between two statistical populations defined by their probability distributions. *Bull. Calcutta Math. Soc.* **1943**, *35*, 99–109.
107. Boixo, S.; Ronnow, T.F.; Isakov, S.V.; Wang, Z.; Wecker, D.; Lidar, D.A.; Martinis, J.M.; Troyer, M. Evidence for quantum annealing with more than one hundred qubits. *Nat. Phys.* **2014**, *10*, 218–224. [CrossRef]
108. Denchev, V.S.; Boixo, S.; Isakov, S.V.; Ding, N.; Babbush, R.; Smelyanskiy, V.; Martinis, J.; Neven, H. What is the computational value of finite-range tunneling? *Phys. Rev. X* **2016**, *6*, 031015. [CrossRef]
109. Navez, P.; Tsironis, G.P.; Zagoskin, A.M. Propagation of fluctuations in the quantum Ising model. *Phys. Rev. B* **2017**, *95*, 064304. [CrossRef]
110. Volkov, A.A.; Artemov, V.G.; Pronin, A.V. A radically new suggestion about the electrodynamics of water: Can the pH index and the Debye relaxation be of a common origin? *EPL* **2014**, *106*, 46004. [CrossRef]
111. Artemov, V.G. A unified mechanism for ice and water electrical conductivity from direct current to terahertz. *Phys. Chem. Chem. Phys.* **2019**, *21*, 8067–8072. [CrossRef] [PubMed]
112. Github Repository with Code. Available online: https://github.com/LuchnikovI/Representation-of-quantum-many-body-states-via-VAE (accessed on 7 November 2019).
113. Kingma, D.P.; Ba, J. Adam: A method for stochastic optimization. *arXiv* **2014**, arXiv:1412.6980. Available online: https://arxiv.org/abs/1412.6980 (accessed on 6 November 2019).

© 2019 by the authors. Licensee MDPI, Basel, Switzerland. This article is an open access article distributed under the terms and conditions of the Creative Commons Attribution (CC BY) license (http://creativecommons.org/licenses/by/4.0/).

Article

# Thermodynamic and Transport Properties of Equilibrium Debye Plasmas

Gianpiero Colonna * and Annarita Laricchiuta *

CNR Istituto per la Scienza e Tecnologia dei Plasmi (ISTP) Bari, via Amendola 122/D, 70126 Bari, Italy
* Correspondence: gianpiero.colonna@cnr.it (G.C.); annarita.laricchiuta@cnr.it (A.L.)

Received: 20 December 2019; Accepted: 18 February 2020 ; Published: 20 February 2020

**Abstract:** The thermodynamic and transport properties of weakly non-ideal, high-density partially ionized hydrogen plasma are investigated, accounting for quantum effects due to the change in the energy spectrum of atomic hydrogen when the electron–proton interaction is considered embedded in the surrounding particles. The complexity of the rigorous approach led to the development of simplified models, able to include the neighbor-effects on the isolated system while remaining consistent with the traditional thermodynamic approach. High-density conditions have been simulated assuming particle interactions described by a screened Coulomb potential.

**Keywords:** Debye plasmas; thermodynamics; pressure-ionization; electrical conductivity

## 1. Introduction

The development of new technologies and experimental techniques has triggered intensive theoretical studies on modeling spatially confined quantum systems [1,2] and also extreme-high-pressure plasmas [3] like in stellar envelopes [4]. The thermodynamics of high-density hydrogen plasmas has been deeply investigated [5–9], due to the necessity of properly accounting for the effects of the multi-body interaction and in principle requiring the reformulation of the statistical mechanics in terms of a global Hamiltonian for the whole gas, instead of the usual separable form of non-interacting chemical species characterized through internal and translational partition functions. The non-ideality also affects the transport properties and in the case of dense, non-ideal, weakly-ionized Debye hydrogen plasma, the electrical conductivity in the non-metal-to-metal transition region at 150 GPa has been measured [10].

The investigation of the thermodynamic and transport properties of highly-dense hydrogen (and its isotopes) and helium plasmas is in fact relevant to many different fields, from astrophysics, for applications to low mass stars and giant planets [11], to inertial confinement fusion for the understanding of the ignition phase. Moreover, hot dense hydrogen and deuterium plasmas can be generated in a laboratory with shock compression, allowing the experimental accurate determination of the molecular-to-atomic transition along the principal Hugoniot to be compared with theoretical first-principle results [12].

It is also worth noting that atomic properties (level ensemble, electrical properties, static polarizability and hyperpolarizability [13–15] and optical oscillator strengths [16]) and the dynamics of collisions (electron impact excitation and ionization [17], symmetric charge exchange [18–20]) change in high-density regimes and are the subject in recent years of an intense investigation focused on the atomic hydrogen system.

In this paper, the thermodynamic properties and the electrical conductivity of weakly non-ideal, high-density partially ionized hydrogen plasma are investigated, accounting for quantum effects due to the change in the energy spectrum of atomic hydrogen when the electron-proton interaction is considered embedded in the surrounding particles. High-density conditions were simulated assuming atomic hydrogen described by a static screened Coulomb potential.

The Debye-Hückel or Yukawa potential, derived from the linearization of the exponential in the Poisson–Boltzmann equation [21,22], is considered suitable for the description of weakly-coupled plasmas, i.e., when the coupling parameter $\Gamma = 1/(ak_BT_e) \leq 1$, where $a = [3/(4\pi N_e)]^{1/3}$ and $N_e$ is the free electron density, and has been used in the literature for the estimation of the effects on collision processes [17,19,23–27]. The conditions explored in the present paper, the electron density ranging from $10^{16}$ to $10^{23}$ cm$^{-3}$ and the temperature from $10^4$ to $5\,10^4$ K, are compatible with weak coupling up to $n_e=10^{22}$, while for higher densities the value of $\Gamma$ is greater than the unity and in principle would require a quantum approach to properly treat the interaction in these strongly-coupled plasmas. In fact, the chemical picture of the interaction offered by the Yukawa potential fails in a strongly correlated quantum regime, where other effects need to be accounted for, such as the ion-ion correlation, the electron exchange, the consistent statistics for electrons and therefore the accurate ab initial molecular dynamics method has to be resorted to [28–30]. Another important issue in both weakly- and strongly-coupled plasmas and neglected in this paper is the dynamical nature of screening, affecting the interaction potential between electrons and ions, and in turn, the transport properties of the plasma and the dynamics of elastic and reactive collisions [21,31,32]. In fact, plasma density fluctuations, due to inter-particle correlation in dense plasmas, produce time-dependent effects in the interaction of electrons and ions, due to the polarization induced by the electron on the surrounding plasma particles, that critically depends on the ratio between the electron velocity and its thermal velocity. The effect of dynamic screening on scattering processes in weakly-coupled plasmas has been investigated [25,33,34], showing that the use of static screening overestimates the shielding, therefore, we would expect an increase of the elastic transport cross-sections reducing the electrical and thermal conductivities.

## 2. Results

### 2.1. Thermodynamics

In weakly non-ideal, partially ionized Debye plasmas, the electron–proton interaction embedded in the surrounding particles can be adequately described by the Yukawa potential, i.e., the static screened Coulomb potential (in atomic units), which is

$$U(r) = -\frac{\exp(-r/\lambda_D)}{r} \tag{1}$$

where

$$\lambda_D = \sqrt{k_B T_e/(4\pi N_e)} \tag{2}$$

is the Debye length, $k_B$ the Boltzmann constant, $T_e$ the electron temperature and $N_e$ the electron density, with severe confined conditions being related to small $\lambda_D$ values.

The atomic hydrogen levels have been calculated by discretization of the radial differential equation and for solving eigenvalues and eigenvectors for different screening conditions, from 2000 to 0.9 Bohr radii [$a_0$], so as to obtain a smooth description of the variation of level energy with the Debye length towards the critical transition to the continuum. In fact, the quantum effects act in modifying the H level structure and lead to a finite number of bound states [7,35]. As the screening increases, i.e., in very high-pressure regimes, the ground state moves towards the continuum, reducing the ionization potential, here estimated through the Koopman theorem, as shown in Figure 1a. Correspondingly, the radial wavefunction of the 1s level, displayed in Figure 1b, becomes more diffuse, describing a physical condition characterized by an electron loosely bound to the nucleus. The system of excited levels also move to the ionization limit, entering the continuum (Mott effect) [7], therefore, the number of bound levels progressively reduces as the Debye length decreases, up to a critical value of last existence of the only 1s state, below which bound states are not admitted and the plasma is fully ionized.

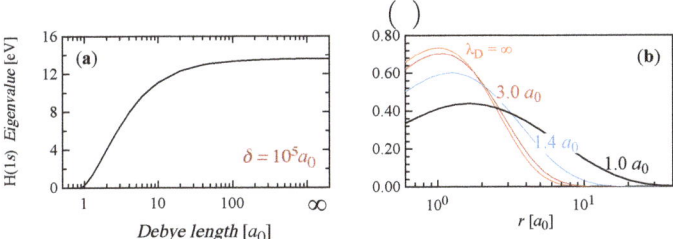

**Figure 1.** (a) Dependence of the ionization potential of atomic hydrogen on the Debye length ($\delta = 10^5 a_0$). (b) Radial wavefunction of the H(1s) ground level for different screening conditions, from isolated atom ($\lambda_D = \infty$) to severe confinement corresponding to very low values of $\lambda_D$.

In the framework of statistical thermodynamics, the state functions are fully determined by the partition function of the system $\mathcal{Q}$ and the ionization equilibrium is governed by the Saha equation

$$\frac{N_{H^+} N_e}{N_H} = \frac{2\mathcal{Q}_e^{tr}}{\mathcal{Q}_H} \exp{-\frac{I_{eff}}{k_B T}} \qquad (3)$$

where $\mathcal{Q}_e^{tr}$ is the translational partition function and is derived for a plasma at pressure $p$ in a continuum approximation, while the $\mathcal{Q}_H$ is the internal partition function of atomic hydrogen.

$$\mathcal{Q}_H = 2 \sum_{n,\ell}^{n_{max}} (2\ell + 1) \exp{[-(\varepsilon_{n,\ell} - \varepsilon_{1s})/k_B T]} \qquad (4)$$

For ideal plasmas, the natural divergency of the internal partition function is avoided, truncating the summation in Equation (4) by using the cutoff criteria, i.e., the minimum value between the Fermi and the Griem cutoff [22]. In Debye plasmas, the finiteness of the number of atomic levels due to the screening presents the very attractive feature of a natural cutoff. In this case, the eigenvalues become dependent on the value of the Debye length that is consistent with the equilibrium in the plasma system, that is $\varepsilon_{n,\ell}^{\lambda_D}$ and in turn

$$\mathcal{Q}_H = 2 \sum_{n,\ell}^{n_{\lambda_D}} (2\ell + 1) \exp{[-(\varepsilon_{n,\ell}^{\lambda_D} - \varepsilon_{1s}^{\lambda_D})/k_B T]} \qquad (5)$$

The mutual dependence of the Debye length and of the equilibrium value of the number density of electrons, $N_e$, makes the determination of $\lambda_D$ an iterative procedure that allows the self-consistency of the values characterizing the plasma at a given temperature and pressure. The non-ideal character of the plasma is usually accounted for, including the Debye-Hückel correction in the calculation of the lowering of the ionization energy [36]. This term for the hydrogen atom corresponds to the so-called self-energy shift, $\Delta = -e^2/\lambda_D$, thus leading to an effective value [7,22] $I_{eff} = I_0 + \Delta$, where $I_0$ is the ionization potential of the isolated, unperturbed hydrogen atom. However, the effect of the modification in the energy level scheme due to the screening also affects the internal partition function, producing an additional lowering that is incorporated in the internal partition function

$$\mathcal{Q}'_H = \mathcal{Q}_H \exp{[-(\varepsilon_{1s} - \varepsilon_{1s}^{\lambda_D})/k_B T]} \qquad (6)$$

It should be stressed that in the present paper, the free electrons are described through the classical Boltzmann statics, but for us to move to strongly non-ideal dense plasmas, the inclusion of the quantum Fermi statistics would be required [37–39].

In Figure 2a,b, the temperature dependence of the internal partition function and of the Debye length is self-consistently determined, following the notation adopted in the literature, for a specific value of the total electron density, $n_e = N_e + N_H$, i.e., electrons bound in an atomic system plus free electrons formed in ionization, $n_e = 10^{20}$ cm$^{-3}$ are reported. The Debye length reported is actually calculated while also considering the shielding of ionic species and not only free electrons [22].

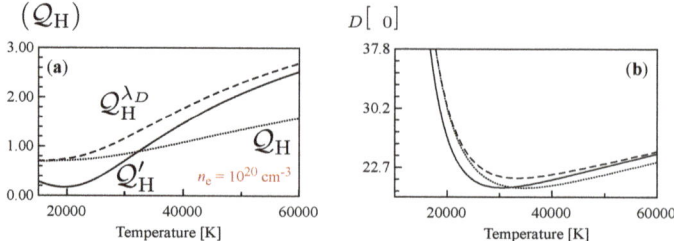

**Figure 2.** (a) Atomic hydrogen internal partition function as a function of temperature at $n_e = 10^{20}$ cm$^{-3}$, calculated with the unperturbed levels with cut-off criteria, $\mathcal{Q}_H$, including all the levels consistent with the Debye length in the plasma and accounting for the lowering of ionization potential, $\mathcal{Q}_H^{\lambda_D}$, and considering the additional ionization lowering, $\mathcal{Q}'_H$. (b) Corresponding temperature behavior of the Debye length, self-consistently determined in the three cases.

The results obtained using the eigenvalues for the unperturbed atom are compared with the partition functions calculated, accounting for the $\lambda_D$-dependent energy levels and of the additional lowering. The partition function in the case of unperturbed levels is actually lower than the values obtained by accounting for the actual levels, as already shown in the literature [40], in fact, the change in the energy-spacing of levels for a screened Coulomb potential allows a larger number of levels to be kept in the summation with respect to what was established with an external cutoff criterion. The inclusion of the additional lowering significantly affects the effective partition function, especially for temperature below 20,000 K. It is also worth noting that the Debye length is also affected in the three different cases attaining values of the order of tens of Bohrs.

The ionization degree $\alpha = N_e/(N_e + N_H)$ has been calculated at a constant total electron density, from $10^{16}$ to $10^{23}$ cm$^{-3}$, over a wide temperature range [15,000–50,000 K]. For higher densities, the theoretical framework is no longer able to deal with the non-ideal effects in the presence of strongly-coupled plasmas and different approaches need to be considered [41].

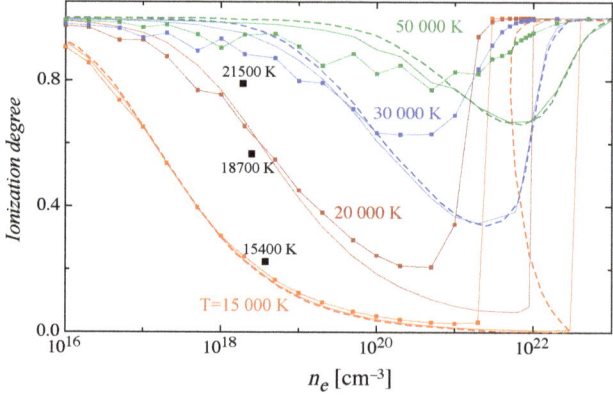

**Figure 3.** Isotherms of the ionization degree of atomic hydrogen plasma as a function of total electron density in the plasma $n_e$, obtained neglecting (dotted lines) and including (markers and lines) the effect of electronic levels, compared with theoretical results in the literature (dashed lines) [7]. Experimental results for a hydrogen arc at a pressure of 10 atm [42] are also reported (squares).

The isotherms are shown in Figure 3, and exhibit the phenomenon of pressure ionization [7,22], i.e., the rapid increase of $\alpha$ in the high-density regime. Contrary to what is expected in ideal plasmas, where the pressure (or density) increase produces a temperature shift in the ionization equilibrium, thus retarding its onset, the non-ideal quantum effects favor the ionization process and produce the observed increase of $\alpha$ merging to the fully ionized case in the limit of high temperature or very-high densities. The results obtained in this work considering only the Debye-Hückel correction to the ionization potential compare well with those reported in the literature [7]. The isotherm at 15,000 K presents a critical behavior around $2 \times 10^{22}$ cm$^{-3}$ that produces an ultra-fast transition to the full ionization and corresponds to the condition of lowest values for the Debye length and to the disappearance of any bound state for the atomic hydrogen. In the same figure, the isotherms are calculated including the effect of additional lowering of the ionization potential, due to the non-ideal effects on the electronic atomic structure, and these show a more pronounced pressure ionization.

Inspection of the isotherms clearly shows the presence of oscillations, more pronounced in the case of additional lowering. These oscillations are due to the fact that the internal partition function $\mathcal{Q}_H$ has a non-regular behavior with the total electron density due to the induced modification in the atomic internal level structure, that introduces discontinuities. This behavior is mirrored on the equilibrium constant $K_P$ that shows a non-regular increase with the density differently from the pressure that increases rapidly and thus producing, as a combined effect, the oscillations in the molar fractions of species and in the ionization degree, representing an ultimate result of the Mott effect of bound levels transitioning to the continuum.

In Figure 3 the ionization degree derived from experiments in a hydrogen arc at a pressure of 10 atm [42] for three different temperature values, approximately corresponding to a total electron density of $10^{18}$ cm$^{-3}$, are also reported, showing a satisfactory agreement with the theoretically predicted values. Unfortunately, there is no available experimental data that could validate the results at higher densities, that is where the non-ideal phenomenon manifests itself.

Concerning the validity of the present approach with respect to theories that can handle the quantum physics of plasmas even in strongly coupled conditions, the results derived for the hydrogen plasma by using the direct fermionic path integral Monte Carlo (PIMC) method [41,43] are compared with present results in Figure 4a,b. In Figure 4c, the temperature behavior of the Helmholtz free energy is also reported for two values of the total electron density.

$$A = -k_B T \sum_s \mathcal{N}_s \left( \ln \frac{\mathcal{Q}_s}{\mathcal{N}_s} + 1 \right) - \frac{1}{12} k_B T \frac{V}{\pi \lambda_D^3} \tag{7}$$

where $V$ is the gas volume, $\mathcal{N}_s$ is the number of particles of the $s$-th species and the last term on the right-side of equation represents the Debye-Hückel correction, contributing not more than 11%.

The PIMC simulations allow for the estimation of the internal partition function from configurational integrals that simultaneously includes the different interactions among elementary particles in the atomic system (electrons and protons) in the frame of a physical picture. The pressure isochors (Figure 4a) for two density values are in very good agreement with PIMC simulation [41]. Figure 4b displays the internal energy of the hydrogen plasma as a function of $n_e$ for a selected value of the temperature and the comparison, limited to the upper value explored in this work, shows again a satisfactory agreement with the PIMC simulation.

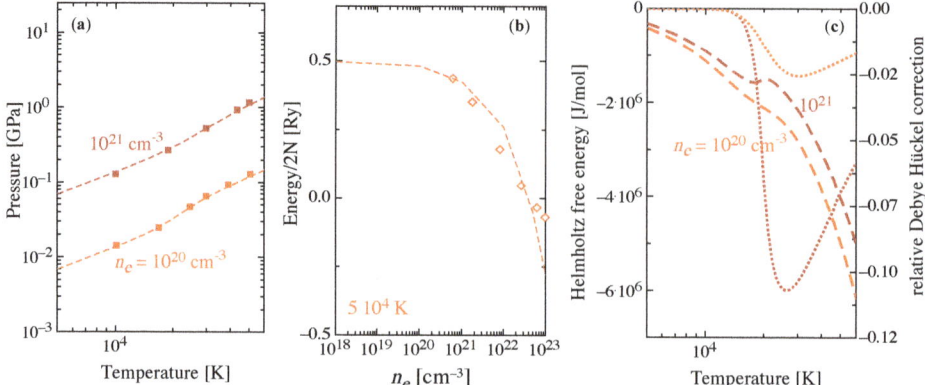

**Figure 4.** (a) Pressure isochors of a hydrogen plasma as a function of temperature for two different values of the total electron density (dashed lines) compared to results in the literature (closed squares) [41]. (b) Internal energy of the atomic hydrogen plasma as a function of the total electron density at the temperature $T = 5 \times 10^4$ K (dashed line) compared with results obtained in path integral Monte Carlo (PIMC) simulation [43]. (c) Helmholtz free energy as a function of temperature for two different values of the total electron density (dashed lines) and corresponding relative Debye-Hückel corrections, $\Delta A / A$ (dotted lines).

## 2.2. Transport: The Electrical Conductivity

The effects of non-ideality on transport properties have been investigated in the frame of the Chapman-Enskog theory [44]. As is well-known, in this theory, the binary interactions are described through the collision integrals, and the non-ideal quantum effects producing a change in the internal level structure of atoms also significantly affects the quantities directly related to the transport cross-sections. However, in this paper we are focused on the effect of the change in the thermodynamic equilibrium, due to the accounting for the additional lowering, on electrical conductivity and therefore the collision integrals for e-H and H-H interactions are assumed to be unaffected, the corresponding screening-independent transport cross-sections taken from the literature [45], while charged-particle interactions, including electron-electron, modeled with accurate Debye-length-dependent collision integrals by Mason [46,47], recently fitted in a wide temperature range in [48].

The electrical conductivity of the atomic hydrogen Debye plasma is displayed in Figure 5 as a function of the total electron density for three values of the temperature, considering the two cases, i.e., neglecting or accounting for the additional lowering of the ionization potential. In Figure 5a the $\sigma$ exhibits a behavior with the increase of the screening in the plasma that is largely dependent on the electron density, and thus mirrors the phenomenon of the pressure ionization in Figure 3: The curves go through a minimum then merge to the fully ionized regime. This first series of results can be compared with the literature, obtained in the frame of different theories. In particular, in [49], the two-term Boltzmann equation is solved including in the collisional terms accurate elastic transport cross-sections for e-e and e-H interactions, re-evaluated so as to account for the additional screened Coulomb potential in the first Born approximation, while in [37] the linear response theory is used for transport. Both [37,49] neglect the contribution of excited levels in the atomic internal partition function, one dealing with the ground-state approximation and the second using the Planck-Larkin approach to avoid divergence which explains the satisfactory agreement found.

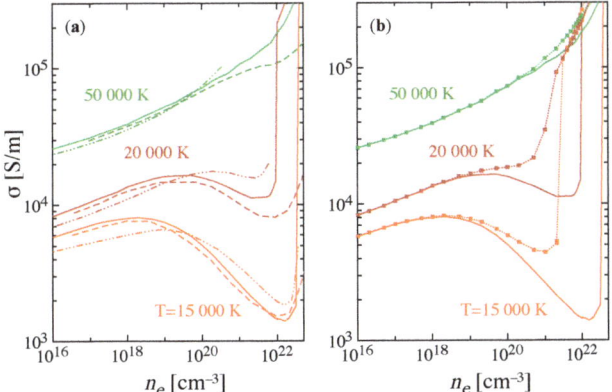

**Figure 5.** Electrical conductivity of an atomic hydrogen plasma for different temperatures as a function of the total electron density. The results (solid lines) obtained neglecting the additional lowering of ionization potential are compared with (**a**) data in literature (dashed lines) [49], (dashed-dotted lines) [37], and with (**b**) calculation including the additional lowering.

Accounting for the effect of the change in the level structure on the effective ionization of atomic hydrogen (Figure 5b) produces significant differences in the isotherms, especially at lower values of the temperature where the dip is pronounced, while the enhancement of the pressure ionization leads to a rapid increase towards the fully ionized case.

It is worth mentioning that for densities >0.1 g/cm$^{-3}$ (for $n_e > 6 \times 10^{22}$ cm$^{-3}$) accurate electrical conductivity results have been obtained with finite-temperature density functional theory molecular dynamics (FT-DFT-MD) simulations [29]. Unfortunately, in this regime of strongly-coupled plasma, the assumption of Debye plasmas is no longer valid and an extension of the present approach to those densities are expected to be unreliable.

## 3. Conclusions

The non-ideal behavior of thermodynamic and transport properties of a partially-ionized, weakly-coupled atomic hydrogen plasma was investigated in the framework of the classical statistical approach and Chapman-Enskog theory, respectively. The approach adopted in literature accounts for the effects of surrounding plasma through the Debye-Hückel correction to the value of ionization potential and disregarding the change in the level ensemble of the H, that are considered in any conditions in those of the unperturbed isolated atom, limited in the internal partition function by different cutoff criteria. The accurate description of the level structure in different screening conditions also correspond to high-density regimes, which allows us to account for all the non-ideal effects on the equilibrium composition, i.e., the natural cutoff of bound levels in $\mathcal{Q}_H$ and the further shift of the ground level to the continuum limit that determines an additional lowering of the effective ionization potential. The most significant result is represented by the emphasized phenomenon of pressure ionization, predicting a more rapid increase of the ionization degree with the total electron density in the whole temperature range considered. These results are expected to also impact the transport properties of the plasma, and in this work, the effect on the behavior of the electrical conductivity is demonstrated. The present results, though relying on the classical theoretical approaches, seem to compare well, at least for the density regime considered, with more accurate methods, reformulating the thermodynamics on the basis of a physical picture and accounting for the modification of transport cross-sections for the relevant interaction in the transport theory. In a future work, the contributions of Fermi statistics and dynamic screening will be investigated.

**Author Contributions:** conceptualization, G.C. and A.L.; methodology, G.C.; software, G.C.; data curation, A.L.; writing—original draft preparation, G.C. and A.L.; writing—review and editing, G.C. and A.L. All authors have read and agreed to the published version of the manuscript.

**Funding:** The research leading to these results has received funding from the ESA GSP Activity No. 07/B78 under CCN1 to ESTECContract 4200021790/08/NL/HE.

**Acknowledgments:** The authors wish to thank Prof Mario Capitelli for the useful discussions in the interpretation of results.

**Conflicts of Interest:** The authors declare no conflict of interest. The funders had no role in the design of the study; in the collection, analyses, or interpretation of data; in the writing of the manuscript, or in the decision to publish the results.

## References

1. Sako, T.; Diercksen, G.H.F. Confined quantum systems: A comparison of the spectral properties of the two-electron quantum dot, the negative hydrogen ion and the helium atom. *J. Phys. B: At. Molec. Opt. Phys.* **2003**, *36*, 1681–1702. [CrossRef]
2. Merkt, U.; Huser, J.; Wagner, M. Energy spectra of two electrons in a harmonic quantum dot. *Phys. Rev. B* **1991**, *43*, 7320–7323. [CrossRef] [PubMed]
3. Dutt, R.; Mukherjee, A.; Varshni, Y.P. Dipole polarizability of hydrogen atom at high pressures. *Phys. Lett. A* **2001**, *280*, 318–324. [CrossRef]
4. Hummer, D.G.; Mihalas, D. The equation of state for stellar envelopes. I - an occupation probability formalism for the truncation of internal partition functions. *Astrophys. J.* **1988**, *331*, 794–814. [CrossRef]
5. Graboske, H.C., Jr.; Harwood, D.J.; Rogers, F.J. Thermodynamic properties of nonideal gases. I. Free-energy minimization method. *Phys. Rev.* **1969**, *186*, 210. [CrossRef]
6. Ebeling, W.; Kraeft, W.D.; Kremp, D. *Theory of Bound States and Ionization Equilibrium in Plasmas and Solids*; Akademie-Verlag: Berlin, Germany, 1977.
7. Kremp, D.; Schlanges, M.; Kraeft, W.D. *Quantum Statistic of Nonideal Plasmas*; Springer: Berlin, Germany, 2005.
8. Capitelli, M.; Giordano, D. Energy levels of atomic hydrogen in a closed box: A natural cutoff criterion of the electronic partition function. *Phys. Rev. A* **2009**, *80*, 032113. [CrossRef]
9. Zaghloul, M.R. On the ionization equilibrium of hot hydrogen plasma and thermodynamic consistency of formulating finite internal partition functions. *Phys. Plasmas* **2010**, *17*, 062701. [CrossRef]
10. Nikolaev, D.; Pyalling, A.; Kvitov, S.; Fortov, V. Temperature measurements and hydrogen transformation under dynamic compression up to 150 GPA. In Proceedings of the AIP Conference, Chicago, IL, USA, 26 June–1 July 2011. [CrossRef]
11. Saumon, D.; Chabrier, G.; Van Horn, H. An equation of state for low-mass stars and giant planets. *Astrophys. J. Suppl. Ser.* **1995**, *99*, 713. [CrossRef]
12. Knudson, M.; Desjarlais, M. High-precision shock wave measurements of deuterium: Evaluation of exchange-correlation functionals at the molecular-to-atomic transition. *Phys. Rev. Lett.* **2017**, *118*, 035501. [CrossRef]
13. Saha, J.; Mukherjee, T.; Mukherjee, P.; Fricke, B. Hyperpolarizability of hydrogen atom under spherically confined Debye plasma. *Eur. Phys. J. D* **2011**, *62*, 205–211. [CrossRef]
14. Qi, Y.; Wang, J.; Janev, R. Static dipole polarizability of hydrogenlike ions in Debye plasmas. *Phys. Rev. A* **2009**, *80*, 032502. [CrossRef]
15. Das, M. Transition energies and polarizabilities of hydrogen like ions in plasma. *Phys. Plasmas* **2012**, *19*, 092707. [CrossRef]
16. Qi, Y.; Wu, Y.; Wang, J.; Qu, Y. The generalized oscillator strengths of hydrogenlike ions in Debye plasmas. *Phys. Plasmas* **2009**, *16*, 023502. [CrossRef]
17. Zammit, M.C.; Fursa, D.V.; Bray, I. Convergent-close-coupling calculations for excitation and ionization processes of electron-hydrogen collisions in Debye plasmas. *Phys. Rev. A* **2010**, *82*, 052705. [CrossRef]
18. Laricchiuta, A.; Colonna, G.; Capitelli, M.; Kosarim, A.; Smirnov, B.M. Resonant charge exchange for H-H$^+$ in Debye plasmas. *Eur. Phys. J. D* **2017**, *71*, 265. [CrossRef]
19. Wu, Y.; Wang, J.G.; Krstic, P.S.; Janev, R.K. Oscillation structures in elastic and electron capture cross-sections for H$^+$-H collisions in Debye plasmas. *J. Phys. B: At. Mol. Opt. Phys.* **2010**, *43*, 201003. [CrossRef]

20. Jung, Y.D. Collective effects on the symmetric resonant charge transfer in partially ionized hydrogen plasma. *Appl. Phys. Lett.* **2005**, *86*, 021502. [CrossRef]
21. Murillo, M.S.; Weisheit, J.C. Dense plasmas, screened interactions, and atomic ionization. *Phys. Rep.* **1998**, *302*, 1–65. [CrossRef]
22. Capitelli, M.; Colonna, G.; D'Angola, A. *Fundamental Aspects of Plasma Chemical Physics: Thermodynamics, in Springer Series on Atomic, Optical, and Plasma Physics*; Springer: New York, NY, USA, 2016.
23. Jung, Y.D. Plasma-screening effects on the electron-impact excitation of hydrogenic ions in dense plasmas. *Phys. Fluids B: Plasma Phys.* **1993**, *5*, 3432–3440. [CrossRef]
24. Yoon, J.S.; Jung, Y.D. Antiscreening channels for ion–ion collisional excitations in dense plasmas. *Phys. Plasmas* **1997**, *4*, 3477–3481. [CrossRef]
25. Yoon, J.S.; Jung, Y.D. Dynamic screening effects on antiscreening excitations for ion–ion collisions in dense plasma. *Phys. Plasmas* **1998**, *5*, 889–894. [CrossRef]
26. Zhang, H.; Wang, J.G.; He, B.; Qiu, Y.B.; Janev, R.K. Charge exchange and ionization in hydrogen atom-fully stripped ion collisions in Debye plasmas. *Phys. Plasmas* **2007**, *14*, 053505. [CrossRef]
27. Jakimovski, D.; Janev, R. Polarization of Balmer alpha radiation resulting from H++ H collisions in Debye plasmas. *Phys. Plasmas* **2015**, *22*, 103301. [CrossRef]
28. Holst, B.; Redmer, R.; Desjarlais, M.P. Thermophysical properties of warm dense hydrogen using quantum molecular dynamics simulations. *Phys. Rev. B* **2008**, *77*, 184201. [CrossRef]
29. Holst, B.; French, M.; Redmer, R. Electronic transport coefficients from ab initio simulations and application to dense liquid hydrogen. *Phys. Rev. B* **2011**, *83*, 235120. [CrossRef]
30. Clérouin, J. The viscosity of dense hydrogen: From liquid to plasma behaviour. *J. Phys.: Condens. Matter* **2002**, *14*, 9089. [CrossRef]
31. Shalenov, E.O.; Dzhumagulova, K.N.; Ramazanov, T.S.; Reinholz, H.; Röpke, G. Influence of dynamic screening on the conductivity of hydrogen plasma including electron–electron collisions. *Contrib. Plasma Phys.* **2019**, *59*, e201900024. [CrossRef]
32. Jung, Y.D. Influence of the dynamic plasma shielding on the elastic electron-ion collision in turbulent plasmas. *Appl. Phys. Lett.* **2012**, *100*, 074109. [CrossRef]
33. Jung, Y.D. Dynamic screening effects on semiclassical ionization probabilities for electron–ion collisions in weakly coupled plasmas. *Phys. Plasmas* **1998**, *5*, 536–540. [CrossRef]
34. Kim, C.G.; Jung, Y.D. Dynamic plasma screening effects on semiclassical electron captures from hydrogenic ions by protons in weakly coupled plasmas. *Phys. Plasmas* **1998**, *5*, 3493–3496. [CrossRef]
35. Rogers, F.; Graboske, H., Jr.; Harwood, D. Bound eigenstates of the static screened Coulomb potential. *Phys. Rev. A* **1970**, *1*, 1577. [CrossRef]
36. Son, S.K.; Thiele, R.; Jurek, Z.; Ziaja, B.; Santra, R. Quantum-mechanical calculation of ionization- potential lowering in dense plasmas. *Phys. Rev. X* **2014**, *4*, 031004. [CrossRef]
37. Reinholz, H.; Redmer, R.; Nagel, S. Thermodynamic and transport properties of dense hydrogen plasmas. *Phys. Rev. E* **1995**, *52*, 5368. [CrossRef]
38. Ott, T.; Bonitz, M.; Stanton, L.G.; Murillo, M.S. Coupling strength in Coulomb and Yukawa one-component plasmas. *Phys. Plasmas* **2014**, *21*, 113704. [CrossRef]
39. Bernu, B.; Wallenbom, J.; Zehnlé, V. On the transport properties of a dense fully-ionized hydrogen plasma. II. Quantum analysis. *J. Phys.* **1988**, *49*, 1161–1171. [CrossRef]
40. Capitelli, M.; Giordano, D.; Colonna, G. The role of Debye-Hückel electronic energy levels on the thermodynamic properties of hydrogen plasmas including isentropic coefficients. *Phys. Plasmas* **2008**, *15*, 082115. [CrossRef]
41. Filinov, V.S.; Levashov, P.R.; Bonitz, M.; Fortov, V.E. Thermodynamics of Hydrogen and Hydrogen-Helium Plasmas: Path Integral Monte Carlo Calculations and Chemical Picture. *Contrib. Plasma Phys.* **2005**, *45*, 258–265. [CrossRef]
42. Radtke, R.; Günther, K. Electrical conductivity of highly ionized dense hydrogen plasma. I. Electrical measurements and diagnostics. *J. Phys. D: Appl. Phys.* **1976**, *9*, 1131. [CrossRef]
43. Levashov, P.R.; Filinov, V.S.; Fortov, V.E.; Bonitz, M. Thermodynamic properties of nonideal strongly degenerate hydrogen plasma. In Proceedings of the AIP Conference, Atlanta, GA, USA, 24–29 June 2001.
44. Hirschfelder, J.O.; Curtiss, C.F.; Bird, R.B. *Molecular Theory of Gases and Liquids*; John Wiley & Sons: New York, NY, USA, 1966.

45. Bruno, D.; Catalfamo, C.; Capitelli, M.; Colonna, G.; De Pascale, O.; Diomede, P.; Gorse, C.; Laricchiuta, A.; Longo, S.; Giordano, D.; et al. Transport properties of high-temperature Jupiter atmosphere components. *Phys. Plasmas* **2010**, *17*, 112315. [CrossRef]
46. Mason, E.A.; Munn, R.J.; Smith, F.J. Transport Coefficients of Ionized Gases. *Phys. Fluids* **1967**, *10*, 1827. [CrossRef]
47. Hahn, H.S.; Mason, E.A.; Smith, F.J. Quantum transport cross-sections in a completely ionized gas. *Phys. Fluids* **1971**, *14*, 278–287. [CrossRef]
48. D'Angola, A.; Colonna, G.; Gorse, C.; Capitelli, M. Thermodynamic and transport properties in equilibrium air plasmas in a wide pressure and temperature range. *Eur. Phys. J. D* **2008**, *46*, 129–150. [CrossRef]
49. Schlanges, M.; Kremp, D.; Keuer, H. Kinetic approach to the electrical conductivity in a partially ionized hydrogen plasma. *Ann. Phys.* **1984**, *496*, 54–66. [CrossRef]

© 2020 by the authors. Licensee MDPI, Basel, Switzerland. This article is an open access article distributed under the terms and conditions of the Creative Commons Attribution (CC BY) license (http://creativecommons.org/licenses/by/4.0/).

*Article*

# Entropy of Conduction Electrons from Transport Experiments

Nicolás Pérez [1,*], Constantin Wolf [1,2], Alexander Kunzmann [1,2], Jens Freudenberger [1,3], Maria Krautz [4], Bruno Weise [4], Kornelius Nielsch [1,2,5] and Gabi Schierning [1]

1. Institute for Metallic Materials, IFW-Dresden, 01069 Dresden, Germany; c.wolf@ifw-dresden.de (C.W.); a.kunzmann@ifw-dresden.de (A.K.); j.freudenberger@ifw-dresden.de (J.F.); k.nielsch@ifw-dresden.de (K.N.); g.schierning@ifw-dresden.de (G.S.)
2. Institute of Materials Science, TU Dresden, 01062 Dresden, Germany
3. Institute of Materials Science, TU Bergakademie Freiberg, 09599 Freiberg, Germany
4. Institute for Complex Materials, IFW-Dresden, 01069 Dresden, Germany; m.krautz@ifw-dresden.de (M.K.); b.weise@ifw-dresden.de (B.W.)
5. Institute of Applied Physics, TU Dresden, 01062 Dresden, Germany
* Correspondence: n.perez.rodriguez@ifw-dresden.de

Received: 17 January 2020; Accepted: 19 February 2020; Published: 21 February 2020

**Abstract:** The entropy of conduction electrons was evaluated utilizing the thermodynamic definition of the Seebeck coefficient as a tool. This analysis was applied to two different kinds of scientific questions that can—if at all—be only partially addressed by other methods. These are the field-dependence of meta-magnetic phase transitions and the electronic structure in strongly disordered materials, such as alloys. We showed that the electronic entropy change in meta-magnetic transitions is not constant with the applied magnetic field, as is usually assumed. Furthermore, we traced the evolution of the electronic entropy with respect to the chemical composition of an alloy series. Insights about the strength and kind of interactions appearing in the exemplary materials can be identified in the experiments.

**Keywords:** electronic entropy; Seebeck coefficient; transport; LaFeSi; FeRh; CuNi

## 1. Introduction

Entropy provides information about the degrees of freedom or ordering of a statistical collectivity, i.e., it is macroscopically seen and treated as an entity. This order directly correlates with changes in the density of states of the respective statistical collectivity. For electrons in crystalline solids, this information is usually extracted from band structure theory assumptions. It is valid in the case that the sometimes quite stringent assumptions of the theoretical model are met. Experimental systems inherently deviate from the ideal solid state model. Due to this, the density of states calculated theoretically is sometimes not enough to describe the electronic properties in real systems. Typical cases where changes in the electronic density of states occur are charge order/disorder phenomena, such as the formation of charge density waves phases, superconducting phases, Fermi liquid systems, or other correlated electron systems. Further systems that are challenging to describe by theoretical solid state considerations are disordered solids, such as alloys, amorphous materials, materials with complex elementary cells, or materials containing a high number of defects induced, for instance, by the fabrication technology.

A usual approach to evaluate the total electronic entropy $S_E$ of a crystalline solid from experimental data is to analyse the low temperature specific heat capacity, $c_p$, measurements under the assumption of a free electron gas [1]. Here the Sommerfeld coefficient is the relevant value, experimentally obtained by fitting the low-temperature $c_p$ data. While this is currently the most widely applied method for an

$S_E$ characterization of crystalline solids, there are some intrinsic drawbacks to this method. These come on the one hand from the assumption of a free electron gas and on the other hand from the fact that the relevant materials properties can only be inspected at low temperature [1]. Both rule out the investigation $S_E$ changes at phase transition, especially those occurring at temperatures above 20 K, and such that induce electronic ordering phenomena.

Within this article, we discuss a recently suggested method for the $S_E$ characterization [2] that overcomes some of the limitations of the low temperature $c_p$ analysis, providing a tool for investigating such mentioned electronic systems by a direct experimental approach. We herein utilize the thermodynamic description of the Seebeck coefficient, $\alpha$, originally described by Onsager [3,4], and later referred to by Ioffe [5] in order to describe the $S_E$ of solids. The inherent advantage of the thermodynamic interpretation of $\alpha$ is that it is not bound to any model, provided the statistical description of the system is significant.

The idea to measure the $S_E$ through the measurement of macroscopic electronic properties like the Seebeck of Thomson effect has been discussed in literature [4–8], and dates, in principle, back to Thomson (Lord Kelvin) who interpreted that the Thomson effect could be seen as the specific heat of electrons, whereas the Seebeck coefficient would be the electronic entropy (divided by the charge of the electrons) [9]. Rockwood [9] pointed out that the measurement of thermoelectric transport properties necessarily only addresses the electrons that participate in the transport. He therefore specified the term "electronic transport entropy" to distinguish from a "static electronic entropy". Furthermore, thermoelectric transport measurement could never be done under truly reversible conditions since the sample needs to be exposed to a temperature gradient and is therefore not under isothermal conditions. Still, he came to the conclusion that the measurement of the thermoelectric coefficients would most likely provide the only practical and generally valid method by which partial molar entropies of electrons could be obtained. Peterson and Shastry construed the Seebeck coefficient as particle number derivative of the entropy at constant volume and constant temperature [8]. Despite this given theoretical framework, examples in which Seebeck coefficient measurements were used to quantitatively deduce $S_E$ are rare and recent but still prove the broad applicability. Our group showed that $S_E$ of a magneto-caloric phase transition could be obtained by thermoelectric transport characterization [2]. Small entities of particles like quantum dots can likewise be characterized [10]. At high temperatures, molten semiconductors and metals were similarly studied [11]. Within this paper, we will discuss the broad applicability of this method. For the following discussion, we refer to the description of the electronic entropy per particle, $S_N$, as derived within a recent review, providing an applied view on the thermodynamic interpretation of $\alpha$ [12]:

$$S_N = \alpha \cdot e \tag{1}$$

where $e$ is the charge of the particle.

In simple metals, a formal expression of $\alpha$ can be derived from band structure arguments as in the case of the Mott formula [13]. Often, a single parabolic band model is assumed. Herein, the relation between $\alpha$ and the density of states becomes evident, thus establishing a direct connection between $\alpha$ and $S_E$. While the general thermodynamic interpretation of $\alpha$ does not rely on any kind of model, the Mott formula already contains simplifications and assumptions. From the description of the quantity $S_N$ as introduced in Equation (1), it is suggested that there exists an absolute value of $S_N$ since $\alpha$ is a quantity that also has an experimentally accessible defined zero-level rather than a relative one where only changes in the quantity can be considered. The case of $\alpha = 0$ occurs, for example, (i) in the superconducting state of matter, where electrons all condense at the state of lowest energy possible and therefore per definition a situation of zero entropy [14] and (ii) in the compensated case that electrons and holes exactly transport the same amount of heat, i.e., intrinsic semiconductors have zero Seebeck coefficients [15]. The latter is an often-seen zero crossing of an n-type conductivity mechanism to a p-type conductivity mechanism. Then, the measured $\alpha = 0$ corresponds to the overall observable $\alpha$ of the material. Naturally, the contributions of the individual bands contain electronic entropy contributions with $S_{E,\text{ individual subband}} \neq 0$. The full evaluation of $S_E$ from $\alpha$ requires a correct description of the collectivity of electrons in the system. This is the point in the complete line of argumentation where assumptions and simplifications necessarily enter the picture. In order to experimentally obtain the entropy of the entity of electrons that participate in the transport, referred to as electronic entropy, $S_E$, the number of electrons contributing to the Seebeck voltage needs to be known. In principle, any experimental procedure to obtain the charge carrier density, $n$, could be used. Herein, it is, as, for instance, suggested in [2,11]:

$$S_E = n \cdot S_N = n \cdot \alpha \cdot e \tag{2}$$

In this work, we measure the ordinary Hall coefficient $R_H$ to obtain $n$, using the relation $R_H = 1/(n \cdot e)$. By doing so, we introduce the strong assumption of a parabolic single-band transport model that is inherent to any Hall measurement. Combining both quantities, we can give a measure of $S_E$:

$$S_E = \alpha / R_H \tag{3}$$

We present examples that highlight the relevance of the entropy interpretation of $\alpha$ and provide insight into the electronic properties: (1) magneto-structural phase transitions of an intermetallic Ni-doped iron rhodium phase, $Fe_{0.96}Ni_{0.02}Rh_{1.02}$ (FeRh) [16], and an intermetallic lanthanum iron silicon phase, $LaFe_{11.2}Si_{1.8}$ (LaFeSi) [17–21]; (2) alloying in the copper–nickel (CuNi) solid solution series.

## 2. Materials and Methods

All samples characterized within this work were obtained by arc melting, and followed by specific temperature treatments to ensure a homogenous microstructure. Details about the fabrication and structural characterization of the samples can be found in [22] and in [23] for $LaFe_{11.2}Si_{1.8}$ (LaFeSi). The samples investigated in the present paper stem from the same batches as the indicated references. In the case of the CuNi alloy series, the processing followed a combination of homogenization (973 K, 5 h) with quenching in $H_2O$, hot rolling (1173 K) and recrystallization (973 K, 1 h).

The transport characterization was performed depending on the temperature range using physical property measurement systems of the Quantum Design DynaCool series and the Versalab series using the thermal transport option for $\alpha$ and the electrical transport option for the Hall characterization in standard Hall bar geometry [24]. For the CuNi alloy series, a Linseis LSR 3 device was used to measure the near-room temperature $\alpha$ (315 K) and electrical conductivity, $\sigma$.

The microstructure of the samples was routinely investigated by scanning electron microscopy and X-ray diffraction.

## 3. Results and Discussion

As briefly discussed above, the entity of carriers needs to be known for the statistical interpretation of $\alpha$. Following Equation (2), we utilize $n$ obtained from a Hall-effect measurement. Herein, one has to be aware of the fact that this evaluation method may be affected by multi-channel transport, induced by multiple bands. However, given a minimal set of regularities, we can compare a homogenous series of samples or one sample under different experimental conditions consistently.

### 3.1. Magneto-Structural Phase Transition

The first example is related to meta-magnetic phase transitions in two magneto-caloric materials, namely Ni-doped FeRh and LaFeSi. They represent examples for a system that can be described with a band magnetism model (FeRh) [25] and a system with a component of localized ionic magnetism (LaFeSi) [26]. General information on the total entropy change in the phase transition of FeRh can be found in Ref. [2] and references therein, as well as a discussion of $S_E$ of this phase transition derived by transport measurements. Additionally, LaFeSi is a well-studied material with respect to magnetic and lattice entropy [26–28]. Due to soft phonon states close to transition, the lattice entropy change is large [29], but a combined contribution of lattice entropy and $S_E$ was suggested [27]. Details on the transport properties of LaFeSi are given in literature with respect to $\alpha$ [28,29] and the anomalous Hall effect [30].

The impact of the applied magnetic field on the transport of the mobile charge carriers shows a clear distinct signature in both materials, which we will discuss in the following. Both magnetic systems behave differently, as best seen in $\alpha$. In the case of FeRh (Figure 1a), it can be seen that the temperature of the phase transition depends on the magnetic field. This is a striking difference of the $S_E$ evaluation by transport experiments and calorimetric measurements that—for intrinsic reasons—do not allow this difference to be unveiled. The $\alpha$ far from the phase transition is independent of the strength of the magnetic field, as emphasized in the inset in Figure 1a that shows an enlarged view of the data in the main panel. In contrast, in the case of LaFeSi (Figure 1c), the $\alpha$ far from the phase transition shows a clear difference in the value depending on the magnetic field. Interestingly, the magnitude of $\alpha$ increases as a magnetic field is applied. The inset to Figure 1c shows the measured Hall coefficient, and the black lines indicate the levels used for the entropy evaluation as was similarly done in [2]. We get a value corresponding to the $\Delta S_E$ at the phase transition, as depicted in Figure 1b,d. In both cases, we see $\Delta S_E$ of a comparable magnitude around 4 J K$^{-1}$ kg$^{-1}$. Moreover, the absolute values of the obtained $S_E$ are also comparable. Furthermore, in both cases, an increase of $\Delta S_E$ is observed when a magnetic field is applied. However, the apparent origin of the increase in $\Delta S_E$ for both materials is different. In the case of FeRh, the first order meta-magnetic transition shifts to lower temperatures as the field is applied (Figure 1a,b). Accordingly, $\alpha$ follows a monotonic trend until the phase transition occurs. In the case of the LaFeSi, the amount of Si ($x = 1.8$) is on the threshold for changing the transition type to the second order [28]. Therefore, the transition temperature does not shift significantly, and only a slight broadening is observed. In this case, it is the change of the over-all entropy level with the applied magnetic field (Figure 1c,d) that causes the increase in $\Delta S_E$. In the case of LaFeSi, this could be an indication of the interaction between itinerant electrons and localized moments, causing the increase of $S_E$ with magnetic field. There is no such interaction in the FeRh case, as magnetism resides to a dominant part within the conduction electrons. Besides minor numerical corrections to the presented results (compare discussion Ref. [2]), it is clear that this method of analysis provides an insight to the interactions relevant to the conduction electrons that go beyond what typical calorimetric experiments can offer.

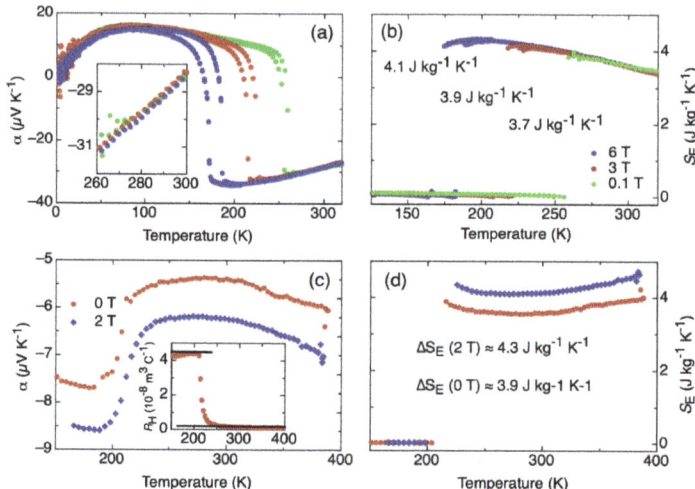

**Figure 1.** Seebeck coefficient and entropy evaluation in Ni-doped FeRh (**a**,**b**) and LaFeSi (**c**,**d**). Inset to (**a**): enlarged view of the high temperature region. Inset to (**c**): Measured Hall coefficient of LaFeSi.

## 3.2. Alloying

The differential evaluation of a systematic series of homogenized CuNi alloys with respect to their $|\alpha|$, $\sigma$, $n$, and $S_E$ at room temperature is shown in Figure 2. Herein, it is the specific situation of alloys that they typically cannot be accurately calculated or predicted by usual band structure models. However, the full alloy series is experimentally accessible. There are no structural phase transitions reported, and, also, all investigated samples were homogenous with respect to their microstructure and composition by scanning electron microscopy and X-ray diffraction. The dependence of $\sigma$ on the Ni content (Figure 2a) presents two minima at around 30 at.%-Ni and at around 70 at.%-Ni, which are better seen in the inset to Figure 2a, where the data of the main panel are presented in logarithmic vertical scale.

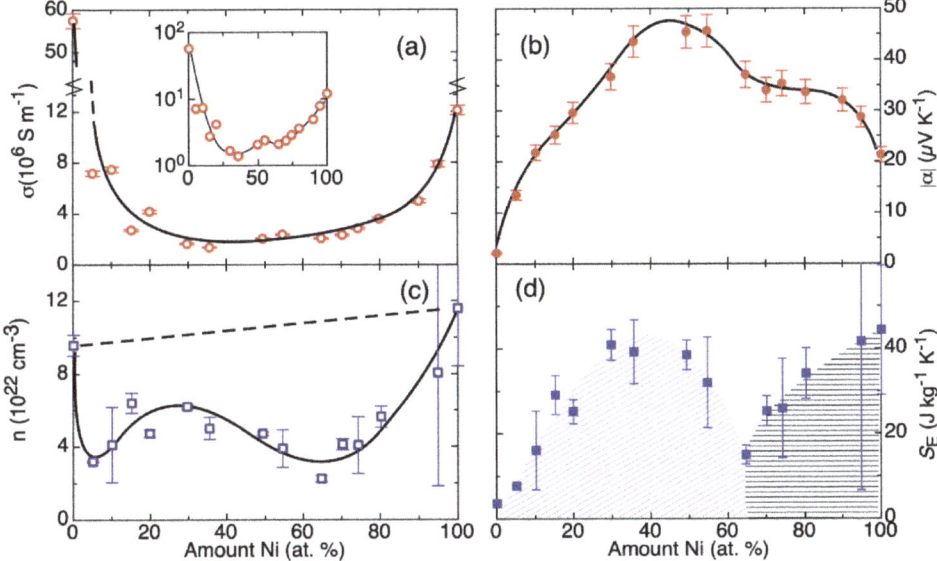

**Figure 2.** Thermoelectric and transport properties across alloy system Cu–Ni at room temperature, alloy composition was obtained with Energy-Dispersive X-Ray spectroscopy: (**a**) electrical conductivity, (**b**) the Seebeck coefficient in absolute values, (**c**) the carrier concentration derived from the Hall coefficient, (**d**) calculated electronic entropy. Lines and shades are guides to the eye.

In the trend of $|\alpha|$ (Figure 2b), a broad maximum can be seen slightly below to the equiatomic composition, close to the composition of the highest chemical disorder, a similar situation to that of other entropic parameters of such alloys [31], but a shoulder at a composition of about 70 at.%-Ni is also evident. This observation of high $|\alpha|$ for a high chemical disorder reflects the general finding that high configurational entropy is a prerequisite for the observation of large $|\alpha|$ [32]. Because of the close relationship between large $|\alpha|$ and high configurational entropy, it was recently suggested to even use configurational entropy as a gene-like performance indicator for the computational search of new thermoelectric materials [33].

The parameters $\sigma$ and $|\alpha|$ follow inverse trends with respect to one another. Additionally, these trends match the description of $\alpha$ under the Mott formula [13]. Consequently, the investigated alloy series represents a good electronic model system. There is no clear trend in the data of $n$. (Figure 2c) The pure metals Cu and Ni have the highest $n$. Different effects superimpose to a more sophisticated dependence of $n$ on the alloy composition: (i) the effect of change in the average lattice parameter by the alloying [31] should create a gradual increase in $n$ as the amount of Nickel increases; (ii) additionally, with the addition of Ni (Ni: $3d^8\ 4s^2$; 2 electrons per Ni atom) into the Cu matrix (Cu: $3d^{10}\ 4s^1$; 1 electron per Cu atom) more charge will also be added [34]. A linear increase is schematically depicted by the dashed line in Figure 2c. The overall result of these measurements is a clear minimum at approximately 65 at.%-Ni. This already indicates that additional degrees of complexity add to this simplified picture.

The combination of $|\alpha|$ and $n$ to extract the $S_E$ allows us to gain additional information compared to the individual transport coefficients. Figure 2d shows a curve in $S_E$ with maximum at approximately 30% of Ni and an additional clear minimum at approximately 65% of Ni. Coming from the Cu-side of the phase diagram, the increase in $S_E$ points out an increase in the available states for the transport electrons, which may be intuitively understood: the disorder in the non-periodic electrostatic potential leads to an increase in the entropy of the transport electrons. This increase in $S_E$ reaches a maximum close to the point where the maximum chemical disorder is expected, following the trend of $|\alpha|$. Coming from the Ni-side of the phase diagram, $|\alpha|$ increases and $n$ decreases. The $|\alpha|$, similar to the Cu-side of the phase diagram, shows higher values because of a higher degree of chemical disorder in the system. But the $|\alpha|$ does not follow a monotonic trend; instead, it has a plateau. This, combined with the reduction of $n$ in the same composition region, results in a sharp minimum of $S_E$. This minimum exactly coincides with the onset of ferromagnetism in the alloy series. Hence, the entropy evaluation provides an insight on how the magnetic ordering mechanism in this alloy affects the localization of charges, possibly due to interactions between d- and s-orbitals. While there is no one-to-one correspondence between the experiment and the microscopic origin, it still provides a meaningful measure of the intensity of correlations in the electronic transport system, which are not easily accessible by usual ab-initio methods.

## 4. Conclusions

In conclusion, this proposed method provides a good instrument for the characterization of electronic interactions or correlations in the material, although the absolute values of $S_E$ or $\Delta S_E$ obtained may, in some cases, need to be corrected (further discussed in [1]). In the case of magnetocaloric materials, the effect of the magnetic field on the electronic entropy change can be traced. In the case of alloys, the effect of the atomic disorder can also be traced on the free electrons. In order to gain deeper insight on the physics of disordered systems or systems with concurring interactions, the goal of future research might be to develop the statistical methods under the point of view of thermodynamics that would allow us to describe the statistical collectivity of electrons. In this way, we could transform the qualitative results of our experiments into quantitative predictions.

**Author Contributions:** Conceptualization: N.P. and G.S.; methodology: C.W., B.W., M.K., A.K., J.F., and N.P.; resources and supervision: K.N.; writing—original draft preparation, N.P., C.W., G.S.; writing—review and editing, N.P., C.W., G.S. All authors have read and agreed to the published version of the manuscript.

**Funding:** This project (Grant No. 100245375) is funded by the European Regional Development Fund (ERDF) and the Free State of Saxony.

**Acknowledgments:** The authors want to thank D. Seifert and R. Uhlemann in IFW Dresden for technical support. Further, G.S. and N.P. want to thank Anja Waske and Sebastian Fähler in IFW Dresden for fruitful discussions. M.K. and B.W. greatfully acknowledge financial support from the Germany Federal Ministry for Economic Affairs and Energy under Project Number 03ET1374B.

**Conflicts of Interest:** The authors declare no conflict of interest.

## References

1. Ashcroft, N.W.; Mermin, N.D. *Solid State Physics*; Brooks/Cole Thomson Learning: South Melbourne, Australia, 2012.
2. Pérez, N.; Chirkova, A.; Skokov, K.P.; Woodcock, T.G.; Gutfleisch, O.; Baranov, N.V.; Nielsch, K.; Schierning, G. Electronic entropy change in Ni-doped FeRh. *Mater. Today Phys.* **2019**, *9*, 100129. [CrossRef]
3. Onsager, L. Reciprocal Relations in Irreversible Processes. I. *Phys. Rev.* **1931**, *37*, 405–426. [CrossRef]
4. Onsager, L. Reciprocal Relations in Irreversible Processes. II. *Phys. Rev.* **1931**, *38*, 2265–2279. [CrossRef]

5. Ioffe, A.F.; Stil'Bans, L.S.; Iordanishvili, E.K.; Stavitskaya, T.S.; Gelbtuch, A.; Vineyard, G. Semiconductor Thermoelements and Thermoelectric Cooling. *Phys. Today* **1959**, *12*, 42. [CrossRef]
6. Rockwood, A.L. Partial Molar Entropy and Partial Molar Heat Capacity of Electrons in Metals and Superconductors. *J. Mod. Phys.* **2016**, *7*, 199–218. [CrossRef]
7. Rockwood, A.L. Partial molar entropy of electrons in a jellium model: Implications for thermodynamics of ions in solution and electrons in metals. *Electrochim. Acta* **2013**, *112*, 706–711. [CrossRef]
8. Peterson, M.; Shastry, B.S. Kelvin formula for thermopower. *Phys. Rev. B* **2010**, *82*, 195105. [CrossRef]
9. Rockwood, A.L. Relationship of thermoelectricity to electronic entropy. *Phys. Rev. A* **1984**, *30*, 2843–2844. [CrossRef]
10. Kleeorin, Y.; Thierschmann, H.; Buhmann, H.; Georges, A.; Molenkamp, L.W.; Meir, Y. How to measure the entropy of a mesoscopic system via thermoelectric transport. *Nat. Commun.* **2019**, *10*, 5081. [CrossRef]
11. Rinzler, C.C.; Allanore, A. Connecting electronic entropy to empirically accessible electronic properties in high temperature systems. *Philos. Mag.* **2016**, *96*, 3041–3053. [CrossRef]
12. Goupil, C.; Seifert, W.; Zabrocki, K.; Müller, E.; Snyder, G.J.; Müller, E. Thermodynamics of Thermoelectric Phenomena and Applications. *Entropy* **2011**, *13*, 1481–1517. [CrossRef]
13. Cutler, M.; Mott, N.F. Observation of Anderson Localization in an Electron Gas. *Phys. Rev.* **1969**, *181*, 1336–1340. [CrossRef]
14. Roberts, R.B. The absolute scale of thermoelectricity. *Philos. Mag.* **1977**, *36*, 91–107. [CrossRef]
15. da Rosa, A. Thermoelectricity. In *Fundamentals of Renewable Energy Processes*; Elsevier: Amsterdam, The Netherlands, 2013; pp. 149–212.
16. Nikitin, S.; Myalikgulyev, G.; Tishin, A.; Annaorazov, M.; Asatryan, K.; Tyurin, A. The magnetocaloric effect in Fe49Rh51 compound. *Phys. Lett. A* **1990**, *148*, 363–366. [CrossRef]
17. Fukamichi, K.; Fujita, A.; Fujieda, S. Large magnetocaloric effects and thermal transport properties of La(FeSi)13 and their hydrides. *J. Alloys Compd.* **2006**, *408–412*, 307–312. [CrossRef]
18. Fujieda, S.; Hasegawa, Y.; Fujita, A.; Fukamichi, K. Thermal transport properties of magnetic refrigerants La(FexSi1−x)13 and their hydrides, and Gd5Si2Ge2 and MnAs. *J. Appl. Phys.* **2004**, *95*, 2429–2431. [CrossRef]
19. Shen, B.G.; Sun, J.R.; Hu, F.X.; Zhang, H.W.; Cheng, Z.H. Recent Progress in Exploring Magnetocaloric Materials. *Adv. Mater.* **2009**, *21*, 4545–4564. [CrossRef]
20. Hu, F.-X.; Shen, B.-G.; Sun, J.; Cheng, Z.; Rao, G.-H.; Zhang, X.-X. Influence of negative lattice expansion and metamagnetic transition on magnetic entropy change in the compound LaFe$_{11.4}$Si$_{1.6}$. *Appl. Phys. Lett.* **2001**, *78*, 3675–3677. [CrossRef]
21. Hu, F.-X.; Ilyn, M.; Tishin, A.M.; Sun, J.R.; Wang, G.J.; Chen, Y.F.; Wang, F.; Cheng, Z.H.; Shen, B.G. Direct measurements of magnetocaloric effect in the first-order system LaFe11.7Si1.3. *J. Appl. Phys.* **2003**, *93*, 5503–5506. [CrossRef]
22. Baranov, N.; Barabanova, E. Electrical resistivity and magnetic phase transitions in modified FeRh compounds. *J. Alloys Compd.* **1995**, *219*, 139–148. [CrossRef]
23. Glushko, O.; Funk, A.; Maier-Kiener, V.; Kraker, P.; Krautz, M.; Eckert, J.; Waske, A. Mechanical properties of the magnetocaloric intermetallic LaFe11.2Si1.8 alloy at different length scales. *Acta Mater.* **2019**, *165*, 40–50. [CrossRef]
24. Haeusler, J. Die Geometriefunktion vierelektrodiger Hallgeneratoren. *Electr. Eng.* **1968**, *52*, 11–19. [CrossRef]
25. Lu, W.; Nam, N.T.; Suzuki, T. First-order magnetic phase transition in FeRh–Pt thin films. *J. Appl. Phys.* **2009**, *105*, 07A904. [CrossRef]
26. Gercsi, Z.; Fuller, N.; Sandeman, K.G.; Fujita, A. Electronic structure, metamagnetism and thermopower of LaSiFe12and interstitially doped LaSiFe12. *J. Phys. D Appl. Phys.* **2017**, *51*, 034003. [CrossRef]
27. Jia, L.; Liu, G.J.; Sun, J.R.; Zhang, H.W.; Hu, F.-X.; Dong, C.; Rao, G.; Shen, B.G. Entropy changes associated with the first-order magnetic transition in LaFe13−xSix. *J. Appl. Phys.* **2006**, *100*, 123904. [CrossRef]
28. Hannemann, U.; Lyubina, J.; Ryan, M.P.; Alford, N.M.; Cohen, L.F. Thermopower of LaFe 13−x Si x alloys. *EPL* **2012**, *100*, 57009. [CrossRef]
29. Gruner, M.; Keune, W.; Landers, J.; Salamon, S.; Krautz, M.; Zhao, J.; Hu, M.Y.; Toellner, T.; Alp, E.E.; Gutfleisch, O.; et al. Moment-Volume Coupling in La(Fe1−x Si x)13. *Phys. Status Solidi (B)* **2017**, *255*, 1700465. [CrossRef]

30. Landers, J.; Salamon, S.; Keune, W.; Gruner, M.; Krautz, M.; Zhao, J.; Hu, M.Y.; Toellner, T.S.; Alp, E.E.; Gutfleisch, O.; et al. Determining the vibrational entropy change in the giant magnetocaloric material LaFe11.6Si1.4 by nuclear resonant inelastic x-ray scattering. *Phys. Rev. B* **2018**, *98*, 024417. [CrossRef]
31. Madelung, O. *Cr-Cs–Cu-Zr*; Springer: Berlin/Heidelberg, Germany, 1994.
32. Gruen, D.M.; Bruno, P.; Xie, M. Configurational, electronic entropies and the thermoelectric properties of nanocarbon ensembles. *Appl. Phys. Lett.* **2008**, *92*, 143118. [CrossRef]
33. Liu, R.; Chen, H.; Zhao, K.; Qin, Y.; Jiang, B.; Zhang, T.; Sha, G.; Shi, X.; Uher, C.; Zhang, W.; et al. Entropy as a Gene-Like Performance Indicator Promoting Thermoelectric Materials. *Adv. Mater.* **2017**, *29*, 1702712. [CrossRef]
34. Hurd, C.M. *The Hall Effect in Metals and Alloys*; Springer Science and Business Media LLC: Berlin, Germany, 1972.

© 2020 by the authors. Licensee MDPI, Basel, Switzerland. This article is an open access article distributed under the terms and conditions of the Creative Commons Attribution (CC BY) license (http://creativecommons.org/licenses/by/4.0/).

Article

# Analysis of Entropy Production in Structured Chemical Reactors: Optimization for Catalytic Combustion of Air Pollutants

**Mateusz Korpyś [1,*], Anna Gancarczyk [1], Marzena Iwaniszyn [1], Katarzyna Sindera [1], Przemysław J. Jodłowski [2] and Andrzej Kołodziej [1,3]**

1   Institute of Chemical Engineering, Polish Academy of Sciences, Bałtycka 5, 44-100 Gliwice, Poland; anna.g@iich.gliwice.pl (A.G.); miwaniszyn@iich.gliwice.pl (M.I.); katarzyna.sindera@iich.gliwice.pl (K.S.); ask@iich.gliwice.pl (A.K.)
2   Faculty of Chemical Engineering and Technology, Cracow University of Technology, Warszawska 24, 31-155 Kraków, Poland; pjodlowski@pk.edu.pl
3   Faculty of Civil Engineering and Architecture, Opole University of Technology, Katowicka 48, 45-061 Opole, Poland
*   Correspondence: matkor@iich.gliwice.pl; Tel.: +48-32-231-08-11 (ext. 124)

Received: 6 August 2020; Accepted: 9 September 2020; Published: 11 September 2020

**Abstract:** Optimization of structured reactors is not without some difficulties due to highly random economic issues. In this study, an entropic approach to optimization is proposed. The model of entropy production in a structured catalytic reactor is introduced and discussed. Entropy production due to flow friction, heat and mass transfer and chemical reaction is derived and referred to the process yield. The entropic optimization criterion is applied for the case of catalytic combustion of methane. Several variants of catalytic supports are considered including wire gauzes, classic (long-channel) and short-channel monoliths, packed bed and solid foam. The proposed entropic criterion may indicate technically rational solutions of a reactor process that is as close as possible to the equilibrium, taking into account all the process phenomena such as heat and mass transfer, flow friction and chemical reaction.

**Keywords:** entropy production; optimization; reactor modelling; irreversible thermodynamics

## 1. Introduction

At the industrial level, optimization of chemical processes, including those based on structured catalytic reactors, is an inherent issue of the design procedure. Process optimization considers the prices of raw materials, energy, products and installations (apparatus); the prices may change rapidly and unpredictably due to market fluctuations, even at the negotiation stage. Therefore, process optimization is usually regarded as being within the engineering domain, it is in fact more connected with business and economic issues. These issues usually exceed the knowledge of an engineer or a scientist and require input from other individuals.

Structured reactors are very important in chemistry and catalysis [1–3]. The process design, i.e., the apparatus and the process conditions, has to secure some economic profitability in spite of potential changes of costs. Regardless of possible economic fluctuations (excluding any collapses), the process has to be profitable during the following years.

A review of the literature provides hints about recommended flow velocities, process temperatures and catalyst carriers. The data originate from the long-standing technical and economic experience of engineers and entrepreneurs. Recently, a new generation of structured catalytic reactors has been introduced into industry, and there is a paucity of knowledge and experience about their optimization. Moreover, the inner-structure design of the reactors is complicated because many geometrical parameters need to be optimized.

In the literature, different criteria can be found, which help identify optimal operating conditions of chemical reactors. "The technical" or "engineering" optimization, with which this work deals, focuses on reactor optimization in terms of fluid velocities, process (reaction) temperature, structured catalyst carrier shape and dimensions. This kind of optimization has begun in energetics due to the introduction of compact heat exchangers that usually exploit a combination of fins, turbulence mixers and other features. In the current literature, even more sophisticated criteria are proposed for multiparameter optimization of different equipment such as heat exchangers. So far, similar criteria for catalytic reactors have been derived. The comprehensive performance evaluation criteria (PEC) use three components: transport coefficients, reaction kinetics and pressure drop [4,5]. Another approach is the comparison of reactor length (or catalyst mass) with the resulting flow resistance as shown in [4,6]. For heat exchanger optimization, there are also evaluation criteria based on entropy production during the process, as presented, e.g., by London [7] and Bejan [8], who also predicted the extension of entropic criteria to chemical reactors. Entropy in economic analysis is treated as trade-off factor and can be a substitute of currency [9]. The application of entropic criterion can also be found in [10–12].

The aim of the study is to propose a highly simplified approach, based on irreversible thermodynamics, suitable for engineering optimization of chemical reactors. The entropic criterion is proposed to optimize structured catalytic reactors. The assumed model process is the catalytic combustion of methane.

## 2. Theoretical Background

To derive the equations governing entropy production, the reactor model must be specified. For the purposes of this paper, the one-dimensional plug-flow model (neglecting axial dispersion) in the steady-state was assumed. Due to the very thin catalyst layer deposited on the structured carrier, the internal diffusional resistance can be neglected.

Mass balance of reactant A, in the flowing fluid, per unit surface area of the reactor cross-section, is as follows:

$$w_0 \frac{dC_A}{dx} + k_C S_v (C_A - C_{AS}) = 0 \tag{1}$$

The initial conditions are: (i) $x = 0$; $C_A = C_{A0}$ and (ii) the reactant A, mass transferred from the gas bulk to the catalyst surface is balanced by the first-order catalytic reaction:

$$k_C (C_A - C_{AS}) = k_r C_{AS}. \tag{2}$$

Deriving concentration of A, at the catalyst surface from Equation (2), Equation (1) becomes:

$$-w_0 \frac{dC_A}{dx} = S_v \frac{k_C k_r}{k_C + k_r} C_A, \tag{3}$$

and, after integration, local concentration $C_{Ax}$ and the reactor length $L$, required for the outlet concentration $C_{AL}$ are:

$$C_{Ax} = C_{A0} \exp\left(-\frac{x}{w_0} \frac{S_v k_C k_r}{k_C + k_r}\right), \tag{4}$$

$$L = \frac{w_0}{S_v} \frac{k_C + k_r}{k_C k_r} \ln\left(\frac{C_{A0}}{C_{AL}}\right). \tag{5}$$

The energy balance may be presented (assuming no heat losses to the environment) as:

$$w_0 \rho c_p \frac{dT}{dx} + \alpha S_v (T - T_S) = 0, \tag{6}$$

the initial conditions: at $x = 0$, $T = T_0$.

The mass and heat transfer in a heterogeneous catalytic reactor are strictly bound up (released reaction heat depends on the reactants mass transferred to the catalyst), thus

$$q = \alpha(T_S - T) = -\Delta H_R J_A = -\Delta H_R k_C (C_A - C_{AS}). \tag{7}$$

The above equations assume an isothermal process. In reality, the process is adiabatic. However, the concentration of organic air pollutants is usually low. For the volatile organic compounds (VOCs), a concentration of very few ppm is typical; for methane, it depends on the kind of source and may be within 1–1000 ppm. The level of concentrations of 100 ppm and higher can be treated by homogeneous combustion in, e.g., reverse-flow reactors due to important reaction heat. Thus, we assumed the concentration of methane at 200 ppm as rational for our analysis. In such a case, the adiabatic temperature rise is about 6 K, so the temperature increase along the reactor can be securely neglected.

Entropy production is an increase of system entropy due only to the irreversible phenomena [13]. This means that there is no entropy production at equilibrium or during a quasi-static process that runs infinitely close to the equilibrium. Any industrial process runs far from the equilibrium, and it produces entropy at irreversible conditions. In irreversible thermodynamics, entropy production is derived as the product of flux $J_i$ and the driving force $\Delta \pi$ (causing the stream) divided by absolute temperature $T$ [13,14]:

$$S_i = \frac{J_i \Delta \pi}{T}. \tag{8}$$

Assuming that the stream $J_i$ is proportional to the driving force:

$$J_i = k_i \Delta \pi, \tag{9}$$

entropy production is proportional to the square of the driving force, thus it increases rapidly with the distance from the equilibrium:

$$S_i = \frac{k_i (\Delta \pi)^2}{T}. \tag{10}$$

In this paper, entropy production is considered due to the following irreversible phenomena:

- heat transfer between the gas phase and the catalyst surface (further denoted as H);
- diffusional mass transfer between the gas phase and the catalyst surface (denoted as D);
- irreversible catalytic reaction (denoted as R);
- flow friction, i.e., work performed against the flow resistance (denoted as F).

Total entropy production (per 1 mole of reactant A consumed in the reactor) is the sum of all the components:

$$S_P = S_H + S_D + S_R + S_F. \tag{11}$$

The above-mentioned components of entropy production are gathered in Table 1.

In the first column, basic equations of local entropy production are presented. In the second and third columns, the equations for the stream and the driving force are presented, respectively, derived using the reactor model. The last column presents reactor-integrated entropy production per 1 mole of substrate A consumed (e.g., burned) in the reactor. Detailed derivations, simple in fact, are not presented for reason of conciseness. The last position in Table 1, *flow friction* needs further comment. The entropy source considered is the volume fluid flow. The stream (flux) is the flow velocity and the driving force is the pressure gradient. The entropy produced is tantamount to viscous dissipation of pumping energy. This approach seems more friendly for engineers than viscous momentum flux often presented by irreversible thermodynamics; the flux is the pressure tensor and the driving force is the velocity gradient [11].

The impact of the reaction rate constant, $k_r$, and the heat and mass transfer coefficients, $\alpha$ and $k_C$, respectively, on the entropy produced by the heat ($S_H$) and mass ($S_D$) transfer is illustrated in Figure 1 for the combustion process and exemplary $k_r$ and $k_C$ values. The heat and mass transfer coefficients are bound by the Chilton–Colburn analogy [15], Equation (12), which allows the influence of mass transport on $S_H$ to be determined.

$$j = \frac{Nu}{RePr^{1/3}} = \frac{Sh}{ReSc^{1/3}}. \tag{12}$$

$$S_D = Rln\left(1+\frac{k_r}{k_C}\right), \tag{13}$$

$$S_H = \frac{k_r}{k_C+k_r}\left[\frac{(-\Delta H_R)^2(C_{A0}-C_{AL})D_A Sc^{1/3}}{2\lambda T^2 Pr^{1/3}}\right]. \tag{14}$$

Table 1. Local and reactor-averaged components of entropy produced.

| Entropy, $\sigma_i$ | Flux, $J_i$ | Driving Force, $\Delta\pi$ | Entropy, Reactor Average Value, $S_i$ (per mol of Substrate A) |
|---|---|---|---|
| Heat transfer (H) $\sigma_H = -\frac{q}{T^2}\nabla T$ | Heat flux $q = -\Delta H_R J_A =$ $= \alpha(T_s-T)$ | Temperature gradient $(T_s - T) =$ $= \frac{k_C(-\Delta H_R)(C_A-C_{AS})}{\alpha}$ | $S_H = \frac{k_C k_r}{k_C+k_r}$ $\frac{(-\Delta H_R)^2(C_{A0}-C_{AL})}{2\alpha T^2}$ |
| Mass transfer (D) $\sigma_D = -\sum_i \frac{J_i}{T}\nabla \mu_i$ | Diffusive mass flux $J_A = k_C(C_A - C_{AS}) =$ $= k_{Cr}C_A$ | Chemical potential gradient $\nabla \mu_A = RT\frac{\mu_A - \mu_{AS}}{S_{ef}}$ | $S_D = Rln\left(\frac{k_C+k_r}{k_C}\right)$ |
| Reaction (R) $\sigma_R = -\frac{A r_A S_z}{T}$ | Reaction rate $r_A = k_r C_{AS} = k_{Cr}C_A$ | Chemical affinity $A = -\sum_i \nu_i \mu_i =$ $= -\Delta G_R^{o,T} - RT\sum_i \nu_i ln y_i$ | $S_R = \frac{A}{T}$ |
| Flow friction (F) $\sigma_F = \frac{W}{TF_cL} = -\frac{w\nabla P}{T}$ | Fluid stream $w$ | Pressure gradient $-\nabla P$ | $S_F = \frac{f}{2T}\frac{w_0^2 \varrho}{\varepsilon^3 k_{Cr}} \frac{ln\left(\frac{C_{A0}}{C_{AL}}\right)}{(C_{A0}-C_{AL})}$ |

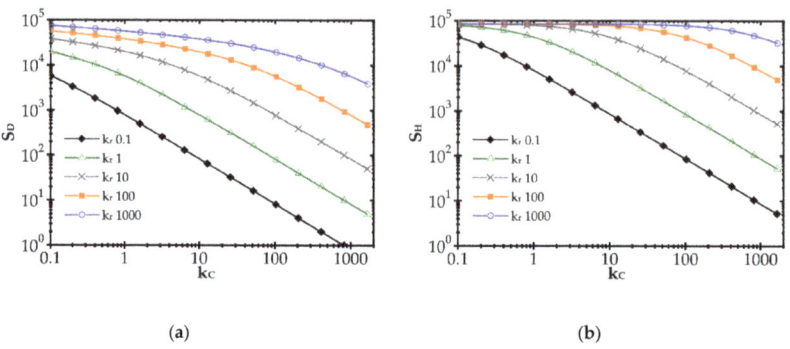

(a)      (b)

Figure 1. Impact of the mass transfer coefficient and reaction rate on entropy production due to: (a) mass transfer and (b) heat transfer.

In Figure 1, a distinct increase of entropy produced with the reaction rate constant, $k_r$, is observed. Conversely, entropy decreases with the mass transfer coefficient, $k_C$ (due to heat, $S_H$, and mass, $S_D$, transfer). A rapid chemical reaction (i.e., high $k_r$) generates intense mass transport of substrates to the catalyst surface and adequate heat transfer in the opposite direction. The faster the reaction, the further the process runs from the equilibrium. When the transfer coefficients are small compared to the reaction rate, the concentration and temperature gradients are large, and even the substrates concentration on the catalyst goes to zero. Entropy production is large, being proportional to the square of the driving force (concentration or temperature gradient, cf. Equation (10)).

The impact of the mass transfer coefficient is opposite. The higher the transfer coefficient for a given reaction rate, the lower the temperature and concentration gradients are and the closer to the equilibrium the process runs. Smaller driving forces lead to lower entropy according to Equation (10). However, when analysing the plots in Figure 1, the impact of mass transfer intensification is distinct only if $k_C$ is close to the $k_r$ value. If $k_C$ is much smaller than $k_r$, slight transfer enhancement will give nothing as the concentration and temperature gradients are still large (zero concentration at the catalyst surface). The gradients start to decrease as the reaction and transfer become comparable.

Obviously, the values of $k_r$ and especially of $k_C$ in Figure 1, may not be found in reality as the plots presented are theoretical, to illustrate the common impact of transfer and reaction rates on entropy production.

## 3. Catalyst Supports Considered

The aim of this study is to show the optimal adjustment of the catalyst carrier geometry, as well as its transfer and friction characteristics to the catalytic reaction kinetics. The catalyst performance (reaction kinetics) is treated as a model parameter only. Therefore, analysed catalyst supports were selected on the basis of similar value of specific surface area. This means that, in all considered cases, approximately, the same area was available for active layer catalyst deposition. For comparison, monolith and packed bed are also examined.

Correlations for the heat transfer and Fanning friction factor were derived experimentally and presented in detail in our earlier papers [4,16]. A photo of catalyst supports considered in the study is presented in Figure 2, and a summary of equations for Fanning friction factor, Nusselt number and Sherwood number of investigated supports are presented in Table 2 and compared in Figure 3.

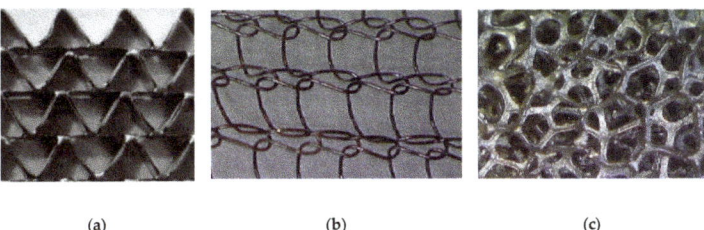

**Figure 2.** Catalyst supports: (**a**) triangular short-channel structure, (**b**) wire gauze, and (**c**) nickel chromium foam.

The kinetic tests were performed experimentally. Two different catalyst deposition methods were applied: (1) for Pd/ZrO$_2$, the incipient wetness (IW) method [20] and (2) for Pd/Al$_2$O$_3$, sonochemical (SC) method [4]. The kinetic studies were conducted in the temperature range of 373–823 K [20]. Kinetic data are presented in Table 3. As was found in [21], the sonochemical method allows higher catalyst activity to be obtained in comparison to the incipient wetness method.

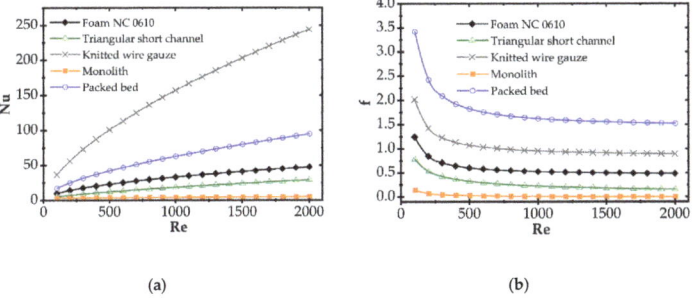

**Figure 3.** (**a**) Average Nusselt number and (**b**) Fanning friction factor for considered catalyst carriers.

Table 2. Correlations used to calculate flow resistance, heat and mass transfer for analysed catalyst supports.

| Structure Description | Correlations |
|---|---|
| Wire gauze [4] | $f = 118.09/Re + 0.836$<br>$Nu = 2.19Re^{0.636} Pr^{1/3}$<br>$Sh = 2.19Re^{0.636} Sc^{1/3}$<br>$S_v = 1355$<br>$\varepsilon = 0.97$ |
| Triangular short channel [16] | $(fRe) = 13.33 + 11.59(L^+)^{-0.514}$<br>$Nu = \left(3.11 + 0.45(L^*)^{-0.61}\right)\left(0.55(PrL^*)^{-0.15}\right)$<br>$Sh = \left(3.11 + 0.45(L^*M)^{-0.61}\right)\left(0.55(PrL^*M)^{-0.15}\right)$<br>$S_v = 1314$<br>$\varepsilon = 0.95$ |
| Nickel chromium foam (NC 0610), Recemat®<br>(Dodewaard, The Netherlands); [4] | $f = 79.9/Re + 0.445$<br>$Nu = 0.96Re^{0.53} Pr^{1/3}$<br>$Sh = 0.96Re^{0.53} Sc^{1/3}$<br>$S_v = 1298$<br>$\varepsilon = 0.89$ |
| Monolith [17] | $(fRe) = 14.23\left(1 + 0.045/L^+\right)^{0.5}$<br>$Nu = 3.608(1 + 0.095/L^*)^{0.45}$<br>$Sh = 3.608(1 + 0.095/L^*M)^{0.45}$<br>$S_v = 1339$<br>$\varepsilon = 0.72$ |
| Packed bed [18,19] | $f = \frac{(\varepsilon-1)[600\eta(\varepsilon-1)-7D_h\varrho w]}{8D_h \varepsilon \varrho w}$<br>$Nu = 2 + 1.1Re^{0.6} Pr^{1/3}$<br>$Sh = 2 + 1.1Re^{0.6} Sc^{1/3}$<br>$S_v = 1240$<br>$\varepsilon = 0.38$ |

Table 3. Kinetic data of tested catalysts.

| Catalyst | Pre-Exponential Coefficient in Arrhenius Equation, $k_\infty$, m s$^{-1}$ | Activation Energy, $E_a$, kJ mol$^{-1}$ |
|---|---|---|
| Slow kinetic, incipient wetness (IW) Pd/ZrO$_2$ | 252.49 | 62.79 |
| Fast kinetic, sonochemical (SC) Pd/Al$_2$O$_3$ | $1.07 \cdot 10^{10}$ | 110.4 |

## 4. Results and Discussion

Plots referring to analysis of entropy production were constructed assuming reactor length required for 90% conversion and show the entropy produced per 1 kmole of methane combusted in the reactor under given process conditions. Entropy production is presented as a function of process temperature and the Reynolds number. Entropy is produced due to the four components denoted as R—reaction, H—heat transfer, D—diffusional mass transfer and F—flow friction. The subscript HDFR means total entropy produced due to the H, D, F and R components.

The components of entropy production (according to Table 1) for the knitted wire gauze are compared for the methane catalytic combustion process vs. process temperature (Figure 4) and the Reynolds number (Figure 5) for the fast (Pd/Al$_2$O$_3$) and slow (Pd/ZrO$_2$) kinetics assuming initial methane concentration of 200 ppm in both cases.

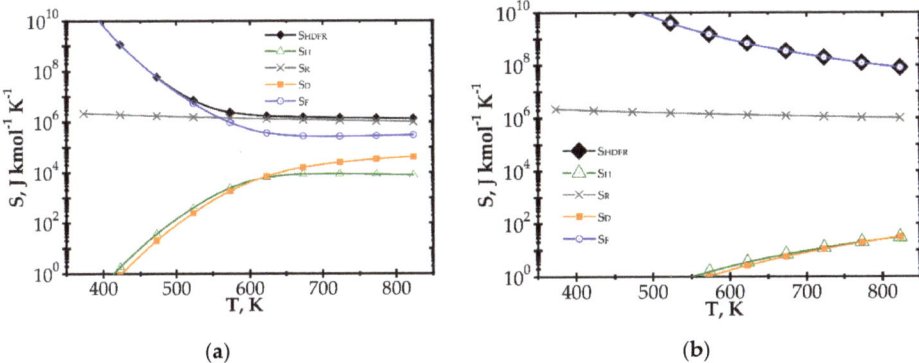

**Figure 4.** Comparison of entropy production components vs. process temperature for knitted wire gauze, Re = 1000, $CH_4$ inlet concentration: 200 ppm: (**a**) fast kinetics, $Pd/Al_2O_3$ and (**b**) slow kinetics, $Pd/ZrO_2$.

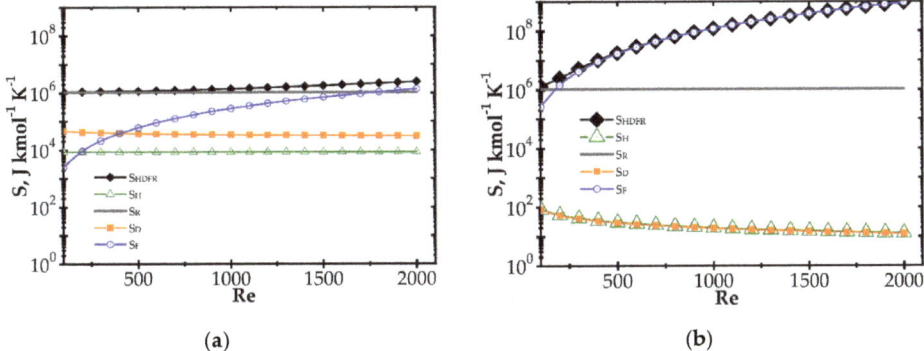

**Figure 5.** Comparison of entropy production components vs. Reynolds number for knitted wire gauze, T = 773 K, $CH_4$ inlet concentration: 200 ppm: (**a**) fast kinetics, $Pd/Al_2O_3$ and (**b**) slow kinetics, $Pd/ZrO_2$.

When analysing the $Pd/Al_2O_3$ catalyst (Figure 4a) within the lower temperature range, entropy due to flow friction, $S_F$, is the major component, and it is close to the total entropy production $S_{HDFR}$. The heat and mass transport components, $S_H$ and $S_D$, play less important roles. However, for higher temperatures, the kinetics become much faster, causing significant shortening of reactor length necessary to attain 90% conversion. The share of flow friction entropy decreases; simultaneously, the entropy components due to heat and mass transport play more important roles. For the highest temperature range analysed, total entropy $S_{HDFR}$ is close to the reaction component $S_R$, while the remaining components are comparable. Increased entropy production due to heat and mass transport at higher temperatures is a result of faster reaction rate. This leads to lower methane concentration on the catalyst surface, and thus to higher temperature and concentration gradients, in consequence of more intense entropy production (cf. Table 1, Equation (10) and Figure 1).

For the $Pd/ZrO_2$ catalyst (Figure 4b), total entropy production is close to the flow friction component in the whole temperature range analysed. The transport component $S_D$, $S_H$ are minor due to low gradients (a result of slow kinetics), and even the reaction component $S_R$ is much lower than the flow friction one, $S_F$.

Figure 5 illustrates entropy production as a function of the Reynolds number for knitted wire gauze assuming a rather high temperature of 773 K. The transport components $S_D$ and $S_H$ are almost constant within the whole Re range analysed. The flow friction component $S_F$ increases with Re, reaching an even higher value than $S_R$, especially in the case of the Pd/ZrO$_2$ catalyst. Moreover, in Figure 5b, the total entropy produced is close to the flow friction component, with a minor role played by the remaining components.

Large entropy production is due to the irreversible reaction of methane catalytic combustion. Moreover, this entropy component is almost the same per mole of reactant, regardless of process conditions (T, Re and catalyst); analysis of the equation for $S_R$ (Table 1) should render this as no surprise. Chemical affinity is close to the standard Gibbs energy of reaction (at the process temperature) $\Delta G_R^{o,T}$, because the sum of the concentration logarithms is minor. For optimization purposes, the place of the minimum total entropy production reflects the process optimum, making the precise value less important. Analysis of Figures 4 and 5 shows that the $S_R$ component is nearly constant within the ranges studied. Note that reaction component, $S_R$, is the lowest possible entropy that can be produced in the chemical reactor. For engineering purposes, such as process optimization, the remaining components are more interesting because they make entropy production higher than that due to chemical reaction ($S_R$) and they are dependent on the physical properties of carriers. For slow reaction, there is no difference between the analysed approaches, because, in this case, flow resistance plays a major role (cf. Figures 4b and 5b) and the minimum is not observed within the considered temperature range. In summarising the catalytic structures displaying close specific surface area $S_v$ (i.e., similar catalyst amount), $S_R$ will be neglected during next analysis.

Analysis of entropy production due to the heat and mass transfer and flow friction (denoted as $S_{HDF}$) is presented in Figures 6 and 7 presents $S_{HDF}$ as a function of the Reynolds number and process temperature for the five catalyst supports considered. In the following figures, minimal entropy production for each support is shown; these points give optimal process conditions for particular catalyst supports.

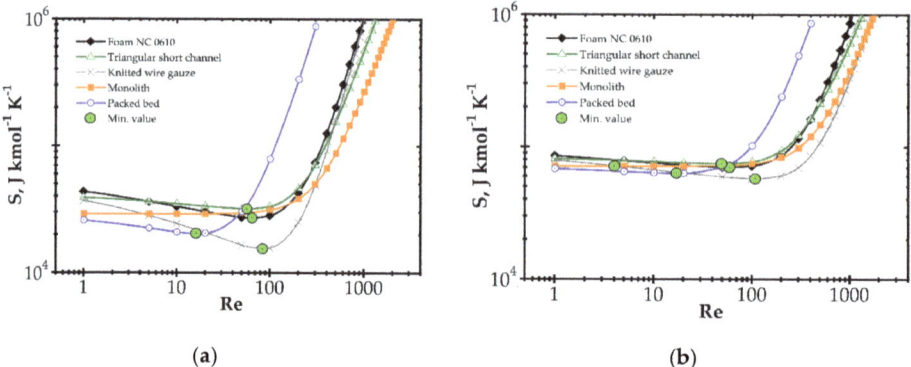

**Figure 6.** Entropy production vs. Reynolds number for different catalyst supports for the fast kinetics, Pd/Al$_2$O$_3$ at temperature: (**a**) 573 K and (**b**) 773 K.

In Figure 6, entropy is presented for two selected temperatures, moderate (573 K) and high (773 K). For the moderate temperature of 573 K (Figure 6a), packed bed seems the best for Re < 20. For Re < 500, knitted wire gauze is optimal (minimum value at Re = 84) in that this results in the lowest entropy production and the most profitable behaviour within this analysis. For a higher Reynolds number, monolith displays the lowest entropy production, undoubtedly due to its lowest flow resistance. For higher temperatures of 773 K (Figure 6b), the impact of transfer properties is more pronounced as a result of faster reaction rate, and knitted wire gauze appears to be the best with classic and short-channel monoliths. Packed bed produces the largest entropy in almost the entire Reynolds range, due to the highest flow resistance. For the higher temperature (773 K), the minima are generally slightly shifted to higher Reynolds numbers and entropy production is several times higher.

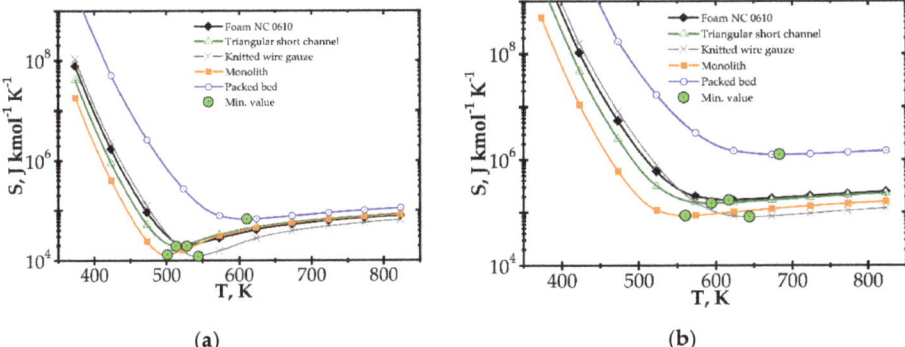

**Figure 7.** Entropy production vs. temperature for the fast kinetics, Pd/Al$_2$O$_3$ for different catalyst supports at Reynolds number: (**a**) Re = 100 and (**b**) Re = 500.

When considering temperature influence on entropy production (Figure 7), the same conclusions may be derived. Low process temperature is favourable for the classic monolith, while for higher temperatures, wire gauze and monolith seem to be the best choice. For Re = 100 (Figure 7a), above 650 K, all the internals display close entropy production. Interestingly, all the structures except packed bed show minima within the narrow range of 500–540 K. For Re = 500 (Figure 7b), entropy produced is higher, especially for packed bed. The minima are shifted towards higher temperatures by 60–100 K. Above 600 K, knitted gauze and monolith are the best.

Analogous plots for slow kinetics (Figures 8 and 9) show quite different behaviour. Here, the reactor is long due to the slow reaction rate. Moreover, slow reaction does not require intense heat and mass transfer. Concentration and temperature differences between the flowing fluid and catalyst surface are very small; entropy production due to transfer is small compared to that due to flow friction. Consequently, entropy produced for the slow kinetics is ordered identically to the friction factors (Figure 3b) vs. the Reynolds number and process temperature. Flow friction is the main entropy source (when neglecting chemical reaction). For slow kinetics, entropy production characteristic considered for all the internals is similar. The shift observed (towards higher or lower entropy produced) results mainly from the flow resistance. All the curves are nearly parallel, and only slight convergence is observed for low Re as a result of different transport properties. The internals displaying the lowest flow resistance (monolith and short-channel structure, cf. Figure 3) offer the lowest entropy production, while those of high flow resistance (packed bed, cf. Figure 3) produce larger entropy, so are less profitable.

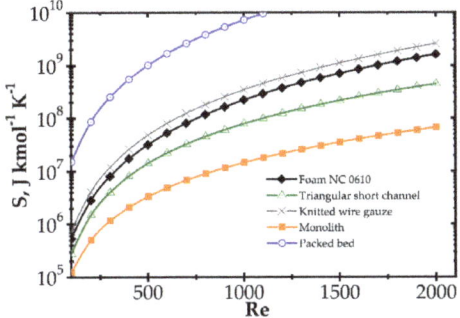

**Figure 8.** Entropy production vs. Reynolds number for different catalyst supports for the slow kinetics, Pd/ZrO$_2$ at temperature 673 K.

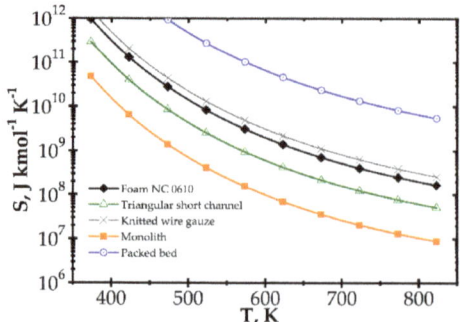

**Figure 9.** Entropy production vs. temperature for the slow kinetics, Pd/ZrO$_2$ for different catalyst supports at Reynolds number 1500.

## 5. Conclusions

The results obtained by entropy analysis indicate that wire gauze is the best choice for the Pd/Al$_2$O$_3$ catalyst and the packed bed is the worst one. In the case of the Pd/ZrO$_2$ catalyst, the best carriers are monolith and short-channel structures, while the worst solution is a packed bed. However, meeting the efficiency criteria cannot be regarded as the ultimate verdict. Any process has its own characteristics and limitations. It is rare for a process to occur separately, as it is usually part of a larger installation. For example, process temperature is limited by catalyst thermal deactivation, and the flow resistance may be limited by the gas pressure available. Therefore, each process needs to be considered individually, and any overall limiting parameters must also be taken into consideration during optimization.

The entropy-based optimization methodology is able to optimize reactor structure (indicating the best geometry, specific surface, etc.), as well as the process temperature and fluid velocity for considered reaction kinetics. The criterion, ensuring the minimum entropy production, ignores the reactor cost and is able to indicate the best structure from among the considered ones, as well as the optimal working conditions of a reactor (e.g., temperature and flow velocity).

Irreversible chemical reaction produces almost the same entropy, per mole of reactant, regardless of the process conditions. Therefore, it can be safely neglected during entropic optimization. The hypothesis is confirmed by analysis presented in Figure 4. For proper results, entropy produced by heat transfer, mass transfer and flow friction should be accounted for.

The gauze structures are assessed as being very effective due to their satisfactory transfer and friction properties. The monolith and short triangular channel display good efficiency for slow kinetics (Pd/ZrO$_2$ catalyst) due to their low flow resistance. The packed bed usually appears as an unsatisfactory solution.

For fast kinetics (Pd/Al$_2$O$_3$ catalyst), the transfer properties of the catalyst support are the most important for low entropy production. The intense transfer properties of, e.g., knitted wire gauze, make the support excellent for such processes. The impact of flow resistance is minor as, for a fast reaction not hampered by insufficient transfer rate, the reactor is very short.

For slow kinetics (Pd/ZrO$_2$ catalyst), the reactor is long. The impact of flow resistance becomes important. In contrast, heat and mass transfer contributions to entropy production are minor. Heat and mass transfer resistance is low, so temperatures (concentrations) gradients between fluid and catalyst surface are low, and the process runs near to the equilibrium.

The optimization methodology presented in this study obviously requires further development, including thorough experimental industrial and economic application. In spite of this, the entropic criterion seems able to indicate technically rational solutions of the reactor process considering the heat and mass transfer, flow resistance and reaction kinetics.

**Author Contributions:** Conceptualization, A.K. and M.K.; methodology, A.K.; formal analysis, A.K. and M.K.; investigation, A.G., M.I., and K.S.; writing—original draft preparation, A.K. and M.K.; writing—review and editing, M.K., M.I., and A.G.; supervision, A.K., A.G., and P.J.J. All authors have read and agreed to the published version of the manuscript.

**Funding:** This work was supported by the Polish National Science Centre (Project No. DEC-2011/03/B/ST8/05455, DEC-2016/21/B/ST8/00496). The financial support is gratefully acknowledged.

**Acknowledgments:** The authors are sincerely grateful for the constructive suggestions and advices provided by Andrzej Burghardt. We regret to inform that A. Burghardt passed away on 1 March 2020.

**Conflicts of Interest:** The authors declare no conflict of interest.

## Nomenclature

| | |
|---|---|
| $A$ | chemical affinity, J mol$^{-1}$ |
| $C_A$ | reagent concentration, mol m$^{-3}$ |
| $c_p$ | heat capacity, J kg$^{-1}$ K$^{-1}$ |
| $D_A$ | diffusivity, m$^2$ s$^{-1}$ |
| $D_h$ | hydraulic diameter, $= 4\varepsilon S_v^{-1}$, m |
| $F_c$ | reactor cross-sectional area, m$^2$ |
| $f$ | Fanning friction factor, $= \varrho w_0^2 L(2\Delta P D_h \varepsilon^2)^{-1}$ |
| $J_A$ | diffusional mass flux, mol s$^{-1}$ m$^{-2}$ |
| $J_i$ | stream (flux) of irreversible process, Equation (9) |
| $k_C$ | mass transfer coefficient, m s$^{-1}$ |
| $k_r$ | kinetic rate constant of the first-order reaction, referred to the catalyst surface area, m s$^{-1}$ |
| $k_\infty$ | pre-exponential coefficient in Arrhenius equation, m s$^{-1}$ |
| $k_{Cr} = k_C k_r/(k_C+k_r)$ | combined transfer-reaction coefficient, m s$^{-1}$ |
| $L$ | bed length, m |
| $\Delta P$ | pressure drop, Pa/m |
| $q$ | heat flux, W m$^{-2}$ |
| $R$ | gas constant, J mol$^{-1}$ K$^{-1}$ |
| $r_A$ | reaction rate, mol m$^{-2}$ s$^{-1}$ |
| $S$ | entropy production rate, J K$^{-1}$mol$^{-1}$ |
| $s_{ef}$ | film thickness, m |
| $S_v$ | specific surface area, m$^2$ m$^{-3}$ |
| $T$ | temperature, K |
| $W$ | pumping power, W |
| $w_0$ | superficial fluid velocity, m s$^{-1}$ |
| $y_i$ | mole fraction |
| $\Delta H_R$ | reaction enthalpy, J mol$^{-1}$ |
| $\Delta G_R$ | reaction Gibbs energy, J mol$^{-1}$ |

**Greek symbols**

| | |
|---|---|
| $\alpha$ | heat transfer coefficient, W m$^{-2}$ K$^{-1}$ |
| $\varepsilon$ | porosity |
| $\eta$ | dynamic viscosity, Pa s |
| $\lambda$ | thermal conductivity, W m$^{-1}$ K$^{-1}$ |
| $\mu$ | chemical potential, J mol$^{-1}$ |
| $\nu$ | stoichiometric coefficient |
| $\Delta\pi$ | driving force of irreversible process |
| $\varrho$ | density, kg m$^{-3}$ |
| $\sigma$ | entropy production per m$^3$ of reactor volume, W m$^{-3}$ K$^{-1}$ |

**Dimensionless numbers**

| | |
|---|---|
| $L^+$ | dimensionless length for the hydrodynamic entrance region, $= LD_h^{-1}\text{Re}^{-1}$ |
| $L^*$ | dimensionless length for the thermal entrance region, $= LD_h^{-1}\text{Re}^{-1}\text{Pr}^{-1}$ |
| $L^{*M}$ | dimensionless length for the mass transfer entrance region, $= LD_h^{-1}\text{Re}^{-1}\text{Sc}^{-1}$ |

| | |
|---|---|
| Pr | Prandtl number, $= \eta c_p \lambda^{-1}$ |
| Re | Reynolds number, $= w0 D h \varrho \eta^{-1} \varepsilon^{-1}$ |
| Sc | Schmidt number, $= \eta \varrho{-1} D_A^{-1}$ |
| Sh | Sherwood number, $= k_C D_h D_A^{-1}$ |

**Subscripts**

| | |
|---|---|
| A | key reactant |
| D | entropy production due to mass transfer |
| F | entropy production due to flow friction |
| H | entropy production due to heat transfer |
| P | total entropy production |
| R | entropy production due to chemical reaction |
| S | catalyst surface |
| x | reactor arbitrary axial coordinate |
| 0, L | reactor inlet, outlet |

## References

1. Vilé, G.; Richard-Bildstein, S.; Lhuillery, A.; Rueedi, G. Electrophile, Substrate Functionality, and Catalyst Effects in the Synthesis of α-Mono and Di-Substituted Benzylamines via Visible-Light Photoredox Catalysis in Flow. *ChemCatChem* **2018**, *10*, 3786–3794. [CrossRef]
2. Amini-Rentsch, L.; Vanoli, E.; Richard-Bildstein, S.; Marti, R.; Vilé, G. A Novel and Efficient Continuous-Flow Route To Prepare Trifluoromethylated N-Fused Heterocycles for Drug Discovery and Pharmaceutical Manufacturing. *Ind. Eng. Chem. Res.* **2019**, *58*, 10164–10171. [CrossRef]
3. Ramirez, A.; Hueso, J.L.; Mallada, R.; Santamaria, J. Microwave-activated structured reactors to maximize propylene selectivity in the oxidative dehydrogenation of propane. *Chem. Eng. J.* **2020**, *393*, 124746. [CrossRef]
4. Gancarczyk, A.; Iwaniszyn, M.; Piątek, M.; Korpyś, M.; Sindera, K.; Jodłowski, P.J.; Łojewska, J.; Kołodziej, A. Catalytic Combustion of Low-Concentration Methane on Structured Catalyst Supports. *Ind. Eng. Chem. Res.* **2018**, *57*, 10281–10291. [CrossRef]
5. Giani, L.; Groppi, G.; Tronconi, E. Mass-Transfer Characterization of Metallic Foams as Supports for Structured Catalysts. *Ind. Eng. Chem. Res.* **2005**, *44*, 4993–5002. [CrossRef]
6. Kołodziej, A.; Łojewska, J. Prospect of compact afterburners based on metallic microstructures. Design and modelling. *Top. Catal.* **2007**, *42–43*, 475–480. [CrossRef]
7. London, A.L. Economics and the second law: An engineering view and methodology. *Int. J. Heat Mass Transf.* **1982**, *25*, 743–751. [CrossRef]
8. Bejan, A. *Advanced Engineering Thermodynamics*, 1st ed.; John Wiley and Sons: New York, NY, USA, 1988.
9. London, A.L.; Shah, R.K. Costs of Irreversibilities in Heat Exchanger Design. *Heat Transf. Eng.* **1983**, *4*, 59–73. [CrossRef]
10. Zimparov, V. Extended performance evaluation criteria for enhanced heat transfer surfaces: Heat transfer through ducts with constant wall temperature. *Int. J. Heat Mass Transf.* **2000**, *43*, 3137–3155. [CrossRef]
11. Kjelstrup, S.; Johannessen, E.; Rosjorde, A.; Nummedal, L.; Bedeaux, D. Minimizing the entropy production of the methanol producing reaction in a methanol reactor. *Int. J. Thermodyn.* **2000**, *3*, 147–153.
12. Nummedal, L.; Kjelstrup, S.; Costea, M. Minimizing the Entropy Production Rate of an Exothermic Reactor with a Constant Heat-Transfer Coefficient: The Ammonia Reaction. *Ind. Eng. Chem. Res.* **2003**, *42*, 1044–1056. [CrossRef]
13. De Groot, S.R.; Mazur, P. *Non-Equilibrium Thermodynamics*; From the Series in Physics; North-Holland Publishing Company: Amsterdam, The Netherlands, 1969.
14. Wei, J. Irreversible thermodynamics in engineering. *Ind. Eng. Chem.* **1966**, *58*, 55–60. [CrossRef]
15. Chilton, T.H.; Colburn, A.P. Mass Transfer (Absorption) Coefficients Prediction from Data on Heat Transfer and Fluid Friction. *Ind. Eng. Chem.* **1934**, *26*, 1183–1187. [CrossRef]
16. Iwaniszyn, M.; Ochońska, J.; Gancarczyk, A.; Jodłowski, P.; Knapik, A.; Łojewska, J.; Janowska-Renkas, E.; Kołodziej, A. Short-channel structured reactor as a catalytic afterburner. *Top. Catal.* **2013**, *56*, 273–278. [CrossRef]

17. Hawthorn, R.D. Afterburner catalysts effects of heat and mass transfer between gas and catalyst surface. *AIChE Symp. Ser.* **1974**, *70*, 428–438.
18. Bird, R.B.; Stewart, W.E.; Lightfoot, E.N. *Transport Phenomena*, 2nd ed.; Wiley International Edition; Wiley: New York, NY, USA, 2007; ISBN 9780470115398.
19. Wakao, N.; Kaguei, S. *Heat and Mass Transfer in Packed Beds*; Routledge: New York, NY, USA, 1982.
20. Jodłowski, P.J.; Jędrzejczyk, R.J.; Gancarczyk, A.; Łojewska, J.; Kołodziej, A. New method of determination of intrinsic kinetic and mass transport parameters from typical catalyst activity tests: Problem of mass transfer resistance and diffusional limitation of reaction rate. *Chem. Eng. Sci.* **2017**, *162*, 322–331. [CrossRef]
21. Jodłowski, P.; Jędrzejczyk, R.; Chlebda, D.; Dziedzicka, A.; Kuterasiński, Ł.; Gancarczyk, A.; Sitarz, M. Non-Noble Metal Oxide Catalysts for Methane Catalytic Combustion: Sonochemical Synthesis and Characterisation. *Nanomaterials* **2017**, *7*, 174. [CrossRef] [PubMed]

© 2020 by the authors. Licensee MDPI, Basel, Switzerland. This article is an open access article distributed under the terms and conditions of the Creative Commons Attribution (CC BY) license (http://creativecommons.org/licenses/by/4.0/).

Article

# Adapted or Adaptable: How to Manage Entropy Production?

### Christophe Goupil *,† and Eric Herbert †

Université de Paris, Laboratoire Interdisciplinaire des Energies de Demain (LIED), UMR 8236 CNRS, F-75013 Paris, France; eric.herbert@univ-paris-diderot.fr
* Correspondence: christophe.goupil@univ-paris-diderot.fr; Tel.: +33-6-0-03-03-50
† These authors contributed equally to this work.

Received: 3 December 2019; Accepted: 20 December 2019; Published: 24 December 2019

**Abstract:** Adaptable or adapted? Whether it is a question of physical, biological, or even economic systems, this problem arises when all these systems are the location of matter and energy conversion. To this interdisciplinary question, we propose a theoretical framework based on the two principles of thermodynamics. Considering a finite time linear thermodynamic approach, we show that non-equilibrium systems operating in a quasi-static regime are quite deterministic as long as boundary conditions are correctly defined. The Novikov–Curzon–Ahlborn derivation applied to non-endoreversible systems then makes it possible to precisely determine the conditions for obtaining characteristic operating points. As a result, power maximization principle (MPP), entropy minimization principle (mEP), efficiency maximization, or waste minimization states are only specific modalities of system operation. We show that boundary conditions play a major role in defining operating points because they define the intensity of the feedback that ultimately characterizes the operation. Armed with these thermodynamic foundations, we show that the intrinsically most efficient systems are also the most constrained in terms of controlling the entropy and dissipation production. In particular, we show that the best figure of merit necessarily leads to a vanishing production of power. On the other hand, a class of systems emerges, which, although they do not offer extreme efficiency or power, have a wide range of use and therefore marked robustness. It therefore appears that the number of degrees of freedom of the system leads to an optimization of the allocation of entropy production.

**Keywords:** out of equilibrium thermodynamics; finite time thermodynamics; living systems

---

## 1. Introduction

The issue of energy conversion is the subject of historical debate. Without going back to its roots, let us mention the work initiated by Glansdorf and Prigogine, which placed at the center the question of entropy production in out-of-equilibrium systems, an issue that is still largely relevant [1,2]. This debate is itself part of an even broader debate that questions the operating points of the systems, considering mainly the maximization of entropy production (MEP), its minimization (mEP), or power maximization (MPP) [3,4]. One of the reasons why these questions do not find a general consensus today is that they are most often considered on very different systems, in particular in the definition of the boundary conditions of the device with its environment, considered immutable. The case of idealized mechanical systems is, from this point of view, much simpler, since, broadly speaking, the absence of any friction process means that the system interacts with its environment via a very limited number of degrees of freedom, which makes variational approaches relevant. On the contrary, it has long been accepted that there is no variational principle that governs the out-of-equilibrium steady state of a thermodynamic system [5]. This can be understood as an impossibility to establish a variational principle when the number of degrees of freedom diverges, which is obviously the case when the system is connected to a thermostat, and when dissipative processes occur. However, it is equally obvious that many out-of-equilibrium systems are perfectly deterministic in their evolution, and have a perfectly defined stationary state, as is the case, for example, for Kirchoff's networks in electronics. As a result, these systems, although not governed by a Lagrangian form and an associated variational principle, have a completely established stationary operating point, without any possible affirmation of an underlying minimization or maximization of the production of the entropy or the power.

These questions of power and finite time performance have been the subject of much work [6] particularly in thermoelectricity [7–11]. Without entering into these debates again, we propose an approach that provides a fairly generic framework for describing a complete thermodynamic system with perfectly established boundary conditions. In this article, we will limit ourselves to the case of locally linear machines, subscribing to Onsager's formalism. This formalism, based on the concept of local equilibrium, makes it possible to consider the thermodynamic potentials of the system, which are the intensive parameters. As a result, it becomes possible to derive a thermodynamics close to equilibrium, with, in particular, a rigorous choice of potentials that allow for obtaining the symmetry of the out-of-diagonal coefficients of the Onsager matrix. The stationary nature also requires that kinetic coefficients and boundary conditions of the system be constant or slowly variable compared to the characteristic relaxation time of entropy production and dissipation diffusion, thus guaranteeing both stationary processes and local equilibrium.

In this article, we consider the transport of energy and matter within a system, where the thermodynamic conversion is produced by coupling the energy and matter currents. By applying the first law of thermodynamics, both of these currents are conservative. By applying the second law, the energy, and sometimes the matter, used during the conversion process is subject to dispersion in the degrees of freedom accessible to the system. As a result, thermodynamics is based on both quantity and quality principles. Since the loss of quality is directly related to dispersion in the degrees of freedom, the search for processes to reduce their number has always been a guideline. It should be noted that, in the case of non-spontaneous processes, it is possible to consider a reduction in the degrees of freedom, but this operation requires the implementation of external processes. These processes offer other opportunities for energy dispersion, in greater proportions than those gained within the system. As a result, any physical process taking place over a finite period of time is the location of a compromise between the total energy used to carry out a process, and the energy actually converted for the needs to be covered. The process efficiency is therefore written as the ratio between the actually converted energy and the total energy supplied. We propose to consider energy conversion processes in a very generic form, in order to establish their main characteristics and constraints. In particular, we address the question of power and entropy production, insisting on the compromises they impose.

The question of adapting a device to the uses assigned to it then arises. In the case of single working point, the system may be designed to be as much adapted as is it possible. However, this single operating working point is a rare configuration, and realistic systems are asked to work in a given range of working points. Then, the concept of adaptability, or flexibility, arises, which enters into competition with the previous adapted concept. This problem of adaptation or adaptability concerns all thermodynamic systems, including, of course, living systems. Indeed, as soon as we define an envelope, we delimit the boundaries of a space occupied by a given device and the interactions of this device with the outside world. Considering the energy and matter budget at the borders of the device, we then characterize the relationship between the device and its environment. Since the processes take place over a finite period of time, it is important to consider an out-of-equilibrium description. In this paper, we consider an out-of-equilibrium thermodynamic description, driven by locally linear equations. We show that the intrinsic characteristics of the device, on the one hand, and the boundary conditions, on the other hand, totally determine the behavior of the system. It appears that the allocation of dissipation largely determines the possible ranges of use of an out-of-equilibrium thermodynamic system.

In terms of boundary conditions, we show that the real coupling conditions of a system with its environment are always located between the Dirichlet and Neumann boundaries, also called "stock" and "flow" boundary conditions. It should be noted that both pure stock and flow are extreme boundary conditions which can never being strictly reached. Between adaptable and adapted, the performances of thermodynamic systems are therefore the result of a compromise between intrinsic performance of a device and the coupling to the environment. This question of coupling to the environment is the subject of the first section of this article. In the following section, we describe the envisaged system in its most general form. The third section concerns the descriptions of the device at the heart of the system, while the fourth section describes its insertion into the complete system. The fifth section considers the different configurations that such a global system may encounter, and the consequences on the production of power, dissipation, and more generally, entropy. The article ends with concluding remarks.

## 2. System Description

### 2.1. Boundary Conditions

As indicated above, the system is composed of two sub-parts: a central zone, which we will call the device, and which is the place of thermodynamic conversion, on the one hand, and the boundary conditions, consisting of the source, and, on the other hand, the sink and the elements connecting it to the device. These elements allow for modifying at will the boundary conditions that condition the coupling of the device with the source and the sink, which is a central question for the optimization. Among the latter, we can distinguish systems whose intrinsic parameters are constant, as is the case for most machines, and systems, whose intrinsic parameters are subject to modification, as is the case for living or societal systems. These latter are subject to potential developments and evolution, which are not possible for the above-mentioned machines. By potential development, we consider the case of living systems, societies or organisms, which can, under conditions of energy and matter supply, develop, maintain, or regress.

In the case of systems under Neumann boundary conditions, the system is somehow fed by a constant current of energy and/or matter, which guarantees the maintenance of the system as much as it constrains its development. Under such conditions, the possible development of the system is limited by the value of the current of matter and/or energy. In the case of Dirichlet systems, there are no restrictions on access to the resource, except for the intrinsic limitations of the conversion device. As a result, the currents of energy and matter may diverge completely, if the characteristics of the device lend themselves to it. The same reasoning applies to the production and rejection of waste to the sink. Access to the resource and waste production are therefore both dependent on these boundary conditions. Let us consider, as an historical illustration, the situation of the industrial revolution, which saw the rise of the use of fossil energy [12]. The latter are by definition stock resources that lead the human societies to find themselves in Dirichlet conditions, as far as access to the resource is concerned. Concerning the waste rejected to the sink, the Dirichlet's condition has been the norm, as long as the planet has been considered a bottomless sink. On the other hand, if we consider the situation before the industrial revolution, it can be noted that the main resource for development, which is the food resource, was dependent on Neumann-type boundary conditions, due to the subjection to solar flux. Without going further into this illustration, which is beyond the scope of this article, we can nevertheless observe the importance of boundary conditions, both on the functioning of systems, but also for their possible evolutions. Indeed, in the case of boundary conditions of the Neumann type, there is no possibility of development, in the sense of increasing the current of energy and matter that feed the conversion device. Consequently, there is no possibility of any increase of the quantities. On the other hand, there are possibilities of increase of the quality because the conditions of coupling between energy and matter may change, as the history of life proved it.

On the other hand, in the case of Dirichlet boundary conditions, there is no limit to the increase in energy and matter currents, which could lead to their possible divergence. It should be noted that the actual Dirichlet conditions for the access to the energy for the human species are quite singular in the history of the living systems. In order to remain explicit and relatively simple to address, these questions need to be modeled in the most compact form possible. This why we propose to describe a generic thermodynamic machine in order to guarantee a general character to the developments of this article. Many extensions and refinements can be added, as for previous systems in the literature [6,10].

## 2.2. Thermodynamic Device

The proposed thermodynamic system is described in Figure 1. It consists of a reservoir providing the resource and a sink receiving the waste, with the respective potentials $\Pi_1^R$ and $\Pi_1^S$ fixed at constant values. Between these two reservoirs is the energy conversion device which is the place of coupling between a current of matter $I_2$, and a current of energy $I_E$. The energy current entering the system is associated with an incoming entropy current, $I_1$, with $\Pi_1$ its conjugated potential. In the case of a thermal system of heat current $I_Q$, temperature $T$ and entropy current $I_S$, we would simply have $\Pi_1 I_1 = I_Q = T I_S$ so $I_1$ would be the classical entropy current. The current of matter is defined by $I_2$ and its conjugated potential $\Pi_2$. The energy currents budget finally writes $I_E = \Pi_1 I_1 + \Pi_2 I_2$. We recognize the fractions of dispersed energy, $\Pi_1 I_1$, and concentrated energy, $\Pi_2 I_2$, which are a generalization of the notions of heat and work extended to the case of non-thermal systems [13,14]. The coupling term between energy and matter is defined, under $I_2 = 0$ condition, as $\alpha = -(\delta \Pi_2/\delta \Pi_1)_{I_2}$. The geometry of the system is given by its length $L$ and its cross-section $A$. The two currents of energy and matter are then associated with two conductivities $\sigma_1$ and $\sigma_2$, which, at the integrated scale, behave like two resistive dipoles $R_{1/2} = \frac{1}{\sigma_{1/2}} \frac{L}{A}$. The connection of the conversion zone with the two reservoirs is defined by the coupling resistors $R_+$ and $R_-$, which allow the boundary conditions to be set, at will, between Dirichlet conditions ($R_+ = R_- = 0$), or Neumann conditions, where $R_+$ and $R_-$ diverge. This type of configuration is not in itself new, and has already been used in specific systems [14,15]. In particular, it has been shown that, under these conditions, the way the system operates is partially governed by the feedback effects induced by boundary conditions. Some of this feedback can lead to the presence of oscillations. It should be noted that these processes do not violate the first principle in that they are not self-sustained oscillations, at least from an energy point of view. They do not violate the second principle either, since these structures are highly dissipative and are only maintained by a continuous supply of energy. It can also be noted that the incoming current of energy is used to produce a potential difference, which, if maintained, allows the circulation of the matter under the action of the thermodynamic force, which is defined from the gradient of the potential. This type of analysis of thermodynamic conversion has been used with success by Alicki in various systems [16,17]. This description of two coupled currents can, of course, be extended to a larger number of coupled currents without changing the spirit of the study.

As it is represented, the system is therefore quite generic. The main determinants of functioning are thus summarized by three terms, the capture of the resource, its conversion into a usable form, and the rejection of waste. It is clear that ideally the target is the one where the output power would be maximum and the amount of energy released would be minimal. The study of the limits to achieving this target is one of the objectives of this article. As the coupling parameter for the conversion, the $\alpha$ parameter is therefore central since it determines the system's ability to convert energy into a usable form. A naive picture may suggest that the largest possible $\alpha$ value necessarily leads to the most efficient system, but this is not correct, as we will see now.

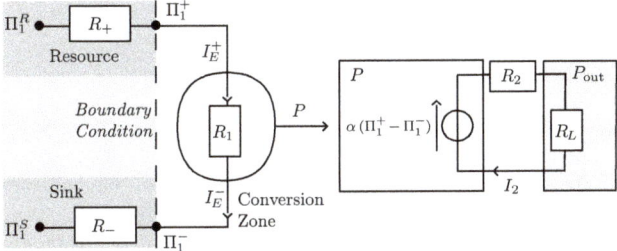

**Figure 1.** Schematic view of the generic system, with a resource and a sink, whose potential $\Pi_1^R$ and $\Pi_1^S$ are constant. The coupling of the conversion zone (circle) with the two reservoirs is ensured by the elements $R_+$ and $R_-$. As a result, the difference potential $\Pi_1^+ - \Pi_1^-$ is less than that between reservoir and sink. Power produced in the conversion zone (circle) is $P = -\alpha \Delta \Pi_1 = \Delta \Pi_2$. The internal resistance $R_2 = \frac{L}{A\sigma_2}$ gives rise to a dissipative contribution $R_2 I_2^2$. The $R_L$ resistance is the output load, and the output power is $P_{out} = R_L I_2^2$.

## 3. Local Energy Conversion

### 3.1. Presentation

At the local level, energy conversion is produced by coupling the energy and matter currents flowing through the device. These currents are generated by the presence of differences between the two thermodynamic potentials $\Pi_1$ and $\Pi_2$. This local modeling is therefore based on the three parameters of conductivity associated with energy transport, $\sigma_1$, conductivity associated with matter transport, $\sigma_2$, and the coupling coefficient between the gradients of the two potentials, $\alpha$. We deduce from this the formulation of local Onsager matrix, where $\nabla = \frac{d}{dx}$ is the spatial gradient, here reduced to 1D in order to simplify the description.

$$\begin{pmatrix} J_2 \\ J_E \end{pmatrix} = \begin{pmatrix} L_{11} & L_{12} \\ L_{21} & L_{22} \end{pmatrix} \begin{pmatrix} -\nabla\left(\frac{\Pi_2}{\Pi_1}\right) \\ \nabla\left(\frac{1}{\Pi_1}\right) \end{pmatrix}. \tag{1}$$

$J_E$ and $J_2$ are the densities of the two currents, and are extensive and conservative quantities. Given the differential form $J_E = \Pi_1 J_1 + \Pi_2 J_2$, the equality of non-diagonal terms $L_{12} = L_{21}$ is insured according to the choice of the correct potentials $-\frac{\Pi_2}{\Pi_1}$ and $\frac{1}{\Pi_1}$ [18,19]. The four terms of the matrix are therefore reduced to three, $\sigma_1$, $\sigma_2$ and $\alpha$, whose correspondences with the coefficients $L_{ij}$ are

$$\sigma_1 = \frac{1}{\Pi_1^2}\left[\frac{L_{11}L_{22} - L_{21}L_{12}}{L_{11}}\right], \tag{2}$$

$$\sigma_2 = \frac{L_{11}}{\Pi_2}, \tag{3}$$

$$\alpha = -\frac{\Delta\Pi_2}{\Delta\Pi_1} = \frac{1}{\Pi_1}\frac{L_{12}}{L_{11}}. \tag{4}$$

In the absence of a matter gradient, the energy conductivity can be defined as $\sigma_{\Pi_2} = \sigma_1\left[1 + \alpha^2\sigma_2/\sigma_1\Pi_2\right]$. The figure of merit is then defined as

$$F_m = \frac{\alpha^2 R_1}{R_2}\Pi_2 = \frac{L_{12}^2}{L_{11}L_{22} - L_{21}L_{12}}. \tag{5}$$

It is known that the ratio $\sigma_2/\sigma_1$, therefore $F_m$, is a direct measure of the intrinsic capacity of energy conversion. $F_m$ can be related to the ratio of the equivalent specific heats by the expression

$\gamma = \frac{C_{\Pi_2}}{C_{I_2}} = F_m + 1$. In their seminal paper, Kedem and Caplan derived the following expression of the coupling parameter between the two fluxes involved in the conversion process [13]:

$$q = \frac{L_{12}}{\sqrt{L_{11}L_{22}}} = \sqrt{\frac{F_m}{1+F_m}} \tag{6}$$

an expression that explicitly includes the kinetic coefficients $L_{ij}$. The figure of merit and the coupling factor $q$ are equivalent in terms of measure of the system performance: the larger their (absolute) values, the better the energy conversion system. This can be evidenced by the derivation of the local maximal efficiency of the conversion process in generator mode, $\eta_{max}$:

$$\eta_{max} = \left(\frac{1+\sqrt{1-q^2}}{q}\right)^2 = \frac{\sqrt{\gamma}-1}{\sqrt{\gamma}+1}. \tag{7}$$

### 3.2. Entropy Production and Efficiency

The volumetric entropy production rate is given by the summation of the force-flow products,

$$\dot{S} = J_2 \nabla\left(-\frac{\Pi_2}{\Pi_1}\right) + J_E \nabla\left(\frac{1}{\Pi_1}\right) = -\frac{1}{\Pi_1}\left[J_2 \nabla \Pi_2 + J_1 \nabla \Pi_1\right]. \tag{8}$$

In the case of a reversible process $\dot{S} = 0$ so does $J_2 \nabla \Pi_2 + J_1 \nabla \Pi_1$. We get $-\frac{J_2 \nabla \Pi_2}{\Pi_1 J_1} = \frac{\nabla \Pi_1}{\Pi_1} = \eta_C$, where $\eta_C$ is the Carnot efficiency. This leads to the general expression of the local efficiency,

$$\eta = -\frac{J_2 \nabla \Pi_2}{J_1 \Pi_1} < \eta_C. \tag{9}$$

Let us define the reduced current as

$$j = \frac{\alpha J_2}{J_1}, \tag{10}$$

which is the ratio between the entropy carried by the transport of the matter, divided by the total entropy transported. In the case of a reversible process, both terms are equal so $j = 1$ [20]. This expression shows three regions for the $\eta(j)$ meaning. For $0 < j < 1$, the device works as a generator. For $j < 0$ and $j > 1$, the device works as a receptor. For reasons of brevity, we will mainly deal with the generator configuration in this article.

Rewriting the Onsager matrix in more suitable form [21], we get

$$\begin{pmatrix} J_2 \\ \Pi_1 J_1 \end{pmatrix} = \begin{pmatrix} \sigma_2 & \alpha \sigma_2 \\ \alpha \Pi_1 \sigma_2 & \gamma \sigma_1 \end{pmatrix} \begin{pmatrix} -\nabla \Pi_2 \\ -\nabla \Pi_1 \end{pmatrix}. \tag{11}$$

Then,

$$j = \frac{\eta \alpha \Pi_1 \sigma_2 - j \alpha \sigma_2 \nabla \Pi_1}{\eta \alpha \Pi_1 \sigma_2 - \frac{j \gamma \sigma_1}{\alpha} \frac{\nabla \Pi_1}{\Pi_1}}. \tag{12}$$

Thus,

$$\eta = \eta_C j \frac{j\gamma - \frac{\alpha^2 \sigma_2}{\sigma_1}\Pi_1}{j\frac{\alpha^2 \sigma_2}{\sigma_1}\Pi_1 - \frac{\alpha^2 \sigma_2}{\sigma_1}\Pi_1}, \tag{13}$$

where $\gamma = \frac{\alpha^2 \sigma_2}{\sigma_1}\Pi_1 + 1$. After a few algebra, we get

$$\eta = \frac{\eta_C}{(\gamma-1)} \frac{\gamma j^2 - (\gamma-1)j}{j-1} \tag{14}$$

$\eta$ presents a maximum for $j_{opt} = 1 + \sqrt{\frac{1}{\gamma}}$ for a receptor mode, and $j_{opt} = 1 - \sqrt{\frac{1}{\gamma}}$ for a generator mode. Both optima reduce to $j = 1$ in the ideal case, when $\gamma$ diverges, where we recover the Carnot efficiency. In this diverging case, the system do not present anymore dissipation production, and the equivalence between the receptor and generator modes is a proof of the absence of causality of the Carnot configuration. This absence of causality is another name for reversibility. We then recover the Kedem–Caplan expression of the maximal efficiency, $\eta_{max} = \eta_C \frac{\sqrt{\gamma}-1}{\sqrt{\gamma}+1}$ for the generator mode, and $\eta_{max} = \eta_C \frac{\sqrt{\gamma}+1}{\sqrt{\gamma}-1}$ for the receptor mode. Let us now plot the efficiency versus the reduced current, as reported in Figure 2.

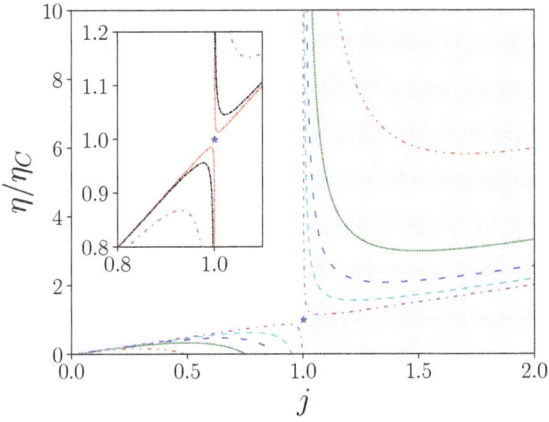

**Figure 2.** Normalized efficiency $\frac{\eta}{\eta_C}$ according to reduced current $j = \alpha J_2/J_1$ with $\gamma = 2$ (red, dot dashed), 4 (green, dots), 8 (blue, loosely dashed), 20 (cyan, dashed), $2 \times 10^2$ (magenta, loosely dot dashed) in main figure, and $\gamma = 2 \times 10^2$ (magenta, loosely dot dashed), $2 \times 10^3$ (black, dot dashed), $2 \times 10^4$ (red, dot dot dashed) in inset. The grey area corresponds to the receptor mode (resp. generator mode). Note that the figure is symmetrical with respect to the Carnot point (blue star), which is never reached. This singular point defines the reversible configuration, where causality is broken.

As expected for the maximum performance achieved, $\eta_{max}$ is an increasing function of the figure of merit. On the other hand, it also appears that the sensitivity to fluctuations in $j$ becomes all the more important as $\eta_{max}$ is important. This is confirmed by estimating the value of the slope in the vicinity of the maximum yield, which is $\partial \eta / \partial(j) \approx -2\eta_{max} F_m$. The larger the figure of merit, the steeper the slope. This local description allows us to conclude that the performance of the device is obtained at the cost of a constraint of stability of the operating points, directly driven by the value of the figure of merit. As an intrinsic quantity, the figure of merit defines the performance ceiling beyond which it cannot be exceeded. It is clear from the figure that the system defined by a high figure of merit exceeds in performance all the systems of lower figure of merit value. However, this result is strongly weighted by the fact that, for excursions of $j$ around the optimal value, the efficiency falls rapidly. Then, it is not necessarily relevant to look for a device with a large figure of merit, without first inventorying the operating range that will be brought to run this device. For simplicity's sake, we have only dealt here with the case where the system works as a generator, which is obtained by $0 < j < 1$. It is clear that the same study can be carried out for the case where the system operates as a receptor, instead of working as a generator. This situation, well known for thermal machines, corresponds to heat pump operation. More broadly, and in the case of non-thermal machines, this case actually corresponds to the operation in recycling mode where the treated quantity undergoes regeneration. It should be noted that the expression of performance refers only to $\gamma$, and therefore to the figure of merit, without specifying any

## 4. Global Conversion System

### 4.1. Presentation

In accordance with the diagram in Figure 1, the device of the conversion zone is connected to its reservoirs via the two resistors $R_+$ and $R_-$, which makes it possible to explore all boundary conditions. The presence of $R_+$ and $R_-$ may lead to the pinching of the potential difference $\Pi_1^+ - \Pi_1^-$ according to the system operating point. More precisely, $R_+$ governs the limitation of access to the resource while $R_-$ reflects possible saturation effects of waste disposal. This global model, although limited, makes it possible to approach the behavior of many systems, including living systems, depending on whether the resource is abundant or scarce, and whether waste disposal, including thermal waste, is easy or not. Living system and non-living systems differ from the fact that the energy current is never zero in living systems, so $R_1$ is always finite, and there is a non-zero resting point. On the contrary, a non-living system may have a zero resting point, with zero energy current, so $R_1$ may be infinite in these systems. Let us consider the set of the four equations that governs the functioning of the system (see Appendix A):

$$I_{E-} = \alpha \Pi_1^- I_2 + (1 - \varphi) R_2 I_2^2 + \frac{(\Pi_1^+ - \Pi_1^-)}{R_1}, \tag{15}$$

$$I_{E-} = \frac{(\Pi_1^- - \Pi_1^S)}{R_-}, \tag{16}$$

$$I_{E+} = \alpha \Pi_1^+ I_2 - \varphi R_2 I_2^2 + \frac{(\Pi_1^+ - \Pi_1^-)}{R_1}, \tag{17}$$

$$I_{E+} = \frac{(\Pi_1^R - \Pi_1^+)}{R_+}. \tag{18}$$

These equations have their origin in the integration of the local form described in the previous paragraph. These developments have been the subject of previous articles [14,22], and will not be re-described here. $\varphi$ controls the dissipation fraction that returned to the source or to the sink. In the following, we will choose $\varphi = 0$. This choice is not critical here since the effect of $\varphi = 0$ is driven by $R_2$, which is equal to zero.

### 4.2. Devices with Zero Resting Point

First of all, we consider that $R_2 = 0$ and $R_1$ diverge, in order to separate the contributions of entropy production and internal dissipation. $R_2$ governs the current of matter, so we therefore consider that this current may not be limited, so there is no intrinsic dissipation within the device. The figure of merit of the device is then infinite and we may expect to reach the ideal conditions and the Carnot efficiency. However, the classical discussion around the Carnot efficiency is based on pure Dirichlet boundary conditions, which is clearly not the case here, so we have to consider the new conditions introduced by the modification of the boundary conditions. In the present configuration of zero resting point systems, the general equations (see Appendix A) can be summarized as

$$I_{E-} = \frac{\Pi_1^S I_2}{\frac{1}{\alpha} - R_- I_2}, \tag{19}$$

$$I_{E+} = \frac{\Pi_1^R I_2}{R_+ I_2 + \frac{1}{\alpha}}, \tag{20}$$

with the output power given by $P = I_{E+} - I_{E-}$.

The plots in Figure 3 summarize the behavior of the global system. The output power presents a maximum and two zero values. The first value corresponds to the case where the efficiency reaches its maximum. This situation is obtained for $I_2 = 0$, so $I_{E-} = I_{E+} = P = 0$. This means that no matter or energy can flow through the system, which is a totally useless situation for a physical system. The second zero power value is reached for a current of matter $I_{2sc}$, named the short-circuit current, by analogy with electronics. In this situation, the produced power is completely re-dissipated inside the system. $I_{2sc}$ is therefore an ultimate operating point for the system, working as an energy generator. For a truly efficient operation, it is therefore necessary to try to push $I_{2sc}$ to large values, which are obtained by getting as close as possible to Dirichlet conditions. In the general case, the approximate expression of this current is

$$I_{2sc} \approx \frac{1}{\alpha} \frac{\Delta \Pi_1}{R_- \Pi_1^S + R_+ \Pi_1^R}, \tag{21}$$

which confirms that Dirichlet's conditions where $R_- = R_+ \approx 0$ are to be sought, if accessible. Since the resting point here is zero, the power curve necessarily intercepts that of $I_{E-}$. Beyond this interception point, the system is in a situation where it releases more waste than it produces output power. We call critical point the point where $P = I_{E-}$, reached for $I_{2cp}$. The fact that power is not a monotonous function of $I_2$ is actually quite unexpected because, to the extent that $R_2 = 0$, the total absence of intrinsic viscosity should not lead to any limit to $I_2$. However, if we carry out a development at the first order of the expression of power we find

$$P \approx \left[ \alpha \Delta \Pi_1 - \left( \Pi_1^R R_+ + \Pi_1^S R_- \right) \alpha^2 I_2 \right] I_2 \tag{22}$$

which clearly indicates the presence of a viscous friction term $R_{fb}$,

$$R_{fb} \approx \alpha^2 (\Pi_1^R R_+ + \Pi_1^S R_-) \tag{23}$$

which reduces the transport of the matter, even though the intrinsic viscosity, i.e., $\frac{1}{\sigma_2}$, associated with the transport of the matter, is zero. This additional dissipation is a pure feedback effect that is due to the presence of boundary conditions at the general limits where $R_+$ et $R_-$ are non-zero. This additional dissipation can only be rendered null if $R_+ = R_- = 0$, i.e., a strict Dirichlet condition, which is, in reality, only very rarely observed. Note that the condition $\alpha = 0$ leads to the same result but it is useless because in this case the transport of energy and matter are fully decoupled, and the device does not convert the energy anymore. The conditions $R_2 = 0$ and $R_1 \to \infty$ determine the performance envelope for a system with an ideal conversion zone. In particular, it is noted that, although $I_{E+}$ and $I_{E-}$ are increasing functions of the current of matter $I_2$, the growth rate of the energy waste current $I_{E-}$ always ends up reaching that of the energy current $I_{E+}$ supplied to the system. In addition, even in the case of a system whose core is composed of an ideal device, ($R_2 = 0, R_1 \to \infty$), the increase in the current of matter inexorably leads to an increase in the current of waste in larger proportions to the rate of supply of resources. The only way out is to limit the current of matter to values below a threshold, which may be that of maximum power, maximum efficiency, minimum waste generation, or below the critical point. In the Figure 3, the response is given for two different values of the coupling parameter $\alpha$. The influence of $\alpha$ is quite surprising. At first we observe that the lower is $\alpha$ and the lower are the output power and efficiencies, as expected for a lower conversion level of the energy. However, in the same time, the short-circuit current is strongly enhanced, opening the way to a large range of $I_2$ working points for the transport of the matter. This is due to the $\alpha^{-2}$ dependency of $I_{2sc}$. This leads to the conclusion that *the search for a very efficient system is in contradiction with the search for a very adaptable system*.

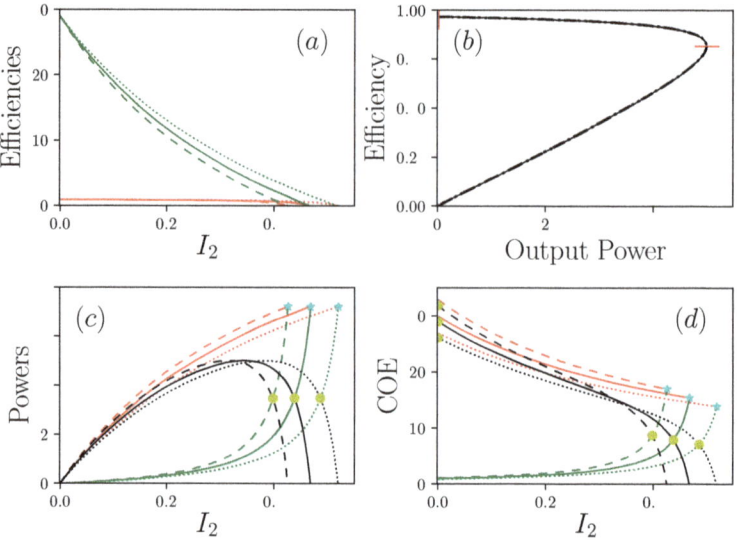

**Figure 3.** Representations of the powers $I_{E+}$, $I_{E-}$, with $R_1^{-1} = 0$, $R_2 = 0$ (and $P = P_{out}$), $R_+ = R_- = 2$, $\Pi_1^S = 1$, $\Pi_1^R = 30$. (**a**) shows efficiencies (resp. in red and green) $\eta_{+/-} = P_{out}/I_{2_{+/-}}$ in function of $I_2$, the current of matter. (**b**) is the efficiency in function of the power produced $P_{out}$. (**c**) show the power (resp. in red, green and black) $I_{E+}$, $I_{E-}$ and $P_{out}$ in function of $I_2$. (**d**) shows (resp. in red, green and black) $COE_{+/-} = I_{E+/-}/I_2$ and $COE_{P_{out}} = P_{out}/I_2$ in function of $I_2$. Dotted lines are $\alpha = 0.9$, solid lines are $\alpha = 1$, dashed lines are $\alpha = 1.1$. In (**c**) and (**d**) cyan stars show short circuit situations $I_{2sc}$, yellow circles are critical points $I_{2cp}$. In (**b**) vertical and horizontal red lines are respectively maximal efficiency and maximal power.

Let us now focus on the issue of the trade-off between power efficiency and waste generation. The Figure 3a represents the curves of the production efficiency $\eta_{prod} = P/I_{E+}$ and the waste efficiency $\eta_{waste} = P/I_{E-}$. Note that $\eta_{prod}$, which is the traditional efficiency, is limited by the Carnot efficiency but $\eta_{waste}$ is not, since it does not refer to the traditional expression of efficiency but is just an extension of the notations. $\eta_{prod}$ is bounded by a zero value, which corresponds to zero power, and a maximum efficiency point, reported in Figure 3b. Between these two values, the system presents a maximum of the power, which absolutely does not coincide with the maximum efficiency. In this configuration the MPP or mEP operations are clearly disjointed as already mentioned [23,24]. Let us now consider the cost of carrying out a unitary process. By unitary process we consider a process standardized by the value of the associated transport of matter, i.e., the ratio between the energy currents and the matter current. We call this quantity Cost Of Energy, i.e., COE. This makes it possible to consider energy expenditures with regard to the associated matter transformation along a unitary displacements. In other words, COE can measure the amount of energy needed to be rejected as a waste, for displacing the matter from a unit length. This quantity is already known in biology as Cost Of Oxygen Transport (COT), where it has made it possible to qualify a unit displacement with regard to the energy released in the form of waste [25,26]. Here, we extend the notion in a more general form where COE is defined by $COE_+$ which is the cost of energy needed to feed the system, and $COE_-$ which is the cost of waste energy that is rejected, so,

$$COE_{+/-} = \frac{I_{E+/-}}{I_2} \qquad (24)$$

Note that the $COE_+$ is a strictly decreasing function of $I_2$ and $COE_-$ is a strictly increasing function of $I_2$. This means that the cost of energy needed for a unitary process decrease when $I_2$ increases but, in the same time the amount of waste always increases. There is therefore no optimum to consider any minimization of the waste. In addition, it is important to note that the $R_1^{-1} = 0$ configuration is the only one that provides the strong coupling conditions, for which the energy and matter currents are roughly proportional [10]. In this case, the Onsager matrix has a zero determinant. This situation is an idealization of the transport of energy entirely achieved by the transport of matter. In other words, it is a question of considering that the behavior of out-of-equilibrium thermodynamics may be equivalently described by pure mechanics. This is obviously never fully encountered unless it is considered that a $\Delta \Pi_1$ difference can persist without an associated current of matter existing. This is the purpose of the following paragraph.

### 4.3. Devices with Non Zero Resting Point

The study of devices with non zero resting points concern the case of all systems for which a shutdown means death. Indeed, unlike a machine, all living systems are never totally shut down, and always keep a minimum operating point value, which we call basal, also known as a resting point. This situation corresponds to the case where $R_1$ has a finite value. While remaining, for the moment in the case where $R_2 = 0$, we can develop the main results from this configuration. The general equations of the system are given in Appendix B. In this situation, the efficiency, nor the power, can reach the previous values, as reported in the Figure 4. At the resting point $I_2 = 0$, the system is in its basal configuration where $P = 0$, so $I_{E+} = I_{E-} = B$ with,

$$B = \frac{\Delta \Pi_1}{R_+ + R_1 + R_-} \quad (25)$$

The typical response of systems with non zero resting points is given in the Figure 4. One can notice that the general shape is not strongly modified from the case of zero resting point configurations, except the presence of a non zero current of energy even at zero $I_2$ and a slight modification of the short-circuit point. Regardless of the reduction in efficiency introduced by the presence of $R_1$, the search for a system with a very low basal point requires to be located in a configuration close to Neumann conditions where $R_+$ and $R_-$ have very large values. This is not problematic except that it requires the system to operate at low values of $I_2$, in order to limit the dissipation due to the term $R_{fb}$. There is therefore a fundamental contradiction between having a system with low resting power consumption and a system that can provide significant power. It is clear that a sober system, in the sense of its consumption at rest, is unsuited to the production of significant power, without leading to significant dissipation at high speed, or equivalently, high $I_2$. If such a power is sought, then it implies that the boundary conditions should be of Dirichlet like with $R_+ \approx R_- \approx 0$. However, in this case the system will have a necessarily high rest consumption. Compared to systems with a zero resting point, it can be seen that the maximum power operating point and maximum efficiency operating point tend to approach each other as $R_1$ increases. In this configuration, as can be derived in [27], the feedback resistance is approximately given by

$$R_{fb} \approx \frac{\alpha^2 \langle \Pi_1 \rangle}{\frac{1}{R_+ + R_-} + \frac{1}{R_1}} = R^* \alpha^2 \langle \Pi_1 \rangle \quad (26)$$

where $R^* = \frac{(R_+ + R_-)R_1}{R_1 + R_+ + R_-}$ and $\langle \Pi_1 \rangle = \Pi_1^R/2 + \Pi_1^S/2$.

Compared to the previous configuration the dissipation introduced by the presence of $R_{fb}$ can now be modified whatever are the boundary conditions because $R^* < Min(R_+ + R_-, R_1)$. More precisely, in the case of Neumann-like boundary conditions, there is a restriction to the value of $R_1$ where $R_1 \ll R_+ + R_-$ is expected. Under Dirichlet-like boundary conditions $R_+$ and $R_-$ are small so there is no condition on $R_1$. Consequently, a system with a very low basal point, with large values of both $(R_+, R_-)$ (Neumann like) and $R_1$ will suffer from a large $R_{fb}$ and is then limited to very low $I_2$ currents. If the boundary conditions are more like Dirichlet conditions, then $R_{fb}$ keeps low but the low basal level now imposes that $R_1$ strongly increases, which reduced both the available power $P$ and the efficiency. Thus, we can see that there is no room for a powerful and efficient system working in all conditions. The main trade-off is between power and efficiency, but it ultimately extends beyond that.

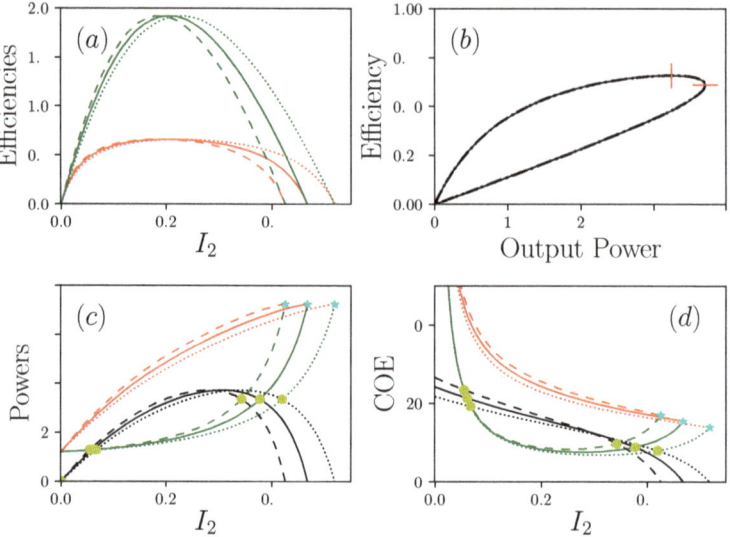

**Figure 4.** Representations of the powers $I_{E+}$, $I_{E-}$ and $P$, with $R_1^{-1} = 0.05$, $R_2 = 0$ (and $P = P_{out}$), $R_+ = R_- = 2$, $\Pi_1^S = 1$, $\Pi_1^R = 30$. (**a**) shows efficiencies (resp. in red and green) $\eta_{+/-} = P_{out}/I_{2+/-}$ in function of $I_2$ the current of matter; (**b**) is the efficiency in function of the power produced $P_{out}$; (**c**) shows the power (resp. in red, green and black) $I_{E+}$, $I_{E-}$ and $P_{out}$ in function of $I_2$; (**d**) shows (resp. in red, green and black) $COE_{+/-} = I_{E+/-}/I_2$ and $COE_{P_{out}} = P/I_2$ in a function of $I_2$. Dotted lines are $\alpha = 0.9$, solid lines are $\alpha = 1$, dashed lines are $\alpha = 1.1$. In (**c**) and (**d**), cyan stars show short circuit situations $I_{2sc}$, and yellow circles are critical points $I_{2cp}$. In (**b**), vertical and horizontal red lines are respectively maximal efficiency and maximal power.

From a rather general point of view, the incoming energy current $I_{E+}$ makes it possible to establish and maintain, thanks to the presence of $R_1$, a potential difference that permits the production of output work. On this point, we join the work of Alicki [16], who considers that the incoming energy current makes it possible to maintain a difference in potential, exactly as a pump would do. This situation is particularly described in the case of photovoltaic structures, with a difference in electrochemical potential [16], or in the case of muscles where the attachment and release cycles of actin and myosin structures lead to the maintenance of a force [28]. It should be noted that, depending on the position of the resting point, the power curve can intercept between zero and twice the $I_{E+}$ curve. It can therefore be seen that, in the case of systems with a relatively low resting point, there may be an area for which the power produced is greater than the power released as a waste. More intriguing, this area can start with a non-zero value of $I_2$. In other words, there may be systems for which the situation $I_2 \neq 0$ leads to a proportionally smaller waste production than at rest. Systems with a non-zero resting point therefore present very different optima than non-living systems, whose zero resting point leads to minimizing power by stopping the machine. By using the definition $COE_- = I_{E-}/I_2$, we can plot its response according to $I_2$. It should be noted that the $COE-$ has a minimum value, which does not coincide with the maximum power point. This defines a new operating point for the system, which characterizes the situation where the system minimizes its production of waste.

An illustration of this can be given if we consider the motion of living systems. Let us consider that the task to be accomplished consists in moving the body over a unit distance, the question arises as to how fast this operation will lead to a minimum of waste, essentially in the form of heat and metabolic degradation products. It is clear that displacement here corresponds to the transport of matter, and is therefore assimilated to $I_2$ proportional to the speed of travel as previously said. There is an abundant amount of literature showing that there exists a minimum of the so-called $COT \equiv COE_-$ point for all animals for which movement appears to be favored when the COT is minimal [25,26]. As expected, see Figure 4, $COE_-$ and $COE+$ curves have a common point at the short circuit point. We previously saw that Dirichlet's conditions, $R_+ = R_- = 0$, were those that minimized the feedback resistance $R_{fb}$ and allowed for considering potentially a divergence of the current of matter and the output power. This simple observation shows that strict Dirichlet's conditions are simply nonphysical. Nevertheless, one can consider that this condition can be approached. However, the presence of $R_+, R_-$ and $R_1$ in series shows that Dirichlet's condition is asymptotically obtained only if the ratios $R_+/R_1$ and $R_-/R_1$ are negligible, which imposes an important value for $R_1$, and therefore a high value of the basal power. We therefore see the emergence of a paradox, which, seeking to minimize the dissipation due to $R_{fb}$ leads to the constraint of high consumption at rest. The same system cannot therefore be both very powerful and very energy-efficient at its resting point. We find here the generalization of a well-known situation, for example for the thermal engines of vehicles, in which the engine's displacement determines its ability to produce power, as well as its efficiency.

### 4.4. Internal Dissipation Devices

Let us now consider the introduction of the dissipative term $R_2$. The output power of the system is now represented by Figure 5. As a thermodynamic engine, the system provides a power $P = \alpha \left( \Pi_1^+ - \Pi_1^- \right) I_2$ as already defined. The efficiency of this part of the system is given by $\eta_2 = \frac{P - R_2 I_2^2}{P}$. Thus, the total efficiency of the system is

$$\eta_{sys} = \eta_1 \eta_2, \qquad (27)$$

with $\eta_1 = \frac{P}{I_{E_+}}$. Compared to the previous configurations, both the power, the short-circuit current $I_{sc}$, and the efficiency are now reduced. The influence of $R_2$ appears to be always detrimental, which was not the case for $R_1$. It is clear that one should look for minimal $R_2$ if possible. In other words, in the expression of the figure of merit, there is a constraint on $R_2$. At first, both $\alpha$ and $R_1$ seem to be non-constrained, and the same figure of merit can be obtained for various values of the couple $(\alpha, R_1)$. Nevertheless, as we have mentioned, the present description shows that $R_2$ is linked in series with $R_{fb}$. Consequently, the constraint on $R_2$ can be relaxed to the condition $R_2 \ll R_{fb}$. According to the expression $R_{fb} \approx \frac{\alpha^2 \langle \Pi_1 \rangle}{\frac{1}{R_+ + R_-} + \frac{1}{R_1}}$, this leads to the condition $1 + \frac{R_1}{R_\Sigma} < \frac{\alpha^2 R_1}{R_2} \langle \Pi_1 \rangle$ where we recognize the figure of merit, so the condition becomes

$$1 + \frac{R_1}{R_\Sigma} < F_m, \qquad (28)$$

where $R_\Sigma = R_+ + R_-$. According to the previous observation, the minimization of the dissipation occurring from the $R_{fb}$ term imposes that $\frac{R_1}{R_\Sigma}$ should be large enough. Thus, we now get a supplementary condition for $F_m$. In this expression, the boundary conditions and the intrinsic performances of the device are considered together. Under Dirichlet conditions, $1 + \frac{R_1}{R_\Sigma}$ diverges so the system keeps its level of dissipation low only in the case of a very large figure of merit, and is forced to work at very low $I_2$ values. Under Neumann conditions, $R_\Sigma$ diverges and then the condition on the figure of merit is then relaxed. Ideally, even when achieved asymptotically, one might want to achieve maximum power, as well as minimal waste production, combined with maximum efficiency. We conclude that looking for maximum efficiency always leads to approaching the Carnot point, which is, even in an out-of-equilibrium description, the point where power production is canceled out.

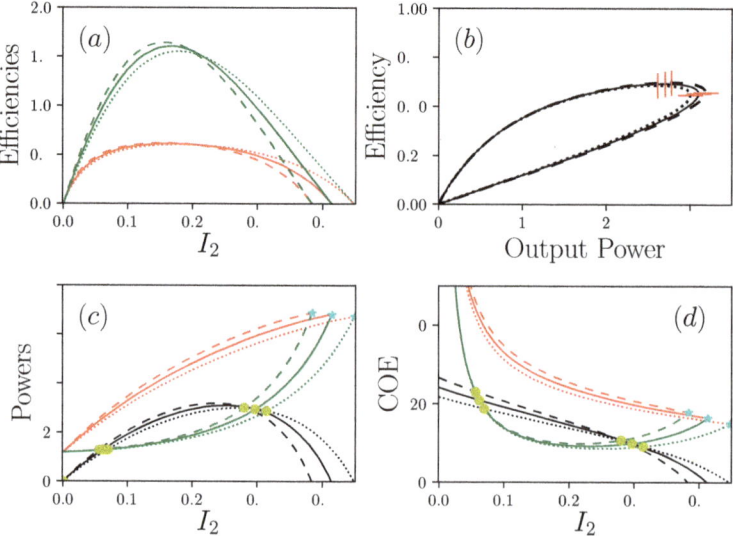

**Figure 5.** Different representations of the powers $I_{E+}$, $I_{E-}$ and $P$, with $R_1^{-1} = 0.05$, $R_2 = 4$, $R_+ = R_- = 2$, $\Pi_1^S = 1$, $\Pi_1^R = 30$. (**a**) shows efficiencies (resp. in red and green) $\eta_{+/-} = P/I_{2+/-}$ as a function of $I_2$ the current of matter. (**b**) is the efficiency in function of the power produced $P$. (**c**) shows the power (resp. in red, green and black) $I_{E+}$, $I_{E-}$ and $P_{out}$ in function of $I_2$. (**d**) shows (resp. in red, green and black) $COE_{+/-} = I_{E+/-}/I_2$ and $COE_{P_{out}} = P_{out}/I_2$ as a function of $I_2$. Dotted lines are $\alpha = 0.9$, solid lines are $\alpha = 1$, dashed lines are $\alpha = 1.1$. In (**c**) and (**d**), cyan stars show short circuit situations $I_{2sc}$ and yellow circles are critical points $I_{2cp}$. In (**b**), vertical and horizontal red lines are respectively maximal efficiency and maximal power.

## 5. Entropic Point of View

The previous power budget analysis highlighted three classes of systems: systems with a zero resting point, systems with a non-zero resting point, and, finally, systems with an additional internal dissipation term $R_2$. Let us consider these three classes again from the entropic point of view.

### 5.1. Devices with Zero Resting Point

The production of entropy from the presence of $R_-$ and $R_+$ is given respectively on both sides of the device by

$$\dot{S}_{E+} = I_{E+} \left( \frac{1}{\Pi_1^+} - \frac{1}{\Pi_1^R} \right) = \frac{\alpha^2 I_2^2 R_+}{1 + \alpha I_2 R_+}, \tag{29}$$

$$\dot{S}_{E-} = I_{E-} \left( \frac{1}{\Pi_1^S} - \frac{1}{\Pi_1^-} \right) = \frac{\alpha^2 I_2^2 R_-}{1 - \alpha I_2 R_-}. \tag{30}$$

The results are given in Figure 6.

There is clearly an asymmetry in the two entropy productions. Indeed, if the two contributions initially increase in a quadratic form with the current of matter, the contribution of the resource side, $\dot{S}_{E+}$, tends to a linear progression independent of the coupling condition $R_+$, while the contribution on the waste rejection side $\dot{S}_{E-}$ tends to diverge as soon as $I_2 \approx 1/\alpha R_-$. It is surprising to see that, in addition, this divergence is more marked as the coupling factor $\alpha$ between energy and matter is important. *There is therefore no other solution than to make $R_-$ as small as possible, and therefore reject all the waste easily. This is an additional constraint for the design of efficient systems.*

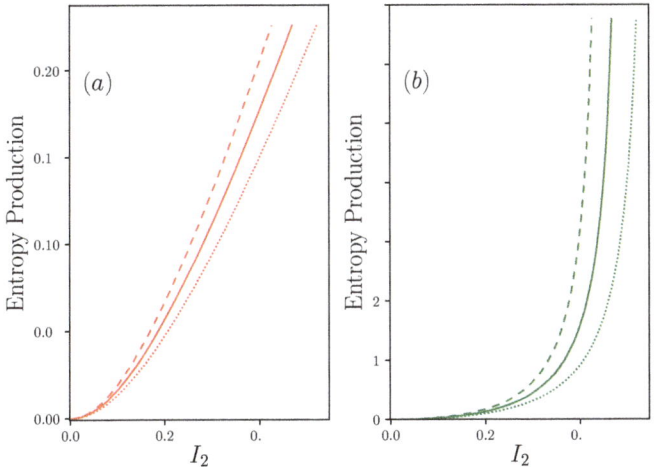

**Figure 6.** Evaluation of the entropy production with the same configuration as in Figure 3, $R_1^{-1} = 0$, $R_2 = 0$, $R_+ = R_- = 2$, $\Pi_1^S = 1$, $\Pi_1^R = 30$. (a) shows $\dot{S}_{E+}$ and (b) shows $\dot{S}_{E-}$, both in function of $I_2$ the current of matter—the same color and line-style code as in Figure 3.

### 5.2. Devices with Non-Zero Resting Points

Let us now look at the configuration of non-zero resting point systems, while keeping $R_2 \approx 0$. In this case, the general expressions become

$$\dot{S}_{E+} = I_{E+} \left( \frac{1}{\Pi_1^+} - \frac{1}{\Pi_1^R} \right) = \frac{R_+ I_{E+}^2}{\left( \Pi_1^R - R_+ I_{E+} \right) \Pi_1^R}, \tag{31}$$

$$\dot{S}_{E-} = I_{E-} \left( \frac{1}{\Pi_1^S} - \frac{1}{\Pi_1^-} \right) = \frac{R_- I_{E-}^2}{\left( R_- I_{E-} + \Pi_1^S \right) \Pi_1^S}. \tag{32}$$

The results are given in Figure 7 where $I_{E+}$ and $I_{E-}$ are defined according to the Appendix B. We can see that the presence of $R_1$ reintroduces a significant symmetry between the two contributions to the entropy production. Moreover, the question of the importance of the quality of the coupling on the resource side, by minimizing $R_+$, or to the rejection side, by minimizing $R_-$, is now of equal importance.

### 5.3. Internal Dissipation Devices

For internally dissipated devices, the term $R_2$ produces a quadratic dissipation $R_2 I_2^2$. We have seen before that the presence of $R_2$ never brings any advantage in terms of energy conversion since it only contributes to lowering the power available at the output of the system. As this dissipation diffuses into the system, it is itself a source of entropy, as shown in Figure 8. At this stage, it is important to know how this dissipation occurs. In the case of some thermal systems, an analytical calculation can be carried out that leads to an equal distribution of this dissipation between the resource and the sink, i.e., $\varphi = 0.5$, see Appendix B in accordance with [22]. In other systems, such as muscles subjected to moderate stress, this dissipation is considered to be completely rejected into the sink ($\varphi = 0$) [14]. For some living systems, including homeothermic species, it is likely that a fraction of this dissipation is partially released, and partially used to maintain the central temperature of the body, leading to a value $\varphi \approx 1$, depending on outdoor conditions. One example is the case of vaso-dilatation and vasoconstriction of peripheral vessels, which is a solution for modulating the value of $R_-$ and consequently reject less, or more, heat outside of the body.

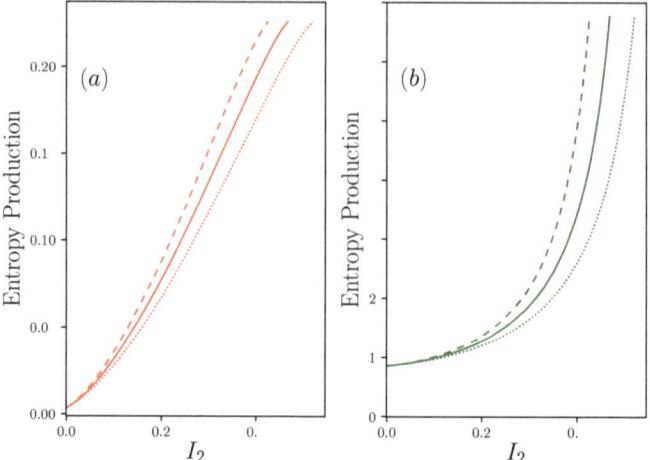

**Figure 7.** Evaluation of the entropy production with the same configuration as in Figure 4 with $R_1^{-1} = 0.05$, $R_2 = 0$, $R_+ = R_- = 2$, $\Pi_1^S = 1$, $\Pi_1^R = 30$. (**a**) shows $\dot{S}_{E+}$ and (**b**) shows $\dot{S}_{E-}$, both in function of $I_2$ the current of matter—the same color and line-style code as in Figure 3.

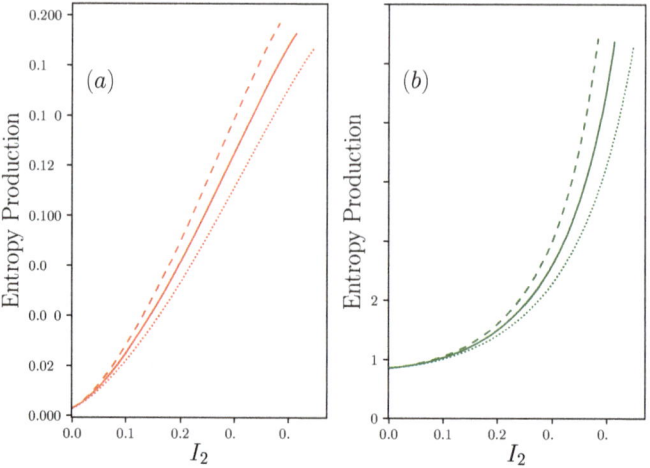

**Figure 8.** Plot of the entropy production with the same configuration as in Figure 5 with $R_1^{-1} = 0.05$, $R_2 = 4$, $R_+ = R_- = 2$, $\Pi_1^S = 1$, $\Pi_1^R = 30$. (**a**) shows $\dot{S}_{E+}$ and (**b**) shows $\dot{S}_{E-}$, both in function of $I_2$ the current of matter—the same color and linestyle code as in Figure 3.

## 6. Adaptable or Adapted?

The study of the behavior of a generic system composed of a conversion device, and the boundary conditions to the reservoirs, now allows us to establish several observations. First, the search for the best device, in terms of power and efficiency, can be summarized by the search for the largest figure of merit $F_m$. However, this result must be modulated by the fact that the value of $F_m$ is determined by the set of the three parameters $R_1$, $R_2$, and $\alpha$ which, at this stage, do not present any constraints. In addition, few thermodynamic devices have a single operating point, but are generally expected to operate over a wide range of uses that principally means large range of $I_2$. In the precedent paragraph, we concluded that the greater the figure of merit, the smaller the effective operating range becomes. Indeed, for such a narrow range, the users must then conform quite strictly to that imposed by the value of the figure of merit of the device. This observation explains quite simply why the consumption observed by vehicle drivers is always larger than that reported by vehicle manufacturers, since the actual conditions of use never coincide with the test conditions. Similarly, the measured performance of equipment in dwellings, as well as the performance of the dwellings themselves, is below the expected performance during construction. This observation leads to the recommendation that devices intended to operate over a wide range of uses should not be designed solely on the basis of their maximum performance in terms of efficiency and power. Beyond this observation, the question arises of determining, within a system, which of the three parameters $R_1$, $R_2$, and $\alpha$ should be optimized as a priority. We can first conclude that, unless there are situations where dissipation is explicitly sought, $R_2$ must be systematically minimized. With regard to $R_1$, we have seen that its choice determines two categories of systems, depending on whether $R_1$ is zero or not. It must be noticed that $R_1 = 0$ is not possible for living systems because a resting point does exist until the death. In the category where $R_1 = 0$, the operating range of the system is limited by the feedback effects that introduce an excess dissipation term $R_{fb}$. Note that this term can be minimized if the boundary conditions are as close as possible to Dirichlet conditions. In this situation, the currents of matter $I_2$ and energy $I_E$ may diverge. This situation has been that of our societies since the beginning of the industrial revolution [12], with coal, followed by an acceleration after the Second World War, due to the rise in oil consumption. The divergence of matter and energy currents is directly linked to an increase in the figure of merit, through an increased facilitation of the circulation of matters and energies, which is produced by a minimization of $R_2$, as well as an increase of $\alpha$, i.e., technological progress that allows thermodynamic potentials to be more strongly coupled. A basic illustration of this increase is the performance of steam machines, which have gradually increased the ratio between outlet pressures and inlet temperatures [29]. The second category of system concerns the case where $R_1 \neq 0$. These systems are particular in that they consume energy, even in a resting situation. We can include living organisms and societies, but also machines, when the latter operate in the idle position, with no other power production than the maintenance of this idle. We have seen that, in this case, there are two categories of systems depending on whether we favor power production or low consumption at rest. These two categories are resolutely distinct and it is illusory to think of a system capable of producing a very high power, while maintaining a very low basic consumption. The choice of $R_1$, i.e., the dissipation at rest, is also decisive in the dissipation produced by feedback. The issues of minimization or maximization of efficiency and power are therefore part of a much broader framework than initially thought.

## 7. Discussion

We proposed a generic thermodynamic system model that allows for considering several situations of coupling of the energy and matter currents, as well as their conversions. At the local level, the intrinsic performance of the device that constitutes the core of the system was studied. It appears that the best intrinsic performance in terms of power and efficiency is obtained for the devices with the largest figure of merit, without specifying the respective contributions of the conductivities associated with the transport of energy or matter. However, the sensitivity of these devices to changes in the reduced current $j$ shows that the intrinsically most efficient devices are also the most constraining because they require precise control of this reduced current, and therefore of energy and matter currents. At the scale of a complete system, the coupling to the external environment very strongly modifies the conclusions compared to the observations made at the local level. It is observed that behavior is mainly governed by the boundary conditions that connect the local system to the resource and the waste. The presence of boundary conditions such as Dirichlet or Neumann leads to a wide variety of behaviors. The ideal Dirichlet conditions are the only ones that do not lead to any feedback, and consequently conduct in the absence of limitations for the energy and matter currents. When the boundary conditions are between Dirichlet and Neumann, many possibilities then arise. The presence or absence of a resting point for the system strongly influences these possibilities in terms of power, but also in terms of waste production associated with the completion of a task. The concept of coefficient of energy cost, COE, is introduced, generalizing the classical COT already established for biological systems. Finally, it is observed that the internal dissipation produced by the presence of $R_2$ is always detrimental for both the efficiency and the power. Its only positive contribution is limited to cases where dissipation and entropy production are explicitly sought, as in the case of homeothermic animals.

**Author Contributions:** Conceptualization, C.G.; methodology, C.G. and E.H.; software, E.H.; validation, C.G. and E.H.; formal analysis, C.G.; writing—original draft preparation, C.G.; writing—review and editing, C.G. and E.H.; visualization, E.H.; supervision, C.G.; project administration, C.G.; funding acquisition, C.G. All authors have read and agreed to the published version of the manuscript.

**Funding:** This research was funded by the CNRS grant METABOLOCOT *Défi Modélisation du vivant*.

**Acknowledgments:** The authors would like to thank Henni Ouerdane for enlightening discussions.

**Conflicts of Interest:** The authors declare no conflict of interest.

## Appendix A

We consider the set of the four equations of the generic model:

$$I_{E-} = \alpha \Pi_1^- I_2 + (1-\varphi) R_2 I_2^2 + \frac{(\Pi_1^+ - \Pi_1^-)}{R_1}, \tag{A1}$$

$$I_{E-} = \frac{(\Pi_1^- - \Pi_1^S)}{R_-}, \tag{A2}$$

$$I_{E+} = \alpha \Pi_1^+ I_2 - \varphi R_2 I_2^2 + \frac{(\Pi_1^+ - \Pi_1^-)}{R_1}, \tag{A3}$$

$$I_{E+} = \frac{(\Pi_1^R - \Pi_1^+)}{R_+}. \tag{A4}$$

The $\varphi$ term defines the fraction of the waste which is respectively rejected to the source and to the sink. This is a well known parameter in some thermal engines [22]. In the case of a living system, $\varphi$ may define the ratio of heat rejected outside of the body and kept inside.

The resolution of the four equations gives

$$\begin{pmatrix} \Pi_1^- \\ \Pi_1^+ \end{pmatrix} = \frac{1}{AD - BC} \begin{pmatrix} D & -B \\ -C & A \end{pmatrix} \begin{pmatrix} \Pi_1^S + (1-\varphi) R_- R_2 I^2 \\ \Pi_1^R + \varphi R_+ R_2 I^2 \end{pmatrix}, \tag{A5}$$

with

$$A = 1 - \alpha R_- I + \frac{R_-}{R_1},$$

$$B = -\frac{R_-}{R_1},$$

$$C = -\frac{R_+}{R_1},$$

$$D = \alpha R_+ I + \frac{R_+}{R_1} + 1.$$

## Appendix B

In the case of a system without dissipation, $R_2=0$, the general equations become

$$I_{E-} = \alpha \Pi_1^- I_2 + \frac{\Pi_1^+ - \Pi_1^-}{R_1}, \tag{A6}$$

$$\Pi_1^- = R_- I_{E-} + \Pi_1^S, \tag{A7}$$

$$I_{E+} = \alpha \Pi_1^+ I_2 + \frac{\Pi_1^+ - \Pi_1^-}{R_1}, \tag{A8}$$

$$\Pi_1^+ = \Pi_1^R - R_+ I_{E+}, \tag{A9}$$

which leads to

$$I_{E+} = \frac{\alpha I_2 \frac{R_1}{R_+} \Pi_1^S + \frac{\Delta \Pi_1}{R_+} + \alpha A I_2 \frac{R_1}{R_-} \Pi_1^R + A \frac{\Delta \Pi_1}{R_-}}{1 + AB}, \tag{A10}$$

$$I_{E-} = \frac{\alpha I_2 \frac{R_1}{R_-} \Pi_1^R + \frac{\Delta \Pi_1}{R_-} - \alpha B I_2 \frac{R_1}{R_+} \Pi_1^S - B \frac{\Delta \Pi_1}{R_+}}{1 + AB}, \tag{A11}$$

with

$$A = \left( \alpha I_2 R_1 \frac{R_-}{R_+} - \frac{R_1}{R_+} - \frac{R_-}{R_+} \right),$$

$$B = \left( \alpha I_2 R_1 \frac{R_+}{R_-} + \frac{R_+}{R_-} + \frac{R_1}{R_-} \right).$$

## References

1. Kondepudi, D.; Kapcha, L. Entropy production in chiral symmetry breaking transitions. *Chirality* **2008**, *20*, 524–528. [CrossRef] [PubMed]
2. Grandy, W.T., Jr. *Entropy and the Time Evolution of Macroscopic Systems*; Oxford University Press: New York, NY, USA, 2008.
3. Martyushev, L.M.; Seleznev, V.D. Maximum entropy production principle in physics, chemistry and biology. *Phys. Rep.* **2006**, *426*, 1–45. [CrossRef]
4. Klein, M.J.; Meijer, P.H.E. Principle of Minimum Entropy Production. *Phys. Rev.* **1954**, *96*, 250–255. [CrossRef]
5. Lebon, G.; Jou, D. *Understanding Non-Equilibrium Thermodynamics: Foundations, Applications, Frontiers*; Springer: Berlin/Heidelberg, Germany, 2008.
6. Esposito, M.; Lindenberg, K.; Van den Broeck, C. Universality of Efficiency at Maximum Power. *Phys. Rev. Lett.* **2009**, *102*, 130602. [CrossRef]
7. Esposito, M.; Lindenberg, K.; Broeck, C.V.d. Thermoelectric efficiency at maximum power in a quantum dot. *EPL* **2009**, *85*, 60010. [CrossRef]
8. Balachandran, V.; Benenti, G.; Casati, G. Efficiency of three-terminal thermoelectric transport under broken time-reversal symmetry. *Phys. Rev. B* **2013**, *87*, 165419. [CrossRef]
9. Benenti, G.; Ouerdane, H.; Goupil, C. The thermoelectric working fluid: Thermodynamics and transport. *CR Phys.* **2016**, *17*, 1072–1083. [CrossRef]
10. Apertet, Y.; Ouerdane, H.; Goupil, C.; Lecoeur, P. Irreversibilities and efficiency at maximum power of heat engines: The illustrative case of a thermoelectric generator. *Phys. Rev. E* **2012**, *85*, 031116. [CrossRef]
11. Schmiedl, T.; Seifert, U. Efficiency at maximum power: An analytically solvable model for stochastic heat engines. *EPL* **2007**, *81*, 20003. [CrossRef]
12. Wrigley, E.A. Energy and the English Industrial Revolution. *Philos. Trans. R. Soc. A* **2013**, *371*, 20110568. [CrossRef]
13. Kedem, O.; Caplan, S.R. Degree of coupling and its relation to efficiency of energy conversion. *Trans. Faraday Soc.* **1965**, *61*, 1897–1911. [CrossRef]

14. Goupil, C.; Ouerdane, H.; Herbert, E.; Goupil, C.; D'Angelo, Y. Thermodynamics of metabolic energy conversion under muscle load. *New J. Phys.* **2019**, *21*, 023021. [CrossRef]
15. Goupil, C.; Ouerdane, H.; Herbert, E.; Benenti, G.; D'Angelo, Y.; Lecoeur, P. Closed-loop approach to thermodynamics. *Phys. Rev. E* **2016**, *94*, 032136. [CrossRef]
16. Alicki, R.; Gelbwaser-Klimovsky, D.; Jenkins, A. A thermodynamic cycle for the solar cell. *Ann. Phys.* **2017**, *378*, 71–87. [CrossRef]
17. Alicki, R.; Horodecki, M.; Horodecki, P.; Horodecki, R. Thermodynamics of Quantum Information Systems—Hamiltonian Description. *Open Syst. Inf. Dyn.* **2004**, *11*, 205–217. [CrossRef]
18. Onsager, L. Reciprocal Relations in Irreversible Processes. II. *Phys. Rev.* **1931**, *38*, 2265–2279. [CrossRef]
19. Onsager, L. Reciprocal Relations in Irreversible Processes. I. *Phys. Rev.* **1931**, *37*, 405–426. [CrossRef]
20. Apertet, Y.; Ouerdane, H.; Goupil, C.; Lecoeur, P. Revisiting Feynman's ratchet with thermoelectric transport theory. *Phys. Rev. E* **2014**, *90*, 012113. [CrossRef]
21. Goupil, C.; Seifert, W.; Zabrocki, K.; Müller, E.; Snyder, G.J. Thermodynamics of Thermoelectric Phenomena and Applications. *Entropy* **2011**, *13*, 1481–1517. [CrossRef]
22. Apertet, Y.; Ouerdane, H.; Goupil, C.; Lecoeur, P. From local force-flux relationships to internal dissipations and their impact on heat engine performance: The illustrative case of a thermoelectric generator. *Phys. Rev. E* **2013**, *88*, 022137. [CrossRef]
23. Novikov, I.I. The efficiency of atomic power stations (a review). *J. Nucl. Energy* **1958**, *7*, 125–128. [CrossRef]
24. Curzon, F.L.; Ahlborn, B. Efficiency of a Carnot engine at maximum power output. *Am. J. Phys.* **1975**, *43*, 22–24. [CrossRef]
25. Tucker, V.A. The Energetic Cost of Moving About: Walking and running are extremely inefficient forms of locomotion. Much greater efficiency is achieved by birds, fish—and bicyclists. *Am. Sci.* **1975**, *63*, 413–419. [PubMed]
26. Hoyt, D.F.; Taylor, C.R. Gait and the energetics of locomotion in horses. *Nature* **1981**, *292*, 239–240. [CrossRef]
27. Apertet, Y.; Ouerdane, H.; Glavatskaya, O.; Goupil, C.; Lecoeur, P. Optimal working conditions for thermoelectric generators with realistic thermal coupling. *EPL* **2012**, *97*, 28001. [CrossRef]
28. Huxley, A.F.; Simmons, R.M. Proposed Mechanism of Force Generation in Striated Muscle. *Nature* **1971**, *233*, 533–538. [CrossRef]
29. Jevons, W.S. *The Coal Question: An Inquiry Concerning the Progress of the Nation, and the Probable Exhaustion of Our Coal-Mines*; Macmillan: London, UK, 1866.

© 2019 by the authors. Licensee MDPI, Basel, Switzerland. This article is an open access article distributed under the terms and conditions of the Creative Commons Attribution (CC BY) license (http://creativecommons.org/licenses/by/4.0/).

MDPI  
St. Alban-Anlage 66  
4052 Basel  
Switzerland  
Tel. +41 61 683 77 34  
Fax +41 61 302 89 18  
www.mdpi.com

*Entropy* Editorial Office  
E-mail: entropy@mdpi.com  
www.mdpi.com/journal/entropy